The Essential Hyland

ALSO AVAILABLE FROM BLOOMSBURY

Academic Discourse, Ken Hyland

Metadiscourse, Ken Hyland

Discourse Studies Reader: Essential extracts, edited by Ken Hyland

Corpus Applications in Applied Linguistics, edited by Ken Hyland, Chau Meng Huat and Michael Handford

Bloomsbury Companion to Discourse, edited by Ken Hyland and Brian Paltridge

The Essential Hyland

Studies in Applied Linguistics

KEN HYLAND

Bloomsbury Academic
An imprint of Bloomsbury Publishing Plc

BLOOMSBURY
LONDON · OXFORD · NEW YORK · NEW DELHI · SYDNEY

Bloomsbury Academic

An imprint of Bloomsbury Publishing Plc

50 Bedford Square	1385 Broadway
London	New York
WC1B 3DP	NY 10018
UK	USA

www.bloomsbury.com

**BLOOMSBURY and the Diana logo are trademarks
of Bloomsbury Publishing Plc**

First published 2018

British Library Cataloguing-in-Publication Data
A catalogue record for this book is available from the British Library.

ISBN: HB: 978-1-3500-3790-8
PB: 978-1-3500-3789-2
ePDF: 978-1-3500-3792-2
ePub: 978-1-3500-3791-5

Library of Congress Cataloging-in-Publication Data
A catalog record for this book is available from the Library of Congress.

Cover design by Olivia D'Cruz
Cover image © gualtiero boffi/Alamy Stock Photo

Typeset by Deanta Global Publishing Services, Chennai, India
Printed and bound in Great Britain

To find out more about our authors and books visit www.bloomsbury.com.
Here you will find extracts, author interviews, details of forthcoming events
and the option to sign up for our newsletters.

Contents

Preface

It seems odd to me, well extraordinary really, that I should be sitting here writing an introduction to a book of some of my collected papers. Something of an accidental academic, I started out as a language teacher pulled towards research by a curiosity about the ways texts work; how did the technical kinds of writing my students had to produce make sense to the people they were written for? It seemed an interesting question to ask and a good idea to try and understand more about these types of writing so I could teach them better. While I haven't always been successful at the teaching part, it turns out that my papers have interested enough people to warrant this book.

Many others have studied writing to support their classroom practices but I was lucky that the beginning of my academic career coincided with an explosion of interest in scholarly discourse. The teaching of English in universities has changed enormously since I first picked up a board marker (or piece of chalk to be more accurate). Something close to a full-blown paradigm shift has occurred during this time with English for Academic Purposes (EAP) programmes taking us towards more targeted and more research-informed language instruction. The last thirty years has been a period characterized by increased specialization, the coming to dominance of genre and corpus analyses, the emergence of teaching paradigms related to social participation, identity and learner experience and the growth of non-Anglo practitioners in research and publishing. This book is one person's experience of these changes and attempts to make sense of them.

While organized thematically rather than chronologically, the pages in this book provide a rough map of a growing sub-discipline, or at least one writer's small part of it. Since the mid-1980s academic discourse analysis, and the written variety in particular, has done a good job of establishing itself as a central field of research interest. With English, for the time being, sweeping away linguistic heterogeneity in the name of globalization and a free market of knowledge, the study of academic English has become something of a cottage industry. Because activities like educating students, demonstrating learning, disseminating ideas and constructing knowledge rely on language to accomplish, English has become of central to concern to everyone in higher education. Textbooks, essays, conference presentations, dissertations, lectures and research articles are indispensable to the academic enterprise while countless students and academics around the world, native and non-native English speakers alike, must now become familiar with the conventions of English academic discourses to understand their disciplines and steer their learning and careers.

The study of academic writing has, then, partly developed under the guise of EAP to better understand and address the needs of students and of academics who are increasingly required to write (and publish) in English. As a result, it has not only influenced what teachers do in the classroom but is itself approaching near-disciplinary status, supported by a burgeoning number of journals, books, conferences and doctoral dissertations and taught in units, departments and centres in almost every university where students need to study in English. But the field of academic writing has also expanded (and perhaps fragmented) because it offers such rich pickings for analysts interested in a diverse array of the twenty-first century's most fascinating and contentious concepts. Here, in the apparently frozen surface of scholarly texts, we find evidence of interaction, interpersonal engagement, community, identity, power and cultural variation. At the same time, these texts reveal the workings of theoretical constructs such as legitimate peripheral participation, genre, agency and the social construction of knowledge.

Something of my own contribution to these issues lies in these chapters, but they must be seen in the wider context of the work of those in the sociology of scientific knowledge, social construction and applied linguistics which inspired and influenced them. I have been standing on the shoulders of giants – among them are the five writers whose commentaries embellish this book: Chuck Bazerman, Brian Paltridge, Vijay Bhatia, Diane Belcher and Ann Johns to whom I am extremely grateful for their contributions. I first became aware that there was more to academic writing than following the prescriptive intuitions of my classroom textbooks when I came across the work of John Swales and Chuck Bazerman in the early 1990s, showing that scientific texts could be studied, much like works of fiction or oratory, for their stylistic choices and rhetorical effectiveness, and that what constitutes an appropriate text largely depends on the social and intellectual activity of the discipline which the text is part of.

These ideas were not new even back then as those working in the sociology of knowledge had long been sceptical of science's view of itself as a dispassionate and neutral observer of reality. But these writers, among others, brought an empirical dimension to the picture. They showed us that scientific persuasion was a result of rhetorical choices which could be unpacked and understood through careful analysis of textual practices.

Readers will quickly become aware that most of the chapters in this book principally focus on texts and less often on how people use them or what they mean for those people. My starting point is usually corpora so it may seem that I am privileging *text* above *practice* and treating writing as primarily a linguistic, even an autonomous, object rather than something which is socially embedded in particular lives, disciplines and contexts. This is not, of course, my intention and where possible I have enlisted the views of disciplinary informants to help me out in understanding what might be gleaned from texts. I believe, however, that texts themselves have a lot to tell us. Academic texts are cultural artefacts which offer evidence of how people interact with others: the ways they present themselves and their ideas to engage with members of their communities and collaboratively construct understandings of what the word is like.

Finally, I should note here that while 'discipline' plays a large part of the analyses in this book, not everyone agrees with the significance I have given it. The emerging importance of local ideologies, of institutional inducements towards interdisciplinary study, of the complexities of research questions, of easier communications and digital technologies have all perhaps undermined disciplinary boundaries in the twenty-first century. However, like Trowler et al. (2012: 246), I believe 'that disciplines have "real" epistemological characteristics, that knowledge structures do condition practices in quite real ways'. Applied linguistics is not electrical engineering and that fact has real consequences for those working in those fields. The training and acculturation involved in becoming a professional scholar largely involves acquiring knowledge in a specialized field, so that the disciplines are not just sources of knowledge but the foundations of a professional identity and the bases for shared communicative practices. Disciplines may be diffuse structures but they are at the interface of academic decision-making and institutional constraints. Academic texts, then, are where individual beliefs and community expectations meet to reveal the everyday workings of disciplines.

All this reinforces the fact that disciplines are not only intellectual but also social, held together by routine interactions among members with investments in a career and vested interests in getting their work accepted by gatekeepers and rewarded with publications, qualifications and advancement. All this involves writing and, in writing, making assumptions about readers to negotiate the successful passage of their work. Through writing, and readers' responses to their writing, individuals learn how things should be done and acquire sets of associations, meanings and tacit beliefs about priorities and values. This knowledge is embodied in the text they produce for each other and forms the basis of the studies printed in this book.

Acknowledgements

Bloomsbury and Ken Hyland are grateful to the following publishers for allowing us to reprint the papers in this book.

Reprinted with the permission of Cambridge University Press:

Chapter 1: Writing in the University: Education, Knowledge and Reputation. *Language Teaching,* 46(1), 53–70 (2013).

Chapter 2: Discipline: proximity and positioning. Chapter 2 of *Disciplinary Identities*. Cambridge University Press (2012), pp. 22–43.

Reprinted with the permission of Oxford University Press:

Chapter 3: Participation: community and expertise. *Chapter 5 of Academic Publishing*. Oxford University Press (2015), pp. 91–112.

Chapter 7: Metadiscourse in Academic Writing: *A Reappraisal. Applied Linguistics*, 25(2), 156–77 (With Polly Tse) (2004).

Chapter 12: Academic Attribution: Citation and the Construction of Disciplinary knowledge. *Applied Linguistics*, 20(3), 267– 341 (1999).

Reprinted with the permission of University of Michigan Press:

Chapter 5: Disciplinary cultures, texts and interactions. chapter 1 of *Disciplinary Discourses: Social Interaction in Academic Writing*. University of Michigan Press (2004), pp. 1–19.

Reprinted with the permission of Sage:

Chapter 4: Community and Individuality: Performing Identity in Applied Linguistics. *Written Communication,* 27(2), 159–88 (2010).

Chapter 6: Stance and Engagement: A Model of Interaction in Academic Discourse. *Discourse Studies*, 7(2), 173–91 (2005).

Chapter 8: Change of Attitude? A Diachronic Study of Stance. *Written Communication*, 33(3), 251–74 (2016).

Chapter 10: Dissertation Acknowledgments: The Anatomy of a Cinderella Genre. *Written Communication,* 20(3), 242–68 (2003).

Chapter 14: Is there an 'academic Vocabulary'? *TESOL Quarterly,* 41(2), 235–54.

(with Polly Tse) (2007)

Reprinted with the permission of Elsevier:

Chapter 9: Constructing Proximity: Relating to Readers in Popular and Professional Science. *Journal of English for Academic Purposes,* 9(2), 116–27 (2010).

Chapter 11: The Presentation of Self in Scholarly Life: Identity and Marginalization in Academic Homepages. *English for Specific Purposes,* 30(4), 286–97 (2011).

Chapter 13: Humble Servants of the Discipline? Self–mention in Research Articles. *English for Specific Purposes,* 20(3), 207–26 (2001).

Chapter 15: As can be Seen: Lexical Bundles and Disciplinary Variation. *English for Specific Purposes,* 27(1), 4–21 (2008).

Chapter 16: Genre–based Pedagogies: A Social Response to Process. *Journal of Second Language Writing,* 12(1), 17–29 (2003).

Chapter 17: Nurturing Hedges in the ESP Curriculum. *System,* 24(4), 477–90 (1996).

Chapter 18: Sugaring the Pill: Praise and Criticism in Written Feedback. *Journal of Second Language Writing,* 10(3), 185–212 (2001). With Fiona Hyland.

Chapter 19: Specificity Revisited: How Far Should We Go Now? *English for Specific Purposes,* 21(4), 385–95 (2002).

PART ONE

Writing, participation and identity

Introduction

The papers in this opening section all address, in different ways, the centrality of writing to the academic enterprise. The perspective is an unashamedly applied linguistic one, looking at what individuals are trying to accomplish in the pages of texts, but it is also a sociocultural one, seeing the ways they write as indicating something about the beliefs, customs and practices of their social groups. Essentially, academic discourse refers to the ways of thinking and using language which exist in the academy. Its significance, in large part, lies in the fact that complex social activities like educating students, demonstrating learning, disseminating ideas and constructing knowledge rely on language to accomplish. Textbooks, essays, conference presentations, lectures, dissertations and research articles are central to what universities are about: they are the very stuff of education and knowledge creation.

But academic discourse does more than enable universities to get on with the business of teaching and research. It simultaneously constructs the social roles and relationships which create academics and students, which sustains the universities, which constitutes the disciplines and which creates knowledge itself. Individuals use language to write, frame problems and understand issues in ways specific to particular social groups and in doing these things they form social communities, personal identities and professional institutions. So while writing involves languages it also involves more than language.

Chapter 1 covers a lot of this ground. This paper was developed out of my professorial inaugural lecture in 2006 at the Institute of Education, University of London (now part of University College London). The audience comprised academics from across the educational spectrum and so the paper was written to engage those outside the field of language education and unfamiliar with the minutiae of discourse analysis. It therefore takes as its topic higher education and the relevance of writing to trends in student recruitment and the strengthening grip of 'publish or perish' on academics. Because my own interests in language stem from an undergraduate background in sociology, and particularly in the sociology of scientific knowledge, it also summarizes some of my work on the ways interpretations of the world are negotiated in becoming facts. Research in the sociology of scientific knowledge seeks to explain why one interpretation rather than another succeeds due to external social and historical circumstances. My work seeks to flesh this out by looking at how this is done at more local levels, exploring the repeated patterns in the surface of texts.

This argument is continued in Chapter 2. Here, however, I attempt to address the problem from a slightly different point of view, exploring how the negotiation of knowledge can never be simply the observance of conformity to disciplinary conventions. It must, to more effectively engage readers and convey a distinctive authorial persona, also provide space for the writer's individual ambitions, character and 'take' on the subject. This is a tension between social belonging on the one hand and individual recognition on the other; between conformity and creativity. Academic writing is largely about persuading others of one's interpretations of data and claims about the world, and this is achieved by displaying one's membership of a discipline and an orientation to one's peers, what I call here *'proximity'*, and simultaneously using the categories of discourse our disciplines make available to *'position'* ourselves as individual academics. In other words, discourses do not form a closed and determining system but are patterns of options which allow us to actively construct a valued academic identity through discourse choices.

Chapter 4 was written earlier than Chapter 2, but was an attempt to work out some of the same ideas. It focuses on the writing of two applied linguistic celebrities, Deborah Cameron and John Swales, and explores how they produce and reproduce distinctive identities through their repeated rhetorical choices. Until this point most of my work had tended to concentrate on what was common in the texts of particular disciplines: how knowledge was constructed through replicating *community*. This paper was really my first effort to try and understand how individuals might manipulate available rhetorical options and to explore the contested concept of 'identity'. This meant looking past the biographical, or narrative, interview which had long been the preferred method studying identity, based on Giddens's (1991) belief that self and reflexivity are interwoven, so that identity is the ability to construct a reflexive narrative of the self. It seemed to me, however, that identity was not the same thing as self presentation and that data generated in low-stakes interviews with a stranger from the local university could never capture the subject's investment in a particular self or its validation in the responses of others.

In contrast, corpora allow us to focus on community practices in negotiating meaning and in so doing they also tell us something of how writers understand their communities – what their readers are likely to find convincing and persuasive. So these repeated rhetorical decisions don't just construct communities; they also construct individuals, they correspond to a view of identity as 'performance' and 'the ways that people display who they are to each other' (Benwell and Stokoe 2006: 6). Identity in this sense, then, doesn't exist *within* individuals but *between* them: *within* social relations. *Who we are*, or rather *who we present ourselves to be*, is an outcome of how we routinely and repeatedly engage in interactions with others on an everyday basis and corpora provide evidence of this. So while it may be a 'performance', and subject to change, it is a performance which is re-inscribed in us over time. Chapter 4 therefore looks at how two accomplished, and rhetorically savvy, academics use the options their disciplinary community makes available to gain credibility as insiders and reputations as individuals.

Chapter 3 comes from my book on academic publishing and looks at some of the ways membership is achieved and academic identities acquired. It therefore sits in contrast to Chapter 4, as it explores the experiences of junior scholars and how they engage in their disciplines, rather than established, well-published professors such as Cameron and Swales who have reached the apex of their discipline. It also differs from Chapter 4 in its methodology. While I have followed the same overall direction, attempting to unpack some of the regulating mechanisms of text production in academic discourse communities, Chapter 3 looks at people rather than texts themselves. This departure from my usual way of working, I hope, helps to offer another view of the ways academic writing is embedded in specific community and personal relationships and illumines something of how scholars learn their trade and engage in a matrix of relationships.

The section closes with a commentary on these chapters by Charles Bazerman, professor of Education at the University of California, Santa Barbara, USA, and one of the world's leading experts on writing and the rhetoric of science.

1

Writing in the university: Education, knowledge and reputation

1 Introduction

The title of this talk has a rather daunting scope but it offers a way of organising my discussion of some key aspects of academic literacy. Essentially, I want to challenge the widespread view that writing is somehow peripheral to the more serious aspects of university life – doing research and teaching students. Instead I want to argue that universities are ABOUT writing and that specialist forms of academic literacy are at the heart of everything we do. Drawing on some of my research over the past ten years, I will explore what writing means in the academy and argue that it is central to constructing knowledge, educating students and negotiating a professional academic career. Seeing literacy as embedded in the beliefs and practices of individual disciplines, instead of a generic skill that students have failed to develop at school, helps explain the difficulties both students and academics have in controlling the conventions of disciplinary discourses. Ultimately, and in an important sense, we are what we write, and we need to understand the distinctive ways our disciplines have of addressing colleagues and presenting arguments, as it is through language that academics and students conceptualise their subjects and argue their claims persuasively.

2 Why this interest?

Writing in the academy has assumed huge importance in recent years as countless students and academics around the world must now gain fluency in the conventions of academic writing in English to understand their disciplines, to establish their careers or to successfully navigate their learning. This is mainly the result of three major developments over the past 20 years.

First, there has been a huge expansion of higher education in many countries around the world, which has meant an increasing ethnic, class and age diversity in the student body. While some groups are still massively under-represented, the fact that almost 40% of the eligible age group now attend university in the UK, for example, means that courses are no longer dominated by white, middle-class, monolingual school leavers in full-time enrolment. This more culturally, socially and linguistically heterogeneous student population means that learners bring different identities, understandings and habits of meaning-making to their learning. This new body of students is very different from those educated in more elitist times, and tutors can no longer assume their students will bring the same understandings, skills and learning experiences that will equip them with the writing competencies they need to meet the demands of their courses.

A second reason for the interest in writing is that universities and other HE institutions around the world are increasingly becoming subjected to 'teaching quality audits' by funding bodies. This has led to universities devoting more attention to the processes of teaching and learning and more resources to the training of teaching staff. Student writing is now often a key area in continuing professional development programmes: here in Hong Kong, for instance, most universities are beginning to require new academic staff to complete a course in teaching. Similarly, in the United Kingdom, following the Dearing Committee's recommendations, writing has become central to the national framework for the training of university teaching staff.

The third reason for this interest in writing is the fact that for most academics, and many students, writing now has to be done in English. English has emerged as the international language of research and scholarship. With half the world's population predicted to be speaking the language by 2050, English seems to becoming less a language than a basic academic skill for many users around the world. Some 1.2 million students now study in English outside their home countries and international students comprise almost 50% of all postgraduates in Britain, contributing £1.5 billion annually to universities and £23 billion to the economy. There is also evidence that many doctoral students studying in overseas universities are completing their Ph.D. theses in English where they have a choice (Wilson 2002).

This also, of course, has consequences for academic publishing. More than 90% of the journal literature in some scientific domains and 70% of the 17,000 titles on the Thompson Reuter Master journal List are published in English (http://science.thomsonreuters.com/). This growth in English medium publications, moreover, is occurring not only in contexts where English is the official language but also where it is used as a foreign language, so that academics from around the world are now almost compelled to publish in English.

Standing alongside these developments, of course, are more enduring reasons for unpacking the black box of academic discourse. Not least of these is the prestige of academic writing and its traditional role as a carrier of what counts as legitimate knowledge and as authorised ways of talking about this knowledge. I want now to look a little more closely at the consequence of these developments in the three domains of my title.

3 Education

Student writing is at the centre of teaching and learning in higher education. While multimedia and electronic technologies are beginning to influence learning and how we assess it, writing currently remains the way in which students both consolidate and demonstrate their understanding of their subjects. The main function of writing, however, often seems to be gate-keeping and assessment. But whatever form writing takes, and it obviously differs by genre, course and discipline, it conforms to a single, institutionalised literacy which differs dramatically from that familiar to students from their homes, schools or workplaces. Trusted ways of writing are no longer valued as legitimate for making meaning when they arrive at university because of the different practices of the academy. Many students, and particularly those who are returning to study later in life, who speak English as a second language, or who have not had a smooth uninterrupted path through the education system, often find these discourses to be alien and privileged ways of writing (Ivanič 1998; Lillis 2001).

3.1 University writing practices

The particular kinds of writing that hold sway in the university have emerged to represent events, ideas and observations in ways that facilitate efficient, even shorthand, communication among insiders. Essentially, the process of writing involves creating a text that we assume the reader will recognise and expect, and the process of reading involves drawing on assumptions about what the writer is trying to do. Hoey (2001) likens this to dancers following each other's steps, each building sense from a text by anticipating what the other is likely to do. But while this anticipation provides for writer-reader coordination, allowing the co-construction of coherence from a text, academic writing disrupts our everyday perceptions of the world and sets up different expectations. Effective writers craft texts in a way that insiders can see as 'doing biology' or 'doing sociology', and this both restricts how something can be said and authorises the writer as someone competent to say it. In other words, students learn what counts as good writing through an understanding of their discipline and the conventions and genres regarded as effective means of representing knowledge in that discipline.

Broadly, the social practices of the academy produce particular configurations of text which cause difficulties for many students. In everyday uses of language, for example, we tend to represent things in a certain way, so that events unfold in a time sequence and agents accomplish actions. This is a 'natural' or congruent representation (Halliday 1994) in that we tend to translate our perceptions of the physical world into the grammatical system of language: we call it as we see it. Academic writing, however, turns our way of expressing meanings on its head through an incongruent use of language. It treats events as existing in cause and effect networks, disguises the source of modality of statements, foregrounds events rather than actors, and engages with meanings defined by the text rather than in the physical context.

These practices often confuse newcomers and force them into roles, identities and ways of writing which run counter to their experiences and intuitions about how language is used, forcing them to represent themselves in certain ways. They require us to change our normal ways of speaking in order to fit in. Ivanič (1998), for instance, found that many of her 'mature' female students felt insecure about their educational identity, as the discourse they were expected to use seemed pretentious and false: it did not let them 'be themselves'. L2 students often experience even greater problems as they encounter writing conventions that can differ considerably from those in their first language. These frequently demand that students are more explicit about the structure and purposes of their texts, more cautious in making claims, clearer in signposting connections, and generally that they take more responsibility for coherence and clarity in their writing (Clyne 1987).

Culture, however, is an extremely controversial notion, with no single agreed definition. One version sees it as a historically transmitted and systematic network of meanings which allow us to understand, develop and communicate our knowledge and beliefs about the world (Street 1995; Lantolf 1999). Language and learning are, therefore, intrinsically bound up with culture. This is partly because our cultural values are carried through language, but also because cultures make available certain taken-for-granted ways of organising our understandings, including those we use to learn and to communicate.

3.2 Modifying claims: Hong Kong versus United Kingdom student exams

The trouble with such broad generalisations, however, is that it is all too easy to overemphasise differences and place the blame on culture for students' writing difficulties. For example, some years ago John Milton and I explored a corpus of 1,800 GCE A level exam papers (a high school matriculation exam) by Hong Kong (HK) and British (UK) school leavers for HEDGES and BOOSTERS (Hyland & Milton 1997). Essentially, these are labels given to the linguistic features we use to mitigate or tone down (hedge) our statements, or to strengthen (boost) them. In that study we found that while both groups of students used the same number of devices overall, about one in every 50 words, the UK students (almost all native speakers of English) used over a third as many hedges and the HK students almost twice as many boosters.

Figure 1.1 shows the percentage distributions among certainty, probability and possibility items in the texts of the two groups, with the dark bars showing the percentage of each used by HK students and the light bars the uses by UK students. We can see from this that the HK students were more likely to be definite and certain, while the UK students expressed statements more cautiously as probabilities. Both groups expressed possibility equally.

While this might seem a small point, such differences are not always viewed as merely preferred alternative ways for expressing ideas in different languages, but have pragmatic consequences. The fact that the HK students' writing seems much stronger can lead English speaking readers, approaching

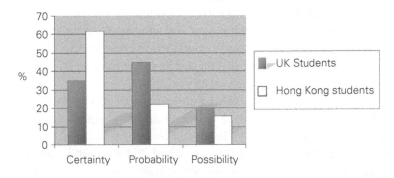

FIGURE 1.1 *Hedges and boosters in A Level corpus.*

a text with another set of expectations, to make negative moral or cultural judgements about the writer. Take this example from the HK students' corpus:

(1) There is <u>strong</u> evidence to <u>demonstrate</u> the relationship between EQ and the academic performance. High EQ is <u>definitely</u> an advantage in any domain of life and <u>we all know</u> that a person with high EQ can <u>certainly</u> manage their own feelings well and deal effectively with others. <u>The fact that</u> the trend from overseas is <u>always</u> affecting Hong Kong people means that schools <u>must</u> now teach boys to be equal to female. They have their right to express emotion.

The generous peppering of boosters in this example (underlined) can make the writer appear rather too assertive, over-confident and perhaps even dogmatic. In contrast, the greater use of hedges (underlined) in the native English speaker sample below is more in line with what we expect, so we attribute positive qualities to the writer, seeing subtlety, circumspection and openness to the views of readers:

(2) Britain is <u>probably</u> one of the few countries in the world where the constitution is not written down. This <u>might seem</u> to be <u>somewhat</u> disorganised and although it has worked <u>fairly</u> well until now, I <u>suggest</u> that this is <u>likely</u> to be unrealistic for much longer and it is <u>possible</u> we <u>may</u> need a Bill of Rights as there is in the USA.

We have to be aware of the possible prejudicial effects of our unconscious expectations, but I don't believe these differences are really a result of differences in culture. On the contrary, contrastive rhetoricians, comparing writing across languages, often expect Chinese students, immersed as they are supposed to be in Buddhist conceptions of face, to be more circumspect and respectful than brash, assertive UK writers (see Hinkel 1997). In fact, proficiency seemed to play a big role in these results, with the higher grade students' uses of these forms more closely approximating the native English speaker uses. Looking more closely for reasons by means of interviews with students and studies of secondary school textbooks, we discovered that these HK writers were over-generalising what they were taught about the need to be direct when writing in English. The main point I want to make here, however, is that examining large numbers of instances and interviewing students can tell us a great deal about students' learning experiences, their understandings of writing, and the ways they respond to writing tasks in English.

3.3 *Seeing literacy as rules*

Overall, though, the idea that university students can't write is central to official and public debate. It forms part of the public discourse of literacy crisis, falling standards, and the collapse of Western civilisation. Unfortunately, the issue is framed in a way that makes it harder to reach a resolution. Public debate on higher education, from radio phone-in programmes in Hong Kong to the *Dearing Report* in the UK, assumes a model of writing which separates language, writer and context. It sees students as identical and isolated, trying to acquire a set of skills independently of their identities, purposes and disciplines.

This echoes traditional linguistic conceptions of language like Saussure's distinction between *langue* – or language as system – and *parole* – language as use. This separation of system and meaning also lies behind the familiar CONDUIT METAPHOR of language, which suggests that others are able to recover our thoughts from our words exactly as we intended: communication simply involves encoding and decoding meanings, with no differences in interpretation or reader positions. Writing is regarded as an autonomous system that we all understand and use in roughly the same way and that is transparent in reflecting meanings, as opposed to the way individuals negotiate and construct meanings. In this view, then, good writing is largely a matter of grammatical accuracy. Literacy is a set of discrete, value-free rules and technical skills, so if we can decode and encode meanings, manipulate writing tools, perceive shape-sound correspondences and so on, all will be well. In higher education this pervasive view has two main implications.

First, by divorcing language from context, academic literacy is misrepresented as a naturalised, self-evident and non-contestable way of participating in academic communities. There is a single literacy, and academic language is all the same: it's just a bit harder than the language we use at home or at school. In response, institutions have invested in language centres to provide learners with the literacy resources they need to cope at university. But these programmes tend to be voluntary rather than compulsory, general rather than specific, and isolated rather than embedded in students' learning experiences. These factors not only limit the effectiveness of such programmes, but contribute to an ideology which transforms literacy from a key area of academic practice – how we construct ourselves as credible linguists, psychologists or whatever – into a kind of add-on to the more serious activities of university life. English for Academic Purposes (EAP), the practice of academic literacy instruction, thus becomes a kind of support mechanism on the margins of academic work.

Second, if we regard academic writing as just a specific kind of top-up to the literacy skills needed in everyday life, any writing difficulties can be seen as a deficiency in the student. Weak writing skills can be attributed to laziness, inattention or poor schooling, problems which can be put right with a few English classes. This diverts attention from a critical understanding of the writing that students are asked to do in their courses, and encourages both students and tutors to see academic writing instruction as a grammar top-up for weak learners. EAP becomes a band-aid measure to fix deficiencies in the students themselves.

3.4 Seeing literacies as practices

The massive expansion of research into academic writing in recent years, however, testifies to the significance of social context and allows us to reconceptualise literacy as a social practice rather than a set of skills (Barton 1994; Anstey & Bull 2004). The concepts of LITERACIES, referring to language use as something people DO when they interact with one another, and PRACTICES, the idea that these language activities are bound up with routine, everyday activities in the real world, provide ways of re-establishing this link between language and context. Moving away from literacy as an individual attribute is a central implication of this view. It helps us to see that words offer broad constraints on understanding but need to be, as Kress (2003: 4) puts it, 'filled in with meaning' which comes from our past experience of them. Basically, then, literacy is a resource for social groups, realised in social relationships and acting to pattern and structure those groups. Studying writing and the activities that surround it therefore become powerful tools for understanding the experiences of everyone in higher education, whether students or tutors.

First, it helps us to see that texts don't exist in isolation but are part of the communicative routines of social communities. This means not only that genres are related to other genres and the texts we read are connected to the texts we write, but that language is intimately related to the different epistemological frameworks of the disciplines and inseparable from how they understand the world. Bartholomae's famous quote captures this well:

> Every time a student sits down to write for us, he has to invent the university for the occasion – invent the university, that is, or a branch of it, like history or anthropology or economics or English. The student has to learn to speak our language, to speak as we do, to try on the peculiar ways of knowing, selecting, evaluating, reporting, concluding, and arguing that define the discourse of our community. (Bartholomae 1986: 4)

Such 'ways of knowing' are not learned by rote but by writing. Teaching writing therefore needs to be embedded in subject learning to provide students with a means of conceptualising disciplinary epistemologies.

Second, this means, in turn, using the analytical tools of applied linguistics to describe the features of academic genres so we can make these explicit to students. This is the domain of EAP, which draws on applied linguistics and an eclectic theoretical base to provide grounded insights into the structures and meanings of texts, the demands placed by academic contexts on communicative behaviours, and the pedagogic practices by which those behaviours can be developed. This means going beyond the teaching of lists of supposedly common grammatical structures and vocabulary which autonomous views of literacy assume, to understanding the choices language users make to most effectively express their intended meanings. It means trying to understand the practices of real people communicating in real contexts by describing and analysing relevant texts; interpreting the processes involved in creating and using these texts; and exploring the connections between disciplinary writing and the institutional practices it sustains (Hyland 2003, 2006).

Third, the study of academic literacy also encourages us to look beyond the immediate context of writing to explore a wider and more abstract notion of culture: the ways that institutional and disciplinary structures impact on language use. Socially powerful institutions, such as education, support dominant literacy practices that are part of organised configurations of power and knowledge embodied in social relationships. Other vernacular literacies in people's everyday lives are less visible, less prestigious and less supported (Barton & Hamilton 1998). In higher education, what Fairclough (1992) calls 'orders of discourse', the arrangement of discourses like seminars, lectures and essays, provide configurations of practices which govern what can be known, who can know it, and how it can be discussed.

Benesch (2001) has argued that teachers need to engage with issues of power by questioning rather than reinforcing conformity to dominant discourses. The view here is that our teaching practices should be less ACCOMMODATIONIST to dominant political and institutional orders, helping students to perform the best they can while 'encouraging them to question and shape the education they are getting' (Benesch 2001: xvii). By exploring these connections between institutions and literacy, both tutors and students are in a better position to understand and critique their disciplines and to identify the role which academic literacy plays in shaping both disciplines and individuals.

4 Knowledge

This brings us neatly to the second part of my talk, as we move on to KNOWLEDGE. If we reject the idea of language as a transparent medium of communication where words correspond unambiguously to meanings, then we problematise writing and the ways that academics report their claims. The fact that there are multiple interpretations of words, and numerous ways of constructing arguments, means that writing can never just simply be a way of neutrally reporting observations made in the lab or in some other research context. It foregrounds writing by giving prominence to how academics reach agreement on claims and the ways disciplines construct knowledge.

4.1 Induction and pursuit of 'truth'

Academic writing clearly enjoys a privileged status in the West, representing what Lemke (1995: 178) calls 'the discourse of "Truth"'. It is perceived as providing an objective description of what the natural and human world is actually like and this, in turn, serves to distinguish it from the socially contingent. We see this form of persuasion as a guarantee of reliable knowledge, and we invest it with cultural authority, free of the cynicism with which we view the partisan rhetoric of politics and commerce. Through induction, experimentation and falsification it gives us direct access to the external world, and the text is just a channel to report observable facts.

But induction does not offer cast-iron proof, and by moving from observations of instances to statements about unobserved cases, we introduce uncertainty. The problem for scientific knowledge, then, is that interpretation always depends on the assumptions scientists bring to the problem. As the celebrated physicist Stephen Hawking (1993: 43) notes:

> . . .a theory may describe a range of observations, but beyond that it makes no sense to ask if it corresponds to reality, because we do not know what reality is independent of a theory.

In other words, there is always going to be more than one possible interpretation of data, and the fact that we can have these competing explanations shifts attention from the laboratory or clipboard to the ways that academics argue their claims. We have to look for proof in the textual practices for producing agreement. At the heart of academic persuasion is writers' attempts both to foresee and address potentially negative reactions to their claims, and to do this they use the discourses of their disciplines.

4.2 The rhetorical construction of truth

This social constructivist position suggests that knowledge and social reality are created through daily interactions between people, and particularly through their discourse. It rejects taken-for-granted knowledge and opposes empiricism and positivism, which have traditionally been the foundation of the natural sciences – the idea that the nature of the world can be revealed by observation. Social constructionism cautions us to be suspicious about the 'facts' of existence and encourages us to recognise that what we see as truth is actually filtered through our theories and our language. It is based on categories which are culturally and historically specific. What this means is that our knowledge of the world is not derived from observation, but constructed by people through their interactions in daily life, acting as members of social groups. Discourse is therefore at the heart of relationships, knowledge and scientific facts, as all are rhetorically constructed. The goal of academic literacy studies is therefore to discover how people use discourse to create, sustain and change these communities; how they signal their membership; how they persuade others to accept their ideas; and so on. Stubbs (1996: 21) succinctly combines these issues into a single question:

> The major intellectual puzzle in the social sciences is the relation between the micro and the macro. How is it that routine everyday behaviour, from moment to moment, can create and maintain social institutions over long periods of time?

This emphasises the community-based orientation to literacy I discussed in relation to education. Just as academic literacies are not just a bolt-on to home literacies, physicists don't write like philosophers nor lawyers like linguists. Writers have to establish a professionally acceptable voice and an appropriate attitude, to both their readers and their arguments. They marshal support, express collegiality, and negotiate agreement by making linguistic choices which connect their texts with their disciplines. They have to balance the claims they make for significance and novelty in their research against the convictions and expectations of their readers. They must make language choices which take account of readers' likely objections, background knowledge and rhetorical expectations (Hyland 2001a; 2004a; 2005a). In other words, the linguistic resources we unreflectively choose when we present our arguments locate us in our disciplines: they present us as competent, credible insiders and allow us to engage with other insiders, anticipating the actual or potential voices and views of our readers. In sum, it is through language that we create a stance and align ourselves with a community.

My research into a range of academic genres over the last decade or so describes some of the ways that writers in different disciplines represent themselves, their work and their readers in very different ways. A corpus of 240 research articles of 1.5 million words, for example, shows that some 75% of all features marking author visibility, such as self-mention, personal evaluation and explicit interaction with readers, occur in the humanities and social sciences (Hyland 2005b). This shouldn't, perhaps, surprise us. After all, empiricism finds its truths by observing the world, so needs a language that represents real events without the mediation of rhetoric. This is what Foucault (1974) characterised as the neo-classical search for a univocal discourse – a one-to-one correspondence between words and categories of things, which linked the rise of science with the eighteenth-century interest in dictionaries, grammars and scientific taxonomies.

But this word–meaning correspondence is an illusion, and legal disputes over the most explicitly written legal documents show how words can have multiple meanings. Simply put, the relative impersonality of scientific discourse is not an absence of rhetoric but a different kind of rhetoric, based, like all persuasive writing, on an assumed consensus about how language can be used within particular communities. Let me illustrate this with some brief examples from my research.

4.3 *Disciplinary variation of lexical meanings*

First of all, our love affair with the idea of a single academic vocabulary seems to have worn off. In a study into an academic corpus of 3.5 million words, for example, Polly Tse and I (Hyland & Tse 2007) found that so-called universal semi-technical items in Coxhead's *Academic Word List* actually have widely different frequencies and preferred meanings in different fields. For example:

- *consist* means 'stay the same' in social sciences and 'composed of' in the sciences.
- *volume* means 'book' in applied linguistics and 'quantity' in biology.
- *abstract* means 'remove' in engineering and 'theoretical' in social sciences.

Similarly, Ward (2009) compared lexical items in textbooks across five engineering fields and found that *gas, heat* and *liquid* occurred almost exclusively in chemical engineering. He also found items like *system, time, value* and *factor,* which were very high across all engineering fields, but they collocated very differently, giving these words different technical meanings (*settling time, critical value, load factor*).

4.4 *Variation in directives*

Another feature which varies across disciplines is the use of directives. These are devices which instruct the reader to perform an action or to see things in a way determined by the writer (Hyland 2002). They are largely expressed through imperatives (e.g. *consider, note, imagine*) and obligation modals (such as *must, should* and *ought*). Overall, they direct readers to three main kinds of activity: textual, physical and cognitive.

- TEXTUAL ACTS direct readers to another part of the text or to another text (e.g. *see Smith 1999, refer to table 1.2*, etc.)

- PHYSICAL ACTS direct readers how to carry out some action in the real world (e.g. *open the valve, heat the mixture*).

- COGNITIVE ACTS instruct readers as to how to interpret an argument, explicitly positioning readers by encouraging them to *note, concede* or *consider* some argument or claim in the text.

Directives represent an explicit intervention in the text by the writer to guide readers to a preferred interpretation. This kind of explicit engagement with readers is a valuable rhetorical resource for writers in the social sciences, as they are unable to rely on the explanatory value of accepted experimental procedures, quantifiable methods or other aspects of scientific proof-making. However, directives are a potentially risky device, as they instruct readers to act or see things in a way determined by the writer (Hyland 2002). As a result, most directives in the soft fields are textual, directing readers to a reference or table rather than telling them how they should interpret an argument. So examples like these are common in social science research articles:

see Steuer 1983 for a discussion of other contingencies' effects.

(marketing)

Look at Table 1.2 again for examples of behavioristic variables.

(sociology)

For transcription conventions please refer to the Appendix.

(applied linguistics)

Two of my social sciences respondents mentioned this in their interviews:

I am very conscious of using words like 'must' and 'consider' and so on and use them for a purpose. I want to say 'Right, stop here. This is important and I want you to take notice of it'. So I suppose I am trying to take control of the reader and getting them to see things my way.

(sociology interview)

I am aware of the effect that an imperative can have so I tend to use the more gentle ones. I don't want to bang them over the head with an argument I want them to reflect on what I'm saying. I use 'consider' and 'let's look at this' rather than something stronger.

(applied linguistics interview)

In hard knowledge fields, in contrast, argument is more highly standardised and less discursive, drawing on semiotic resources which are graphical, numerical and mathematical rather than simply textual. Arguments are formulated more succinctly and papers are often less than a third the length of those in the social sciences and humanities, as both editors and scientists struggle with the continuing expansion of knowledge in the sciences. Cognitive directives are far more frequent here, as they help writers cut directly to the heart of complex arguments, explicitly positioning readers by leading them through an argument or emphasising what they should attend to:

What has to be recognised is that these issues. . .

(mechanical engineering)

Consider the case where a very versatile milling machine of type M5. . .

(electrical engineering)

A distinction must be made between cytogenetic and molecular resolution.

(biology)

My informants noted this in their interviews:

I rarely give a lot of attention to the dressing, I look for the meat – the findings – and if the argument is sound. If someone wants to save me time in getting there then that is fine. No, I'm not worried about imperatives leading me through it.

(electrical engineering interview)

I'm very conscious of how I write and I am happy to use an imperative if it puts my idea over clearly. Often we are trying to work to word limits anyway, squeezing fairly complex arguments into a tight space.

(mechanical engineering interview)

4.5 *Variation in bundles*

Finally, lexical bundles, or frequently occurring word sequences. These are a key way that particular disciplines produce community specific meanings and contribute to a sense of distinctiveness and naturalness in a register.

So collocations like *as a result of* and *it should be noted that* help identify a text as belonging to an academic register, while *in pursuance of,* and *in accordance with* mark out a legal text. Their frequent use helps to identify competent language ability among individuals and to signal membership of a particular academic community. Using a corpus of 120 research articles and 120 postgraduate dissertations in four disciplines, I found that the most common bundles in this academic corpus of 3.5 million words were *on the other hand, at the same time* and *in the case of,* all of which occurred over 100 times per million words (Hyland 2008a & b).

There are considerable disciplinary variations in the use and frequencies of different bundles, with the top 20 most common shown in Table 1.1. The extent to which specific fields make use of particular strings is shown by items that occur in all four disciplines being marked in bold and those in three disciplines shaded (Hyland 2008a). Just two forms occur in all four disciplines (*on the other hand* and *in the case of*) and a handful in three fields. In fact, it turns out that over half of all items in the top 50 bundles in each discipline don't occur in the top 50 of any of the other disciplines studied here.

TABLE 1.1 Most frequent four-word bundles (bold = in four disciplines; shaded = in three disciplines)

Biology	Electrical engineering	Applied linguistics	Business studies
in the presence of	**on the other hand**	**on the other hand**	**on the other hand**
in the present study	as shown in figure	at the same time	**in the case of**
on the other hand	**in the case of**	in terms of the	at the same time
the end of the	is shown in figure	on the basis of	at the end of
is one of the	it can be seen	in relation to the	on the basis of
at the end of	as shown in fig	**in the case of**	as well as the
it was found that	is shown in fig	in the present study	the extent to which
at the beginning of	can be seen that	the end of the	the end of the
as well as the	can be used to	the nature of the	significantly different from zero
as a result of	the performance of the	in the form of	are more likely to
it is possible that	as a function of	as well as the	the relationship between the
are shown in figure	is based on the	at the end of	the results of the
was found to be	with respect to the	the fact that the	the Hang Seng index
be due to the	is given by equation	in the context of	the other hand the
in the case of	the effect of the	is one of the	in the context of
is shown in figure	the magnitude of the	in the process of	as a result of
the beginning of the	at the same time	the results of the	the performance of the
the nature of the	in this case the	in terms of their	Hong Kong stock market
the fact that the	it is found that	to the fact that	is positively related to
may be due to	the size of the	in the sense that	are significantly different from

Predictably, perhaps, we find the greatest similarities between cognate fields, so that business studies and applied linguistics share 18 items in the top 50 with *on the basis of, in the context of, the relationship between the,* and *it is important to* exclusive to these two fields. Similarly, biology and electrical engineering have 16 bundles in common, with *it was found that, is shown in figure, as shown in figure, is due to the,* and *the presence of the* not in the social science lists at all.

The electrical engineering texts contained the greatest range of high frequency bundles with 213 different four-word strings, and also the highest proportion of words in the texts occurring in four-word bundles. Biology, on the other hand, had the smallest range of bundles, the fewest examples, and the lowest proportion of texts consisting of words in bundles. Thus, the electrical engineering texts were most dependent on such common prefabricated bundles as 'it can be seen' and 'as shown in figure', and used many sequences not found in the other disciplines, perhaps because technical communication is relatively abstract and graphical. This supports analyses in systemic functional linguistics which suggest that language in science continually constructs conceptual objects that are arrayed in various syntactic relations of coexistence and causality (Halliday & Martin 1993). Arguments might therefore function by linking data or findings in routinely patterned, formulaic ways with the same forms used repeatedly.

The functions of these sequences also reveal disciplinary specificity, with a broad split across the hard and soft fields. Table 1.2 shows that RESEARCH-ORIENTED BUNDLES, which refer to real world activities, make up about half of all bundles in the sciences and TEXT-ORIENTED BUNDLES, which focus on argument, make up half those in the social sciences. PARTICIPANT-ORIENTED BUNDLES, which focus on the writer or reader of the text, provide a structure for interpreting a following proposition, conveying the writer's individual

TABLE 1.2 Distribution of functions of four-word bundles

Discipline	Research-oriented	Text-oriented	Participant-oriented	Totals
Biology	48.1	43.5	8.4	100
Electrical engineering	50.4	40.4	9.2	100
Applied linguistics	31.3	49.8	18.9	100
Business studies	35.0	48.4	16.6	100
Overall	**41.2**	**45.5**	**13.2**	**100**

stance and personal engagement with readers, and these are twice as common in the more discursive social science fields.

Once again, these uses reflect the argument patterns in the two domains. So we can see how text-oriented examples function to connect aspects of argument (a) while those in the sciences point to research and findings (b):

(a) The term 'linguistics' might be too narrow *in terms of the* diverse knowledge-base and expertise that is required.   (applied linguistics)
The *purpose of this paper* is to investigate the perceptions of consumers in the Hong Kong market toward fast food.   (business studies)

(b) *The structure of the* coasting-point model can be divided into three areas.   (electrical engineering)
The DNA was precipitated *in the presence of* 2.5 volumes of ethanol and 0.1 volume of 3.0 M sodium acetate pH.   (biology)

The relatively substantial presence of these items in the hard sciences reflects the fact that these disciplines place considerable emphasis on precision, particularly to ensure the accurate understanding of procedures and results. In contrast, the text-oriented strings which were particularly marked in the soft fields reflects a more discursive and evaluative pattern of argument, where persuasion is more explicitly interpretative and less empirical.

5 Reputation

Clearly, writing has enormous relevance to the ways individuals construct themselves as competent academics, build professional visibility and establish reputations. I've mentioned the importance of disciplinary discourse conventions and the need for writers to project an appropriate, community-defined stance and to engage with their readers in ways they are likely to find familiar and persuasive. In this sense the importance of academic literacy to academics cannot be overestimated. But writing has become even more important as the institutionalised system which both creates knowledge and distributes rewards: the system of publication. The system of institutional incentive, in fact, has transformed academic endeavour into a machine for the manufacture of papers. Researchers find that their work has little value for them or their university without publication. A paper is judged as a contribution to the disciplinary literature if colleagues can find it and make use of it. If editors, referees, proposal readers, conference attendees or journal readers regard a submission as original and significant, if it is published in prestigious places, and if it is cited and taken up by others in future papers, then the writer receives the reward of recognition.

5.1 Recognition and influence

Recognition is not simply an intangible benefit to the holder: it brings the reward of influence and the ability to wield influence in their field. Academics who excel in publishing their research achieve promotion, tenure and access to senior administrative and gate-keeping roles from which they can oversee research agendas, appointments, funding and project applications. They therefore form an elite which exercises influence in setting standards, directing strategies and determining what is considered good work or an important topic. They may also gain greater influence as spokespeople for their colleagues, both in the media and as members of government committees, thereby helping to shape perceptions of their field and the direction it moves in. This prestige and recognition is what Bourdieu (1991) refers to as symbolic capital, which bestows power on the holder. Latour & Woolgar (1979) express this using a market metaphor, which sees academics as engaged in a cycle of activity which involves them in converting different kinds of 'credit' to enhance their credibility. Thus, a series of successful publications bestows credit on researchers, which they may be able to convert into a research grant to finance equipment and attract colleagues to conduct further research. This, in turn, will produce more data which can be converted to arguments, new publications, and so on through another cycle. Credibility thus helps academics to progress along an upward mobility path in their careers. Latour & Woolgar put it like this:

> For example, a successful investment might mean that people phone him, his abstracts are accepted, others show interest in his work, he is believed more easily and listened to with greater attention, he is offered better positions, his assays work well, data flow more reliably and form a more credible picture. (Latour & Woolgar 1979: 204)

Success is seen as largely measured by recognition and, in turn, the process of acquiring recognition as dependent on the capacity to write papers valued by one's colleagues.

5.2 Publishing and competition

Academics are, of course, motivated by curiosity and driven by a need to understand the issues they research, although most admit that recognition is an important source of professional reward, so they publish in order to further their careers. But because reputation is translated into concrete consequences, and because both material and symbolic capital are scarce, academic publication is fiercely competitive. This institutionally sanctioned competition is generally supposed to be a spur to advancing knowledge, but it is now inseparable from the process by which prestige and credibility are assessed. Publication comes to equal 'productivity' and is used as a crude measure of worth, with institutions conferring promotion and tenure on the length of personal bibliographies. The emergence of Research Assessment Exercises and publication of university league tables in many countries act to fan these flames.

Writing is both the stick and the carrot that propel us on the academic treadmill. As James Watson, Nobel laureate and a member of the biology establishment, spells out:

> It starts at the beginning. If you publish first, you become a professor first; your future depends on some indication that you can do something by yourself. It's that simple. Competitiveness is very dominant. The chief emotion in the field. (Cited in Judson 1995: 194).

Competition is increasingly important with the growth of commercial incentives which, in technological fields in particular, may be even stronger than intellectual ones.

So the pressure has been increased on academics – not only is the need to publish greater than ever, but publication must be in English. Research shows that academics all over the world are increasingly less likely to publish in their own languages and find their English language publications cited more often. There were over 1.1 million peer-reviewed research articles published globally in English in 2005; this number has been increasing by 4% annually, despite falling library budgets. This has meant that the numbers of non-native English speaking academics publishing in English language journals is increasing dramatically.

References to English language publications, for example, have reached 85% in French science journals, and English makes up over 95% of all publications in the *Science Citation Index*. There were, in addition, over 4,000 papers in English language social science journals in 2005, compared with only 400 just eight years earlier. With publishers increasingly encouraging libraries to subscribe to online versions of journals, the impact of English becomes self-perpetuating, since it is in these journals where authors will be most visible on the world stage and receive the most credit. Many European and Japanese journals, for instance, have switched to English, journals in Swedish, Dutch and German being extremely hard hit.

Clearly a lingua franca facilitates the exchange of ideas far more effectively than a polyglot system, but there is a real danger this will exclude many L2 writers from the web of global scholarship, so depriving the world of knowledge developed outside the Anglophone centres of research. One consequence has been the development of 'writing for publication' courses for academic staff at universities around the world. In Hong Kong, staff were extremely appreciative of these courses, but also very secretive about actually attending them. This is perhaps another feature of academic writing that I haven't mentioned: not only does it have to be engaging, interesting and persuasive, it must also be effortless.

6 Conclusions

It might seem that as an applied linguist I've attached too much importance to the object of my trade and been guilty of over-emphasising the impact of writing in the academy. After all, being a successful student, publishing academic, influential researcher or whatever involves other competencies – craft skills, wide reading, analytical and critical abilities and so on. But while diligence and brilliance will get you a long way, we are ultimately defined and judged by our writing. It is important to stress that this writing isn't just an abstract skill, but a core aspect of the epistemological frameworks of our fields and of our identities as academics. We are what we write, and we need to understand the distinctive ways our disciplines have of identifying issues, asking questions, addressing a literature, criticising colleagues and presenting arguments.

To sum up: knowledge, disciplines and the professional careers of academics themselves are ultimately constructed through the ways we write. What I have been arguing is that we need to understand the full complexity of writing as a situated activity and to recognise its central place in our practices. To do this is to both improve our practice and ensure that the teaching of academic discourse is not artificially divorced from the teaching of academic subjects.

2

Discipline: Proximity and positioning

We produce and reproduce an identity through our repeated interactions as members of social groups. Identity is a person's relationship to his or her social world: it is a joint, two-way production, and language allows us to create and present a coherent self to others because, over time, it ties us into webs of common sense, interests and shared meanings. *Who we are* and *who we might be* are therefore built up through participation and linked to situations, to relationships and to the language we use in engaging with others on a routine basis.

Communities based around occupation, recreation, family and so on thus provide meaningful reference points and so help shape collective definitions of identity within their frameworks of understandings and values. In academic contexts, disciplines provide such reference points. They are places where participation involves communicating, and learning to communicate, in appropriate ways, and membership depends on displaying the ability to communicate as an *insider*: using language to represent ourselves as legitimate members. But while individuals enact who they are only in their dealings with others, they all bring different experiences, inclinations and proclivities to their performances as academics, teachers and students. Disciplinary membership, and identity itself, is a tension between conformity and individuality, between belonging on the one hand and individual recognition on the other. In this chapter I introduce the terms *proximity*, to refer to the first of these, the relationship between the self and community, and *positioning* to talk about the second, the relationship between the speaker and what is being said, as a way to begin unpacking these tensions. First, though, I want to establish the idea of discipline as a rhetorical community and site of identification.

2.1 What is a *discipline?*

Discipline is a common enough label, used to describe and distinguish topics, knowledge, institutional structures and individuals in the world of scholarship. While emergent multi-focused and practitioner-based fields may challenge conventional notions of academic disciplines, students and academics themselves typically have little trouble in identifying their allegiances (Becher and Trowler, 2001).

Definitional difficulties

Although less controversial than the idea of identity, *discipline* is, however, no easier to define. Disciplines, in fact, have been seen in numerous ways: as institutional conveniences, networks of communication, political institutions, domains of values, modes of inquiry and ideological power-bases. Kuhn (1977) identifies them according to whether they have clearly established paradigms or are at a looser, pre-paradigm stage; Biglan (1973) and Donald (1990) draw on faculty perceptions and Kolb (1981) on learning-style differences to provide categories which distinguished hard from soft and applied from pure knowledge fields; Storer and Parsons (1968) oppose analytical to synthetic fields; and Berliner (2003) distinguishes 'hard' and 'easy-to-do' disciplines in terms of the ability to understand, predict and control the phenomena they study. Others, writing from post-modern positions, predict their imminent demise, arguing that the fragmentation of academic life has resulted in the death of disciplines (e.g. Gergen and Thatchenkery, 1996; Gilbert, 1995).

Clearly, the term needs to be treated with caution. As a gloss on the complexity of how research and teaching is socially organised and conducted, the concept of *discipline* has its limitations. Some writers, in fact, see it as little more than a convenient shorthand for practices that are less distinguishable and stable than we usually suppose (e.g. Mauranen, 2006). It is certainly true that the boundaries of scholarship shift and dissolve, aided by institutional changes such as the emergence of practice-based and modular degrees and the fact that research problems and investigations often ignore disciplinary boundaries. New disciplines spring up at the intersections of existing ones and achieve international recognition (biochemistry, gerontology), while others decline and disappear (philology, alchemy). Similarly, there are cultural and geographic variations among disciplines as different education systems, levels of economic development and political ideologies come into play (Podgorecki, 1997). It is, however, important to recognise that centripetal as well as centrifugal forces are at work, with increased global information flows and resource networks counteracting the influence of nation-states and local cultural practices.

Divergences and differences

Nor do all disciplines exhibit the same degree of cohesiveness and internal agreement. Some (such as modern languages and biochemistry) are convergent and closely knit in terms of their ideologies, awareness of a common tradition, research methodologies, allowable topics, standards of quality and so on. Others (perhaps geography and literary criticism) are more divergent and loose, where members lack a clear sense of group cohesion, where disciplinary borders are ill-defined, and where ideas cross boundaries more readily. Disciplines also often embrace a spectrum of specialisms, of theory, methods or subject matter, which seem to fragment cohesiveness. As Becher and Trowler (2001: 65) observe, 'there is no single method of enquiry, no standard verification procedure, no definitive set of concepts that uniquely characterizes each particular discipline'. In some contexts, in fact, it is more accurate to see communities in the overlapping subsidiary domains within disciplines (Hyland, 2009). Instead of uniformity, then, we find fluid and permeable entities impossible to pin down with precision.

There are also difficulties in reconciling how large, amorphous and dispersed groups of individuals defined by common bonds of discourse practices and conventions can be matched up with groups of individuals who typically work together and subscribe to common work practices and patterns of communication (e.g. Killingsworth and Gilbertson, 1992). Everyday local practices of teaching, supervision, research, marking and committee work can often seem removed from a wider involvement in publishing, networking and conference attendance. Killingsworth and Gilbertson themselves argue that membership in 'global communities' now dominates the activities of Western academics and that this 'tends to be regulated exclusively by discourse-governed criteria (writing style, publication in certain journals, presentations at national conventions, professional correspondence, and so forth)'. Local communities, on the other hand, have largely been characterised in terms of shared practices and the engagement of individuals around some project which fosters particular beliefs, ways of talking and power relations (Swales, 1998).

In fact, how far realms of knowledge come to be accepted as *disciplines*, rather than remaining as, say, *approaches*, seems largely a matter of institutional recognition. International significance is a key measure, particularly the extent to which leading universities recognise the independence of an area and give it the status of a department with professorial chairs, budgets and degrees; whether a distinct international community has appeared around it with the professional paraphernalia of conferences, learned societies and specialist journals; and whether the wider populace generally see it as having academic credibility and intellectual substance. All this points to the contextually contingent nature of disciplines and the fact that they are not monolithic knowledge-generating apparatus but are very much dependent on local struggles over resources and recognition. In other words, a discipline is as much determined by social power and members' categorisations as it is by epistemological categories.

Disciplines as sites of identity

Disciplines, however, are not mere chimera. They have a very real existence for those who work and study in them, who attend conferences, supervise students, manage programmes, read journals, write books, sit through lectures and take exams. Although they work day to day in local *institutions,* members have a sense of being part of a *discipline* and of having a stake in something with others. In fact, one often has more to do with people one meets occasionally at conferences or corresponds with by email than with those on the next floor of the same building (e.g. Swales, 1998). It is in disciplines, rather than particular physical sites, that the important interactions in a professional's life occur, bringing academics, texts and practices together into a common rhetorical locale. Group membership, in other words, implies some degree of group identification and this, in turn, presupposes that members will see themselves as having some things in common and being, to some extent, similar to each other (e.g. Turner, 1984).

Identity is about who the individual is, and an academic identity is who the individual is when acting as a member of a discipline. This clearly involves a range of practices, understandings and values in which familiarity with a common literature, knowledge of accepted and topical theories, fluency in arcane research practices, and awareness of conferences, journals, leading lights, prestigious departments and other paraphernalia of daily academic life are important. However, individuals identify themselves, and are identified by others, largely through their use of a language; a language which interprets and reports these activities and beliefs in specific ways and which simultaneously helps distinguish insiders from others. There is an essential and integral connection between identity and the cultures of disciplinary groups, which is mediated by distinctive patterns of language so that, to put it crudely and no doubt a shade polemically: we are what we write. An engineer is an engineer because he or she communicates like one and the same is true for biologists, historians and linguists.

Because social groups objectify in language certain ways of experiencing and talking about phenomena, these come to influence the self-perceptions and presentations of group members. Conditions of homogeneity tend to be created within communities defined by that language so that texts help evoke a context where the individual activates specific recognisable and routine responses to recurring tasks. Whether this context is a face-to-face lecture, an online supervision, an exam board meeting, or a research report to anonymous others around the world, it involves actors constructing themselves in terms of how they understand reality and want to be seen. All this, in turn, increases the social distance between members and outsiders and works against cross-community cooperation:

> As the work and the points of view grow more specialised, men in different disciplines have fewer things in common, in their background and in their daily problems. They have less impulse to interact with one another and less ability to do so.
>
> (Clark, 1987: 273)

The idea of interdisciplinarity, almost universally held to be a positive aspect of knowledge and research, actually often amounts only to the accumulation of knowledge gleaned from more than one field. Relentless diversification and specialisation appear to be more common, leading to the growing mutual incomprehensibility of academic communities.

The concept of discipline thus helps explain how individuals identify themselves and act to persuade others of that identification. Discipline helps apparently heterogeneous individuals form first-year undergraduate classes, departmental committees and research groups. It helps concretise, in specialised textually encoded knowledge, the 'invisible colleges' which span the globe through webs of journals, symposia, conferences, Skype chats, pre-print circulations and email exchanges. In short, the notion puts each individual's decision-making and engagement at centre stage and underlines the fact that academic discourse involves language users in constructing and displaying their identities, both moment by moment and over extended periods, as members of *disciplines*. This membership, however, implies both similarity and difference, conformity and individuality, and I turn to these issues now, recasting existing work on academic discourse to show its relevance to the study of disciplinary identity.

2.2 Proximity: Hegemony, solidarity and convention

I have been arguing that actions that we perform routinely, and particularly routine *rhetorical* actions, are central to the performance of self. This is particularly the case in the performance of a *disciplinary* self where language ties actors into webs of association and allows them to become part of the collective, to identify with others and, as Mead has it, to 'take on the role of the other'. Seeing others as similar or different allows members, for practical purposes, to create a sense of self through consistent engagement with those like them. Over time, they construct a recognisable and valued identity through competent participation in the common genres and discourse forms of a discipline. I use the term *proximity* to refer to a writer's control of these practices to manage themselves and their interactions with others; it is *the use of a disciplinary-appropriate system of meanings.*

An orientation to readers and discipline

Proximity implies a receiver-oriented view of communication and is closely related to Sacks et al.'s (1974: 272) notion of 'recipient design', or how talk is shaped to make sense to the current interactant. In fact, talk has figured far more prominently than writing in identity research as analysts have established that gender (Butler, 1990), ethnicity (Day, 1998), family role (Wortham and Gadsden, 2006) and so on are performed through verbal interaction. Writing is no less important in constructing identity, however, as lexical choice, topic selection, conventions of argument and so on also display an orientation and sensitivity to co-participants.

The particularities of any situation therefore connect with wider norms and practices. This is not because members rationally decide these norms are sensible, but because constant exposure to this discourse leads them to work out what norm the group favours. Choices are narrowed to the point where we do not have to decide on every option available. In academic contexts this means that the process of writing involves creating a text that we assume the reader will recognise and expect, and the process of reading involves drawing on assumptions about what the writer is trying to do. Skilled writers are therefore able to create a mutual frame of reference and anticipate when their purposes will be retrieved by their audiences. This is not to claim anything new, of course: a considerable literature supports the fact that individuals are socialised into disciplinary membership. But the use of preferred discourse patterns has not usually been seen in terms of proximity or identity construction.

Proximity ties the individual into a web of disciplinary texts and discourses, recalling Bakhtin's (1981) influential notion of *dialogism,* mentioned in Chapter 1. This suggests that all communication reveals the influence of, refers to or takes up what has been said or written before and at the same time anticipates the potential or actual responses of others. Utterances exist 'against a backdrop of other concrete utterances on the same theme, a background made up of contradictory opinions, points of view and value judgements, (Bakhtin, 1981: 281). Most obviously, this means talking to others by explicitly responding to a disciplinary literature, and in some cases with long-dead thinkers (examples from my corpus of 240 research articles in eight disciplines):

(1)
The first is derived from Durkheim's (1938) notion that there is a general . . .

(Soc)

However, both Piagetian and Vygotskian thinking involve constructivist . . .

(AL)

Wittgenstein insists that what is true or false is what people say . . .

(Phil)

Matthei's equations [17, 19] were first used as a starting point in the scale model.

(EE)

But it also involves anticipating other viewpoints and positions, as yet perhaps unstated, talking to readers and drawing on a knowledge of what particular interlocutors might find contentious, unexpected or limited:

(2)
Some readers will want to argue that this is a comparative analysis of neighbourhood associations more than social movements.

(Soc)

These results were surprising in that they showed that a $Ca2+$ influx apparently is not required for a low level of increase in the rate of transcription from flageltar genes following acid shock.

(Bio)

Readers who feel sceptical about this argument should consider its pedigree, or perhaps find better examples of their own.

(AL)

More generally, individuals present and represent themselves by locally managing their use of argument conventions to engage in such dialogues, immersing themselves in the reservoir of meaning options that constitute a distinctive culture. Who we identify with therefore contributes to who we are and says something to others about us, but this reliance on a common, authoritative discourse also points to the fact that a lot of what happens in the field of identity is done by others, not by oneself, as those we interact with can validate who we claim to be, refute it or impose an alternative. In other words, as I shall return to below, power and authority are critical in determining whose definition counts. Socially competent individuals therefore learn to bring their self-presentation into line with the ethos of the group, both to be understood and to be recognised. Proximity, then, involves both identification *with* and *by* others.

Hegemony and ideology

An individual's proximity to his or her discipline is built on language, but our displays of who we are and what we are doing as members of particular groups always involve more than language. Following Foucault (1972), Gee (1999), for instance, talks of Discourse (with a big 'D') to refer to a hegemonic system of thinking and seeing the world specific to particular groups which encompasses an entire interlocking network of practices, structures and ideologies. This is the language we use to enact our identities and get things done in the world:

> Such socially accepted associations among ways of using language, of thinking, valuing, acting, and interacting in the 'right' places and at the 'right' times with the 'right' objects (associations that can be used to identify oneself as a member of a socially meaningful group or 'social network'), I will refer to as 'Discourses' with a capital 'D'.
>
> (Gee, 1999: 17)

This, then, is discourse as form-of-life: the stuff of our everyday world of activities and institutions which is created by our routine uses of language, together with other aspects of social practices. It is through Discourses, for example, that we build meanings for things in the world such as meetings, shopping and drugs, and in academic contexts such as lectures, journal impact factors and research. Discourse is how we foster group mythologies, solidarity and social control; how we distribute prestige to people and value to ideas; and the ways we make connections to the past and to the future. Discourses consist of the basic assumptions, vocabularies and values of group members which form the shared and standardised frameworks for acting in and making sense of social settings and interactions. Displays of proximity thus help to ring-fence disciplines by identifying their users as insiders and excluding others. They create both group and individual identities.

Identities are therefore anchored around sets of norms that regulate values and behaviour, and particularly language-using behaviour, so that identity construction necessarily involves perceptions of 'correctness' and 'appropriateness'. But proximity does not imply regimented conformity or slavish rule observance. Just as gender identities allow degrees of masculinity and femininity, it is also possible to negotiate nearness or distance from a discipline. The neophyte and the Nobel prize-winner have different commitments to the epistemological values and social practices of a discipline and different investments in creating and sustaining a disciplinary identity. Individuals, in other words, can be more or less 'proximate'.

Disciplinary ideologies and collective practices

Proximity underlines the fact that appropriate language choices do not just offer a view of the world but reveal what writers and speakers see as important, how they believe they should select and present material, and what these selections imply about disciplinary values and practices. In other words, specific linguistic realisations represent disciplinary ideologies and help individuals to identify themselves as members.

Rhetorical strategies which highlight a gap in knowledge, present a hypothesis related to this gap, and then detail experiments and findings to test this hypothesis, for example, construct a scientific 'objectivity' and the view that knowledge emerges in a linear step-wise fashion, each new finding building on the last and leading inexorably towards the truth (Prelli, 1989; Rorty, 1979). This appearance of objectivity is further strengthened by suppressing any reference to the fact that experimental observations depend on personal interpretation or that experimenters are committed to specific theoretical positions (e.g. Gilbert and Mulkay, 1984). Reporting excludes the role of the interpreting scientist and the symbolic construction of papers as Introduction – Methods – Results – Discussion scaffolds an ideology of inductive procedure. Truth, in Knorr-Cetina's (1981) words, is 'theory impregnated'. Form is ideological.

This abstract from a letters' journal in physics is unremarkable, but in reporting experimental procedures in a dispassionate, impersonal and expository style, it mimics inductive research practices and so helps to reinforce an empiricist ideology:

(3)
   The influence of the quantized centre-of-mass motion on the dynamics of a one-atom laser with a quantized field is investigated. Both the atom-field coupling and the decay channels depend on the motional state. It is demonstrated that sideband cooling can prevent the atom from heating up while laser action is maintained. Furthermore, the discrete nature of the motion is reflected in multiple vacuum Rabi splitting.

   (Lage, et al., 1990: 6550)

Writing as a physicist means accepting certain scientific precepts, or at least behaving in that way. It means acting as if truth were built on non-contingent pillars of impartial observation, experimental demonstration and replication and as if the text is merely the channel which relays them.

Research in the humanities employs equally formal conventions although they are generally more explicitly interpretive and less abstract, with less 'exact' data collection procedures. Here we find discourses which recast knowledge as sympathetic understanding in readers through an ethical rather than cognitive progression (Dillon, 1991; Hyland, 2004a). Readers expect that authors will display more of a 'reflexive self, unable to write with the classic detachment of positivism' (Starfield and Ravelli, 2006). Cultural studies is just one discipline where we expect to find this kind of reflexivity, where the involvement of the writer is an essential element in the credibility of the account:

(4)

Blurry-eyed, what can and can't we say about Hong Kong now? Well, mostly can't. Not yet anyway. Administratively, our political destiny has been settled. However, the politics of administration has so far been most unsettling. I share Fred Chiu's view that leading up to 30 June 1997, the intense debates over Hong Kong's future that were undertaken in a hothouse atmosphere may have been 'overplayed'.

(Erni, 2001: 390)

Arguments thus involve a more personal stance and exploration of specific cases, but disciplines vary both in the kinds of problems they address and in the forms of argument they adopt. All humanities disciplines are *reiterative*, for example, in that they are obliged to revisit and reinterpret material already studied, but historians and literary critics persuade readers to accept their reinterpretations of the familiar in very different ways. One difference worth mentioning is that literary critics use the *it is* ADJ *that* pattern (e.g. *it is clear that*) to include previously introduced aspects of the writer's argument far more than historians, who are more likely to use it to emphasise the empirical grounding of claims (Groom, 2005). Proximity is also evident in the extent to which knowledge is attributed to individual scholars, schools of thought or conventional wisdom, or whether it is expressed in a non-attributed canonical form. Here economics textbooks are apparently more like those in physics than sociology, with the latter containing far more references to social actors and processes, perhaps because sociology has failed to establish clear agreement among its members about how the world is seen and how research issues are to be tackled (Moore, 2002).

The most obvious expression of proximity to a disciplinary Discourse, however, is an orientation to a topic recognised as current and relevant. Topics, in fact, are more than just a research focus as they represent resources of joint attention which coordinate activities and mark out co-participation in disciplines. Selecting a topic and arguing for its novelty and significance are critical in securing colleagues' attention and in displaying the speaker's right to be taken seriously as a disciplinary member. Whether discussing xylem embolisms, moth-erese or Collective Consequentialism, academics are staking a claim to be a certain kind of person, drawing on the themes of their fields and the vocabularies of their trade in the careful management of identity.

Situating research and constructing novelty

Topics, however, must also be handled appropriately so that local research is situated in the broader concerns of the discipline in order to establish novelty and value. The pursuit of novelty, in fact, is often seen as the driving force of disciplinary activity. As Foucault (1981: 66) observes, 'for there to be a discipline, there must be the possibility of formulating new propositions, ad infinitum', while Kaufer and Geisler (1989: 286) refer to academic communities as 'factories of novelty, encouraging members to plod towards their yearly quota of inspirational leaps'. Creating novelty is, however, also a demonstration of identity as novelty is seen differently across disciplines.

As my corpus data suggest, in the sciences constant innovation and progress are expected, and practitioners look for new results to develop their own research (5), while engineers construct a practical, applied orientation by combining novelty with the utility of their research to the industrial world (6):

(5)
Analyzing a Xenopus liver CDNA library, we identified a CDNA encoding a new HNF4 gene.

(Bio)

In this paper a new coherent transient effect is reported: a two-pulse stimulated echo observed via the angular distribution of anisotropic gamma-radiation from oriented nuclei under axial GE.

(Phy)

(6)
In this work, a novel inductive sensor system has been realized using an arrangement of planar microcoils that allows for a precise measurement of the wheel position, independently of speed and target distance.

(EE)

To handle design problems with conflicting objective functions, a novel procedure has been developed for pruning the design space iteratively that is based on identification of relative significance of design factors on different objectives.

(ME)

To write in the humanities, on the other hand, means situating novelty as having a disciplinary-internal relevance, so it is new only as a contribution to current theories and issues (Hyland, 2004a):

(7)

It offers a new way of theorising ageism itself, as a contingent and negotiated interactional practice.

(AL)

On this new account, we have a pluralist story, not just about morality, but also about benevolence.

(Phil)

Novelty has to be sold to peers as a valid contribution, and this is most obviously achieved by establishing explicit intertextual links to existing knowledge through citation and the marketing of the newsworthy in the structure of research papers. It is apparent, for instance, in abstracts where writers seek to gain readers' attention and selectively highlight what they are likely to find new. Similarly, the introduction assembles claims for novelty by reference to what the discipline knows and believe is worth knowing. This is more carefully elaborated in the literature review, which seeks to justify the value of the current research and persuade the reader that it links into a coherent chain of disciplinary activity. Finally, writers urge the value of their research onto the reader in results sections through a series of rhetorical moves designed to justify the methodology and evaluate their findings (e.g. Ruiying and Allison, 2003).

In short, persuasion can be effective only if the author has correctly identified the opinions and expectations of his or her audience and the culture of his or her community. The semiotic resources which our disciplines put at our disposal are so rich and complex that we have only a limited conscious awareness of them, and we can never second-guess colleagues with absolute precision. Through conversations with peers, attending conferences, reading the research and by generally engaging with the interdiscoursal paraphernalia of the discipline, we discover not only which topics and theorists are currently hot, but also the kind of arguments and warrants that are ideologically appropriate. Proximity to the conventions of a discipline thus allows individuals to engage with peers to construct effective knowledge claims while performing a competent insider identity.

Proximity and identity

Observations on disciplinary differences have been made before, but their links to identity have not been previously established. The idea of proximity, however, not only suggests how individuals display an alignment with the group, making connections to the cultural models which are the discourse of a discipline, but also draws attention to the symbolisation of this discourse. Proximity helps reinterpret this earlier work to indicate that discourse can be seen as both an attitude *toward the another* and *of the other,* building and occupying a body of dispositions in the context of a social group. The idea underpins the constructivist conviction that scientific papers are not persuasive because they communicate independently existing truths about the external world, but because they appeal to community-grounded and subjectively shared beliefs of members. Focusing on the language of the disciplines thus reveals how the epistemological and social assumptions of the author's disciplinary culture helps to form academic identities discursively.

Claiming a disciplinary identity involves collusion, at least to some extent, in community Discourses. This is not to say resistance is impossible or that constraints are immutable. Hegemonic Discourses are blueprints which are always changing and changeable; they are not determining or controlled by any particular individual or group. In fact, they are constantly recreated in every act of communication so that the identity claims that individuals make through a disciplinary performance simultaneously re-make these Discourses. To study the social interactions expressed through academic talk therefore reveals something of the sanctioned social behaviours, beliefs and institutional structures which underpin both disciplinary and individual identity.

2.3 Positioning: Diversity, appropriation and stance

The idea of proximity highlights the importance of disciplinary Discourses in performing identity work by focusing on the shared social representations which provide broad templates for how people see and talk about the world, but it is not the whole story. The stress it places on what is 'shared' among members tends to emphasise a structuralist notion of discipline and a deterministic view of identity, downplaying personal creativity and what the individual brings to the context. While processes of identity formation draw heavily on these disciplinary schema to both shape and enable particular 'speaking positions' and disable others (Baynham, 2006), how individuals actually occupy these positions is more variable.

Academics, like anyone else, pursue individual as well as collective goals and interests. They want to 'be somebody', to make a name for themselves and stand out from the crowd, and while this involves engaging with others in ways they understand and value, it also means staking out a distinctive territory. So while we become who we are only in relation to others, adopting the modes of talk that others routinely use, identity also means assembling the elements of a communicative performance from our backgrounds, which shapes our interpretations of these discourses. Positioning refers to the fact that while we might recognise 'the ways things are done', we also see these as the enabling conditions for individuality.

Positions and positioning

Positioning is a bridge between identities and Discourses, referring to how people locate themselves in discourse when interacting with others. Essentially, I am using it here to refer to how we experience and construct identities by appropriating the discoursal categories of our communities as our own.

The literature on positioning has become extremely diverse, embracing post-structuralist feminism (Norton, 2000), cultural theory (Hall, 1996) and Critical Discourse Analysis (Fairclough, 1992), however, it concerns the fact that an individual emerges in interaction not as a relatively fixed product, but as one continually reconstituted thorough the various discursive practices in which they participate. Davies and Harré (1990) argue that our experiences and sense of self can only be expressed and understood using the categories made available by discourse, which acts as a kind of dynamic conceptual schema through which meanings are actively achieved. Central to acquiring a sense of self and interpreting the world from that perspective is learning the categories which our communities make available (male/female, student/supervisor) and then participating in the practices that allocate meaning to those categories. We position ourselves as belonging to certain categories and in relation to the 'story lines' that are articulated around those categories (student not teacher, good student not plodder).

Thus, disciplinary contingencies guide language use through the positions they make available to speakers and writers, and individuals come to invest in particular positionings so that they act like someone in that position and 'become like that kind of person' (Wortham, 2001: 9). Foucault (1972) sees this as the potential denial of agency, but while actors *are positioned* in terms of what disciplinary Discourses allow, individuals can also *position themselves* in terms of personal stance and interpersonal alignments. Davies and Harré (1990: 46) capture the dynamic like this:

> Once having taken up a particular position as one's own, a person inevitably sees the world from the vantage point of that position and in terms of the particular images, metaphors, storylines and concepts which are made relevant within the particular discursive practice in which they are positioned. At least a possibility of notional choice is inevitably involved because there are many and contradictory discursive practices that each person could engage in.

Identifying ourselves as an undergraduate student, research assistant or lecturer, for example, does not involve stepping into a pre-packaged self. It always entails negotiating overlaps with other simultaneously held aspects of identity (such as being a single, working mother postgraduate student) and making meaning in interaction with the people and the events around us (a celebrated, high-status professor). The ensemble of selves influences how we play any one of them and thus helps shape a particular identity. For Davies and Harré (1990: 47), it is the experience of contradictory positions which provides the dynamic for understanding both ourselves and the world because we can negotiate new positions by posing alternatives. So while we all 'ventriloquate' (Bakhtin, 1981) the voices we encounter in a community to demonstrate our claims to membership, the demands of dominant discourses do not form a closed and determining system. Instead, they can be seen as patterns of options which allow us to actively and publicly accomplish an identity through discourse choices. There is, in other words, always potential for transformation as well as reproduction in academic discourses.

Collective identity and possible selves

Positioning is made possible by proximity and the cognitive routinisation, or habitualisation of behaviour which regular participation in a community enables (e.g. Bourdieu, 1977; Berger and Luckman, 1967). Not only do familiar patterns of discourse, values and other practices get laid down though repeated experiences so that individuals display membership through their proximity to them, but also this routinisation opens up space for deliberation and innovation so that every situation does not need to be reconsidered as novel. Proximity thus creates a secure and relatively enduring context of the 'way things are' and so facilitates positioning and reflexivity, while positioning contributes to the intersubjective agreement which sustains proximity.

Clearly texts can be read as a desire to shape an individual identity which resonates with the writer's perceptions of the collective identity of a discipline. Actors fashion genres using conventional rhetorical forms to claim membership, but these language choices are also ways of making sense of personal experiences and understandings of themselves. We claim affiliation to any number of groups at different times and with varying degrees of commitment, and so who we portray ourselves as being changes with interactants, settings and life-stage. Importantly, however, we also bring residues of these diverse experiences of group memberships to our participation in each of them. Gender, social class, religion, race, ethnicity, age and geographical region are the most obvious of these; but other communities like school, family and the workplace also shape our perceptions and understandings (e.g. Bondi, 2004). Collective identifications are not just masks we put on and take off; they are real for us and have real consequences.

Helpful here is the idea of *possible selves* (Markus and Nurius, 1986), which provides a conceptual link between cognition and motivation by representing individuals' ideas of 'what they might become, what they would like to become, and what they are afraid of becoming' (ibid., 954). An individual is able to create any variety of possible selves, yet the available pool derives from the categories made salient by the individual's particular sociocultural and historical context and from the models and symbols provided by immediate social experiences. Possible selves are therefore personalised, but they are also distinctly social as they result from previous comparisons to salient others and so provide an evaluative and interpretive context for the current view of self. We develop an idea of what is possible, a motivation to *be* and *become* someone.

Every discipline is therefore composed of individuals with diverse experiences, backgrounds, expertise, commitments and influence, all differing in how far they subscribe to its various goals and methods, participate in its activities and identify with its values. Collective identity emphasises similarity but not at the expense of difference. The idea of positioning helps reveal the creative, local nature of identity construction as part of the shaping forces of social processes and ideologies, what I have called proximity. Together they offer a way of seeing how language acts as both a constraint and a resource: how discursive practices represent people in particular ways and, at the same time, how they offer the means for people to negotiate new positions. In speaking or writing, we are not just mimicking others in a similar position but imprinting our own unique take on it as a result of the experiences we bring to it.

Positioning in spoken interaction

Positioning is perhaps most evident in spoken contexts where the formal normative 'defaults' are more easily overridden by individual proclivities and personal style. The constraints of the essay or research paper are relaxed in the lecture hall and seminar room as the immediacy of a co-present audience and the need to impart information in an engaging and effective way gives individuals the licence to take explicitly personal positions towards their material.

This extract from an undergraduate lecture on the Holocaust, for instance, shows the extent of individual positioning while displaying proximity to the conventions of an academic genre:

(8)

> and um, I want to start with, um, with the essay that I recommended you read and and uh which you can, refer to, after the class if you, um if your interest, has been, um peaked. um, the essay, from the, the uh the book *The Imaginary Jew,* in which he examines his, the development of his own, Jewish identity it's a it's an essay that um, I find particularly resonates with my, my own experience and I I find it um, really um, thoughtful and um, and um, and thought provoking. um. . .I I must say that I share myself um his own development from a, um an unexamined, uh pride of um, of being, um, of descending from one from a family that was affected by the Holocaust, to a more, reflexive stance and and um, a more self-conscious, stance on, on uh the place of the Holocaust in my own identity and that's one of the reasons why, um, I told you, that it was with mixed feelings that I, that I had um, revealed to you where, where I come from. but um, Alain Finkielkraut, he's a French philosopher, writes about this, um, this identity. and uh I just wanna quote to you, from page seven in the, in the essay. um he's talking about this this um, new Jewish identity that's uh that's a privileged, identity a privileged identity as, as a victim. um and he says 'I Inherited a suffering to which I had not been subjected. for without having to endure oppression the identity of the victim was mine. I could savor an exceptional destiny, while remaining completely at ease. without exposure to real danger I had heroic stature. to be Jewish was enough to escape, the anonymity of an identity indistinguishable from others, and the dullness of an unevent-uneventful life'. and I think the observation is um is is uh accurate.
>
> (MICASE: LEL542SU096)

Within this enactment of an academic register there is an explicit positioning of the speaker. The numerous framing constructions (*I want to start with, I just wanna quote to you*), organising metadiscourse (*I must say that, he's talking about*), evaluative commentary (*I find it really thoughtful, I think the observation is accurate*) and quotation from a source text are typical of lecture discourse (Mauranen, 2001; Swales and Malczewski, 2001). The extract also contains the colloquial features typical of much impromptu lecturing such as hesitations, false starts, fragments, repetitions, contractions and so on (Biber, 2006). Most striking, however, is the identity work going on here. This is not simply the way the speaker positions himself in relation to the quote, how he agrees with and values the idea it expresses, but how he positions himself as a person: *'the place of the Holocaust in my own identity'*. So while conveying proximity to the conventions of an academic lecture, he takes a very personal position on the issues and moves beyond the role of lecturer to present a personal and intimate characterisation of himself to his students.

Positioning in academic writing

The ability to take a position is also apparent in the collaborative negotiations of written academic arguments. While some observers see disciplinary discourses as agonistic and aggressive (e.g. Belcher, 1997; Salager-Meyer et al. 2003), explicit confrontation and the destruction of an opponent's standpoints tend to actually be quite rare in published texts in most fields. The systematic avoidance of conflict is an important aspect of disciplinary competence, and while fast-moving science fields may be more competitive and potentially more confrontational than the individual philosopher working to his or her own rhythms (Becher and Trowler, 2001), antagonism tends to be kept below the surface. Debate involves negotiating difference in disciplinary-defined ways. Of course, academic reputations are based on saying something new, by staking a claim to a novel idea or seeing things in an original way, and this often involves disputing previously held truths and challenging positions perhaps cherished and nurtured by others. Effective argumentation requires a display of proximity to the community's collaborative practices, so while it often involves genuine disagreement, it also maximises opportunities to find common ground.

This is clear in the fact that it is crucial for writers to establish a common position at the outset, recognising the importance of a topic and the value of exploring it. The starting point of an academic paper is typically the identification of a gap or shortcoming in earlier research (Swales, 1990), but in establishing this gap, writers largely attempt to respond to a general body of more-or-less impersonal literature (9) or particular theories (10) rather than individual authors:

(9)

> Some commentators have, on the basis of the emerging survey data, argued that drug use among young people is no long regarded as a deviant activity by them. They claim that drug use is becoming normalized. We wish to challenge this view and contend that far more has been read into the survey data than is warranted.
>
> (Soc article)

> Some authorities believe the permanent conidiogenous cells in Ascochytato to be annellidic, but this is difficult to interpret in many species, especially with light microscopy.
>
> (Bio article)

(10)

> Critical linguistic studies have tended to be 'fragmentary' and 'exemplificatory'. Analyses have generally focused on isolated segments of texts to demonstrate how linguistic theory and categories provide a method for examining the constitutive and ideological character of texts.
>
> (AL article)

> Detailed two-dimensional (2-D) physical models have been reported (for example [4] [5]) which provide a valuable insight in to the physics behind the operation of HBT's. However, there are a number of issues that still need to be addressed to obtain good quantitative correlation between these physical models and measured data.
>
> (Phy article)

What emerges from academic argument is a reformulation of the original position in a way that brings members onside rather than results in head-to-head confrontation.

Positioning of this kind enables writers to project an identity which is both disciplinary-oriented and personally committed. Within the constraints of this collaborative environment, the use of various evaluative and engagement features enables writers to construct a distinctive stance, or disciplinary performance.

Stance and engagement

In academic writing, positioning is largely accomplished through stance and engagement (Hyland, 2005b). *Stance* refers to the writer's textual 'voice' or community-recognised personality, an attitudinal, writer-oriented function which concerns the ways writers present themselves and convey their judgements, opinions and commitments (Biber, 2006; Conrad and Biber, 2000). This includes the use of hedging and boosting devices to express an epistemic attitude, conveying either tentativeness and possibility or assurance and certainty (Hyland, 1998a), authorial self-mention to give prominence to the role of the author in the text (Hyland, 2001b) and attitude markers, which indicate the writer's affective attitude to propositions, conveying surprise, agreement, importance, frustration and so on. This example, from a philosophy paper in my corpus, illustrates how clear such a stance can be:

(11)

> **No doubt** there are a number of criticisms that adherents to the justice-based paradigm <u>might</u> make of the moral model Dworkin proposes. Still, **I believe that** Dworkin's investment model has **remarkable** resonance and **extraordinary** potential power. **The worry I have** about Dworkin's proposal arises from inside his model. **It is interesting** right off the bat to notice that. . .
>
> (Phil article)

Similarly, writers can position themselves towards other views through their choice of reporting verbs (Hyland, 2004a; Thompson and Ye, 1991) and other references to external sources (Hyland, 2005b). Verb selections, for example, can ascribe an epistemic status to information, evaluating it as true, doubtful or false (Hyland, 2004a) or marking it as 'received knowledge' (Hunston, 1993). Tadros (1993) argues that some choices allow writers to detach themselves from a proposition and so position themselves to deliver their own viewpoint. In this example, the writer frames his argument in relation to a currently accepted view, drawing on a range of intertextual markers in a way which most effectively supports his own argument. His position thus emerges from the contrast he makes with earlier studies:

(12)

> **Her reading** of Twiggy as an oppressive icon **is consistent with** many popular press analyses **that render** the Twiggy phenomenon as a culturally important manifestation of the 'ideology of thinness' that is, in turn, **widely associated with** eating disorders (see Lague 1993; Leland 1996); hence, the 'waif' look exemplified by supermodel Kate Moss and the so-called postwaif look embodied by the newly controversial model Trish Goff are **commonly characterized** as a regressive turn toward to this oppressive ideal (e.g., Goodman 1996). **This revisionist interpretation** of Twiggy **overlooks** historical research **indicating that** the Twiggy phenomenon represented an intersection of class and gender politics.
>
> (Mk article)

Engagement, on the other hand, is more of an alignment function, concerning the ways that writers rhetorically address their readers to actively pull them along with the argument, include them as discourse participants and guide them to interpretations (Hyland, 2001a; Martin and White, 2005). These include the ways authors try to involve readers in the communication process through the use of addressee features such as questions and second-person pronouns (Hyland, 2001a, 2002d). These identify the reader as someone who shares similar understandings to the writer as a member of the same discipline and invite dialogic involvement, encouraging curiosity and bringing interlocutors into an arena where they can be led to the writer's viewpoint:

(13)

In other words, **we** are facing a dilemma: Wilson's natural theory fails to explain many observations, while cultural theories provide mere verbal explanations of them. **Is it, in fact, necessary to choose between nurture and nature?** My contention is that it is not. A very simple example shows that **we** do not need to evoke any moral sense, or cultural conditioning, to explain familiar moral reactions.

(Phil article)

A different and slightly more risky strategy is for writers to position themselves in relation to their readers with directives (Hyland 2002a). Mainly expressed through imperatives and obligation modals, these instruct readers to act or see things in a certain way:

(14)

It is important to recognize the complexity of mycorrhizal systems, and address the appropriate scale when assessing mycorrhizal function.

(Bio article)

It <u>must be</u> understood, however, that there are wide variations in applications that describe themselves as 'interactive multimedia'.

(Comm article)

Together these resources enable writers to adopt positions, both towards their material and towards others who hold positions on the topic. They contribute to the writer's ability to take a stand on something and dialogically refer to the positions of potential readers, signalling competent academic engagement and performing a disciplinary identity.

Disciplinary arguments are therefore both an affiliative resource and a mechanism for professional recognition, a display of proximity to community norms and the performance of individual positioning. Academics use these resources to chart a course between critique and collegiality, minimising personal threat while demonstrating a critical expertise. Identity is a representation of ideas as a cooperative accomplishment, presenting a viewpoint while engaging with others' ideas in agreed-upon ways. Ultimately, disciplines are the contexts in which disagreement can be deliberated and identities constructed.

2.4 Conclusions

In this chapter I have tried to flesh out the idea of discipline and its connections to identity, exploring the view that members' participation in academic discourses contributes to both disciplinary cohesion and individual identity. The difficulties involved in identifying disciplines in terms of purely academic criteria mean that we have to see them as the collective engagement of individuals in certain discourses and practices so that identity and discipline can be understood only by reference to each other. Each is emergent, mutable and interdependent. The key feature of these processes is language, for while we do not simply parrot the words of our disciplines, we index who we are in the ways we selectively draw on and negotiate the linguistic repertoires they make available.

Individual identity is a unique complex of different group and personal experiences. It is locally constructed along two main discoursal paths: through relationships between the self and community, what I refer to as *proximity,* and through relationships between the speaker and the message, or *positioning.* The former concerns the interactive character of the talk represented through the social and discursive conventions of a discipline and the underlying power relations these draw upon; it means using language to adopt a disciplinary voice to establish the writer as someone competent to engage as an insider, but also creates the potential for positioning and agency through broadly agreed meanings and correspondences of expectations. *Positioning,* on the other hand, means adopting a point of view both to the issues discussed in the text and to others who hold points of view on those issues, including actors' individual evaluations and stances towards what they discuss.

Together these terms help to show how disciplines can be seen as cultures and not merely institutional designations. They are enacted and constructed by social actors. At the same time these actors are also enacting and constructing identities for themselves as disciplinary members and individual academics. An orientation to proximity and positioning is therefore an orientation to disciplines and those we encounter within them, a context where individuals shape a writer–reader dialogue which situates both their research and themselves. The academic is not a victim of his or her discipline or a puppet of its discourses, for while these routinely patterned uses of language help make the world predictable and interactions relatively straightforward, they do not eliminate choice. Our particular biographies, with experiences of different relationships, understandings and interactions, and with varied social, educational and professional affiliations, mean that we fill the positions we occupy with unique experiences of other positions and discourses.

The constructs of proximity and positioning also inform methodology and suggest how we might investigate identity empirically. These features can be studied systematically by investigating the everyday discoursal activities of academics, students and teachers.

3

Participation: Community and expertise

Academic publishing presents authors with a number of choices and challenges. Decisions about whether to publish locally or aim for an international journal, to write in their mother tongue or take on the demands of writing in English, to collaborate with others or to work alone, are just a few of these. We have also seen, however, that these are rarely completely free options, as faculty members are largely required, by carrot or stick, to research and publish internationally.

The concept of 'participation' is useful here as a way of understanding a scholar's involvement in community practices and everyday patterns of activity. It is a key metaphor in work on discourse communities and situated learning and can be applied to how individuals go about writing for publication as members of social groups. Clearly *discipline* is one such influential group, shaping how we conceive and construct research for publication, but less abstract and more immediate contexts are also important. Authors have a commitment to the local as well as the global, to their colleagues in the here-and-now as well as to those in cyberspace across the world. This chapter explores the ways researchers develop community-located expertise in academic text production, first examining the notion of *community* and its relevance to the local writing context.

3.1 Global communities, local interactions, and personal positions

The processes of text production, reader commentary, and text revision put text and context into a reciprocal relationship, so that context affects text production and, equally, acts of writing change that context. We participate, and learn to participate, in academic publication by negotiating with others who have similar interests and this, in turn, changes our texts and how we interact with those others. In this way we can see published papers as community artefacts, characterized by the assumptions, preferred argument styles, and epistemological beliefs of these communities. But while academic publications encode the conventions of disciplinary practices and values, they also carry the traces of more local interactions. The meaning of any text is both the outcome of activity within a global community and the product of 'the real-time, contexted discourse of individual conservants' (Nystrand, 1992: 162). In this section I look at the interlayered contexts of participation in academic text production suggested by notions of 'community'.

Global communities and rhetorical practices

The notion of *community* has been widely used to characterize an author's projection of a shared professional context in academic writing. In Chapter 4 I discussed how published texts evoke a social milieu where the writer activates specific recognizable and routine responses to recurring tasks. Texts, in other words, are constructed in terms of how their authors understand reality and these understandings are, in turn, influenced by their membership of social groups which have objectified in language certain ways of experiencing and talking about phenomena. Assumptions about what can be known, how it can be known, and how certainly it can be known all help shape research and writing. But while informing a great deal of discussion of academic text production, the notion of community is by no means a settled and accepted concept.

Bakhtin's (1981) notion of *dialogism* has been influential in discussions of community. This stresses that all communication, whether written or spoken, reveals the influence of, refers to, or takes up, what has been said or written before, while simultaneously anticipating the potential or actual responses of others. All utterances exist 'against a backdrop of other concrete utterances on the same theme, a background made up of contradictory opinions, points of view and value judgements' (Bakhtin, 1981: 281). In writing for publication authors enter into community-located system of beliefs concerning what is interesting or worth discussing and, through their language choices, align themselves with, challenge, or extend what has been already been said on the topic. The notion of community therefore offers a framework within which authors can situate their actions and so position themselves towards others.

Communities based around occupation, recreation, family, and so on, provide meaningful reference points and so help shape collective definitions of identity and belonging within their frameworks of understandings and values. In academic contexts, disciplines qualify as such communities, as participation involves communicating and learning to communicate as an *insider*: it means using language to represent ourselves as legitimate members. But while the notion of community occurs frequently in studies of academic authorship, it lends itself to many different readings. On one hand it is possible to see communities as real, relatively stable groups whose members subscribe, at least to some extent, to a consensus on certain ways of doing things and using language; on the other they can be regarded metaphorically, as collections of practices and attitudes (Kent, 1991). Swales (1990 and 1998), discussing disciplines more specifically, emphasizes their heterogeneous, socio-rhetorical nature, focusing on collectivities which share occupational or recreational goals and interests and which employ particular genres. These genres 'orchestrate verbal life' by linking 'the past and the present, and so balance forces for tradition and innovation' (Swales, 1998: 20).

The most concrete expressions of a community's conventions and conversations are found in its genres. Insiders have implicit knowledge of these genres as a result of their participation in the community, as Berkenkotter and Huckin point out:

> Our knowledge of genres is derived from and embedded in our participation in the communicative activities of daily and professional life. Such genre knowledge is a form of situated cognition (Brown, Colins, and Duguid, 1989) that continues to develop as we participate in the activities of the ambient culture.
>
> (Berkenkotter and Huckin, 1995: 7)

Berkenkotter and Huckin describe this process of learning as a 'cognitive apprenticeship' conducted through participation in the community so that novices gradually become members over time.

While 'discourse community' offers a useful way of exploring the discourse associated with a specific disciplinary group, especially its texts and genres, the idea also has its critics. One problem is that we don't know how to identify such communities or get any sense of where they begin or end so that the idea often reduces to a nebulous assortment of conventions. More seriously, the concept is viewed by some as altogether too structuralist, static, and deterministic, giving too much emphasis to a stable underlying core of shared values which removes discourse from the actual situations where individuals make meanings (see Canagarajah, 2002b; Prior, 1998). Nor is it readily apparent how communities come into being or how they develop and change; how they shed redundant genres and practices to take up new ones; and how they replace established members with young blood. Such criticisms alert us to the dangers of viewing communities as stable, rule-conforming groups which adhere to a collection of values and uphold a consensus concerning ways of doing and communicating.

Local communities and communities of practice

One difficulty with the concept of community has been reconciling what might be seen as the local and the global. That is, how large, amorphous and dispersed groups of like-minded individuals spread across the planet and defined by common bonds of discourse practices and conventions can be squared with groups who typically work together and subscribe to common practices of work and patterns of communication. The ghostly 'invisible colleges' that influence how we encode our specialized knowledge in books, journal articles, and pre-prints certainly influence language choices, but can't explain how we learn these conventions or participate in their maintenance and development. Because of this, Harris (1989) argues that we should disregard such 'discursive utopias' which fail to state either their rules or boundaries and focus instead on specific local groups. Writing can then be seen as a social action in communities constructed by the interactive work of their members.

Discourse communities only have a reality in lived experience, expressed through the local practices which carry traces of their orientations, methodologies, and beliefs (see Porter, 1992). This has led some authors to distinguish local groups of speakers and writers, who habitually interact in laboratories or research units, from global communities 'defined exclusively by a commitment to particular kinds of action and discourse' irrespective of where they work (Killingsworth and Gilbertson, 1992: 162). While 'global communities' may now dominate the lives and activities of Western academics, putting individuals under pressures to publish and attend conferences, the engagement of individuals around some project often fosters particular beliefs, ways of talking, and power relations in local contexts. This shifts the focus from language and social structure to the situated practices of individuals acting in 'affinity spaces' (Gee, 2004) or 'Place Discourse Communities' (Swales, 1998), where people interact in a mutual endeavour and with a sense of their common roles, purposes, and discourses.

These conceptions obviously have similarities with that of 'Community of Practice' (CoP), a group of people who engage in a joint endeavour of some kind (Lave and Wenger, 1991). This idea brings us down from abstract rhetorical conventions and closer to acting individuals, an inner circle of context which refocuses on the immediate relationship of texts and their creation and use. In other words, discourse communities are not *sui generis* realities but part of the social context of text production, reception, and revision. A CoP denotes a group of active practitioners in some project; it is a way of being in the world and a means of participation and learning. Unfortunately, Real groups are less utopian than Lave and Wenger (1991) suggest, with the potential for disharmony, prejudice, fragmentation, and power games (cf Wasko and Faraj, 2000). The idea does, however, serve to emphasize how members have an opportunity to develop themselves personally and professionally through the process of sharing information and experiences within the group.

This idea of learning as situated activity is a powerful one and has clear implications for the everyday activities of writing for publication. The kind of understandings a novice author needs in order to be accepted into a community of scholars is rarely acquired entirely in the classroom but through experience: by engaging in the process of writing itself. Local communities are places where experts can assist learners to bridge the gap between knowing *what* and knowing *how* (Duguid 2005), bringing their tacit knowledge of the publishing process to help novices avoid mistakes and shorten their learning curve. Conversely, they are also sites where experienced authors can acquire social capital themselves, both through sharing their expertise and so constructing themselves as experts (Jacoby and Gonzales, 1991), and in benefitting from the advice of junior researchers when writing in a new topic area (see Myers, 1990).

Personal contexts and positioning

The metaphor of 'community', both as a global structuring frame for the ways academics construct worlds of meanings through discourse or the local work practices of specific groups, is a useful way of conceptualizing participation in academic writing, but does not capture the whole picture. The innermost layer of the onion is the context writers construct for themselves when writing, a personal and perhaps more idiosyncratic context which touches the self more directly. This is a context, as Casanave (1995: 88) observes, comprised of local factors which 'reside both outside the writer (people, settings, assignments) and inside the writer (intentions, intellectual histories, interests)'.

While individuals enact who they are only in their dealings with others, this context reminds us that they all bring different experiences, inclinations, and proclivities to their performances as academics. Being a sociologist or physicist means talking like one, but individuals do this in different ways, as disciplinary membership, and identity itself, is a tension between conformity and individuality, between belonging on one hand and individual recognition on the other. Academic text production involves what I have referred to as *proximity,* the relationship between the self and community, and *positioning*, the relationship between the speaker and what is being said (Hyland, 2012a).

In other words, texts, and academics themselves, are locally constructed through both the social and discursive conventions of a discipline and how these are used to adopt a point of view to both the issues discussed in the text and to others who hold points of view on those issues.

Proximity, and participation in disciplinary authoring practices, does not imply regimented conformity or slavish rule observance. Just as gender identities allow degrees of masculinity and femininity, academics also negotiate nearness or distance from a discipline. Academics fashion texts for publication using conventional rhetorical forms in order to claim membership, but these language choices are also ways of making sense of personal experiences and of themselves. We claim affiliation to any number of groups at different times and with varying degrees of commitment, so that who we portray ourselves as being changes with interactants, settings, and life-stages. Importantly, however, we also bring residues of these diverse memberships to our participation in each of them. Gender, social class, religion, race, ethnicity, age, and geographical region are the most obvious of these, shaping our perceptions and understandings. Collective identifications are not just masks we put on and take off, but are real for us and have real consequences. The neophyte and the Nobel prize winner have different commitments to the values and practices of a discipline, as well as different investments in creating and sustaining a disciplinary identity.

Academics, like anyone else, pursue individual as well as collective goals and interests. They want to 'be somebody', to make a name for themselves and stand out from the crowd, and while this involves engaging with others in ways they understand and value, it also means staking out a distinctive territory. As authors we recognize disciplinary conventions and 'the ways things are done', but we also see these as the enabling conditions for individuality. So while we all 'ventriloquate' (Bakhtin, 1981) the voices we encounter in a community to demonstrate our claims to membership, the demands of dominant discourses do not form a closed and determining system. Instead they can be seen as patterns of options which allow us to actively and publicly accomplish an identity through discourse choices. There is, in other words, always potential for transformation as well as reproduction in the discourses of academic publishing.

So personal factors of various kinds are likely to mediate an individual author's participation in a disciplinary community as well as in more local networks and academic research groups. Perhaps most important among these factors is how far authors need to mobilize such resources to support their participation in academic publishing as a result of either inexperience or their efforts to write in a foreign language. The next two sections explore these issues, beginning with expertise.

3.2 Experts and newcomers

The process of authoring is an important encounter with the regulating mechanisms of a disciplinary community and a crucial site of disciplinary engagement (Hyland, 2004a). Crafting an article, negotiating revisions, and then incorporating the feedback of editors and reviewers is also a process where insiders are able to make use of their experience and cultural knowledge to gain advantages over novices and where novices are socialized into an academic community: it is the route to insider status. Through publishing their work they become familiar with the rhetorical and ideational expectations of their discipline and gradually develop an academic persona. In this section I look at the role of expertise in the ways individuals participate in writing for publication.

Expertise: What is it and does it matter?

Expertise refers to the demonstration of special skills or knowledge which enables the performer to zero in on a problem without wasteful consideration of a large range of alternative solutions. It is 'knowing in action' and involves the exercise of professional judgement and person-sensitive performance based on knowledge and experience (Sarangi and Roberts, 1999). While both general and local knowledge seem necessary to account for writing expertise, the more individuals become familiar with the genres and expectations of their target communities, the greater the accumulated store of experiences they can draw on to meet those expectations. The strategic deployment of discursive resources is, in fact, a key marker of expertise in writing for publication, as it allows experts to orient to specific features of the activity and to tailor both information and interpersonal aspects of their messages to the needs and knowledge of disciplinary gatekeepers.

It certainly seems that experts have the edge over novices in academic text production, with both a greater command of the discipline's rhetorical resources (Hyland, 2011b) and understanding of how a paper might best be steered through the review process (see Myers, 1990). Pagel et al. (2002), for example, found that post-doctoral fellows in a leading medical research university had greater difficulty with writing and publishing than faculty members. Berkenkotter and Huckin (1995) show that one reason for this is because graduate students struggle to imagine an appropriate audience in constructing arguments which readers find persuasive. Several of my novice researcher informants noted that their unfamiliarity with a community's discourse was a significant problem for them:

Structuring is pretty complicated for me. I still spend a lot of time trying to think where I should put information and the structure of each section. Let's say with the introduction, how you organize it in a way that makes sense. They don't want too much information but they want enough and in a way they can use.

(Biology student)

I never really thought much about writing or style as an undergrad – we wrote reports but describing the experiments was important, not how we wrote it. Now I struggle. The right way of putting it is really important. I've had reviewers criticize me for it.

(Engineering student)

A lot of our post docs need a lot of help, a lot of spoon-feeding. Not just what to do in the lab but how to write. They really seem to be at a loss about how to organize the information for journals.

(Chemistry professor)

Coming to terms with appropriate rhetorical conventions is just one obstacle for novices, as they must also become familiar with what Bazerman (1994) refers to as the 'conversations of the discipline' or the issues, problems, and questions that are current in the field. These conversations demand awareness of the main disciplinary paradigms and the community zeitgeist, as writers must address research topics which will interest colleagues and also frame their research claims within these topics. It means knowing which journals are likely to be most sympathetic to the particular methodologies, arguments, and theoretical perspective of their paper and being able to incorporate successfully the feedback of editors and reviewers. All this is challenging for novice writers and responding to editorial changes in particular can be a fraught one for them, as my informants noted:

Getting 'major revisions' is a real problem. I mean there was one case that the reviewer asked me to change the whole introduction and discussion of my paper. They said the results are over-interpreted and so they asked me to change the whole story: to redo the statistics and find a bigger sample. I had no way to improve it because I mean the study was already done. It was wasted time.

(Biology student)

Sometimes you don't know what they want. The reviewers might say that it is interesting and relevant, a useful study, but then want you to change so much it becomes something new. It's frustrating. Like learning a new language.

(Linguistics student)

I think the main thing you learn through publishing is when to argue with reviewers and when to keep quiet and accept what they say. I remember finding it incomprehensible at first that someone could find fault after all that work I put in! I still get bad reviews and rejected but have learnt to roll with the punches.

(Economics professor)

Experience thus counts heavily in navigating peer review. In a detailed study of two established biologists, for example, Myers (1990) traced how they drew on their expert knowledge of the review process and understandings of reviewers' expectations in revising grant proposals in response to reviewers' comments. While both were successful researchers with long publication lists, one of them was seeking entry to a new research area and the rhetorical construction of their texts reflected these different contexts. Thus 'the cautious tone adopted by Bloch, appropriate for his situation as a newcomer, would be disastrous for Crews, who is well established in his specialized field'. As Myers points out, the rhetoric of an appropriate text varies with each discipline and with the writer's relation to the discipline, rather than conformity to any ideal list of stylistic criteria.

More generally, novices are often shocked by the time it takes for papers to pass through review, by the bluntness of some reviewers, and by the need to revisit, often several times, what seemed to be finished work. In sum, expertise not only refers to the expert use of specialist language which defines someone as belonging to a disciplinary elite, but also to the know-how to negotiate the publication process. It establishes the authority of certain individuals and reinforces the legitimacy of certain practices, representing capital in an asymmetrically structured symbolic marketplace (Bourdieu, 1991).

Developing expertise and negotiating identities

Novices gain expertise through engagement with individuals and texts in their communities, and one of the main things that develops is a sense of who they are as academic writers. They construct identities for themselves through their 'negotiated experiences' with others; identities which change over time as they change their trajectories—where they have been and where they are going (Wenger, 1998). The expert–novice metaphor tends to support professional privilege and widen the distance between those who know and those who do not, but novices also play a role in socializing mentors into expert knowers. They contribute to an interactive process in which both novices and competent experts are transformed (see Jacoby and Gonzales, 1991). The local, concrete interactions of everyday engagement in writing and talking about writing for publication thus contribute both to the production of texts and to the growth of academic authors. Most importantly, such interactions shift the identities of novices from student to scholar.

One of the most significant aspects of this development is the ability to balance a personal stance or identity against the demands and expectations of the professional discourse community, as authors relinquish some of their textual ownership to incorporate others' views (see Lee and Norton, 2003). Novice writers often attempt to strive for an original voice in order to be noticed, but this makes them vulnerable to criticism and fails to grasp the essentially social nature of writing for publication. Several of my informants raised this issue:

> I want to put *me* in a paper. To say what I think about what others are saying but you can't always do this. Or at least you can't always do it in the way you want to. It's not a question of just putting down the findings or whatever. You have to do it in a certain way otherwise they won't bother with it. You've got to write it in the right way.
>
> (Sociology student)

> Reviewers seem to be very conservative. While they definitely help improve a paper – I've had a lot of help from them – they make it something different from what I wanted to say. It's not how I wanted to say it.
>
> (Business Post Doctoral respondent)

Experience helps authors come to see that 'originality' is not the expression of an autonomous self but of writing which is embedded in and built on the existing theories, discourses, and topics already legitimated in the community. Kubota, an EAL academic working in the USA, puts it like this:

> In my experience, the more the publishing community recognises the credibility of my work, the more I feel empowered to explore alternative ways of expressing ideas. Thus it is advisable for a writer to follow closely the conventions at least in the initial stages of writing for publication in order to gain the cultural capital that will facilitate her or his initiation into the academic community.
>
> (Kubota, 2003: 65)

The process of reflecting on, becoming aware of, and trying out the persuasive options the discipline makes available is achieved through reading and by following the example of others, although the new identities that novices try on may sometimes feel uncomfortable. Ivanic's (1998) case studies of mature-age students in higher education, for example, suggests how the semiotic resources available in academic conventions can present novice writers with dilemmas about the ways they are presenting themselves. The possibilities they make available for the expression of self may not always mesh with the experiences individuals bring with them to the academy.

> A lot of academic writing is awful. Dense, obscure stuff that you have to fight your way through to find the point. I taught myself to write like that but I've also learnt I don't need to. You get published by reaching them, the reviewers.
>
> (Junior Sociology researcher)

> Science is not really a human thing but I am human. Writing like a science paper doesn't feel like me writing it.
>
> (Biology graduate)

Developing expertise is, at least in part, where novices learn to put the rhetorical conventions of the field to work for them, expressing a voice or identity that suits them. In other words, they come to see that the linguistic resources of their disciplines are 'fundamentally *co-owned*, so that any form of "personal expression" is perforce also a social one' (Atkinson, 2002: 169).

Exercising expertise

The production of texts is always the production of self, but negotiating a representation of self from the standardizing conventions of disciplinary discourses is a skilled accomplishment, involving both recognizing and exploiting community constraints. This is the tension between positioning and proximity mentioned above. While novice authors may be overawed by the cultural authority of convention and the demands of disciplinary gatekeepers, established figures are more likely to have the confidence and expertise to exploit the wriggle room which these conventions afford for individual agency. Reputation is established by marking out and occupying a specialization, but it also involves establishing a rhetorical persona, carving out an academic identity from the rhetorical options a discipline makes available. A comparison of the single-authored output of two celebrated applied linguists, Deborah Cameron and John Swales, suggests how experienced academics can consistently position themselves in relation to their colleagues and their material (Hyland, 2012a).

Deborah Cameron is a sociolinguist known most widely for her work on gender, workplace discourse, and globalization and language, and for her contributions to public debates in both print and broadcast media in the UK.

The most striking feature of Cameron's published work is her willingness to engage in head-on debate with alternative positions, thus projecting a confident, combative identity, while at the same time aligning herself with her disciplinary colleagues. One way she does this is through her significantly above average use of *is*. This is the fifth most frequent keyword in her corpus, meaning that it occurs statistically far more often in her work than in applied linguistic writing more generally. She often uses this to avoid modality and specify an unambiguous relationship between referents:

1 Verbal hygiene is used to affirm a particular view of the ideal social order.
(Cameron, 1994: 385)

The upskilling of talk at work . . . is mainly a means for managers to exert stricter control over the talk produced by service workers.
(Cameron, 2004: 56)

More often, however, we find other collocational patterns with *is* in her writing. *It is* co-occurs most frequently (370 times) with a particularly high use of *it is* + *adj.*+ *to infinitive* (161 times):

2 Yet *it is difficult to* believe the results are advantageous to women.
(Cameron, 1994: 392)

It is important to distinguish between the ideological representations of gender found in texts like conduct books and the actual practice of real historical gendered subjects.
(Cameron, 2006: 18)

Thematic *it* introducing an embedded clause as subject is a highly directive strategy: it asserts the writer's opinion and recruits the reader into it, and Cameron uses the pattern some four times more frequently in her style of argument than in applied linguistics more generally.

Another way which Cameron deploys the linguistic resources of the discipline to construct a distinctive identity is through the use of rebuttal and counter-argument, with *not* (904 times), *but* (572), and *though* (144), all in her top 20 keywords. Once again this is a forceful and dialogistic means of engaging with others' views and Cameron uses it far more than is common in applied linguistics, responding to possible viewpoints through direct challenge, as in this example:

3 The idea that access to higher education should be widened, that degree courses should be for the many and not just the few, has attained the status of received wisdom, and it is hard to dispute it without appearing

snobbish, reactionary or simply out of date. What lies behind it is not, however, a desire to democratise the 'life of the mind', but a set of ideas about the changing nature of work.

(Cameron, 2004: 54)

Here Cameron appears to concur with the policies promoting wider university access but then steps back to question the assumptions which underpin it, presenting her own position that 'knowledge work' is actually a skill acquired for the benefit of employers. The reader is not bludgeoned by her argument but construed as potentially vulnerable to a pervasive ideology, which she then disputes.

John Swales, the doyen of the ESP movement and influential champion of *genre* in English language teaching and research, projects a very different identity to Cameron. By conveying clear personal attitudes and a strong interpersonal connection to his readers, we get an impression of a cautious and inquiring colleague exploring the mysteries of the ways people use language. Frequent use of the first person is perhaps the most striking feature of Swales' discourse, with both *I* and *my* occurring in the top 10 keywords. Self-referential *I, me,* and *my,* in fact, occur 9.1 times per 1000 words in the Swales corpus, almost twice as often as in an applied linguistics reference corpus, imparting a clear authorial presence and a strong sense of personal investment to his writing. An unusual aspect of this is the extent self-mention is used in a self-deprecatory way, breaking with convention to admit his uncertainties and failures:

4 The account presented there was incomplete and somewhat misleading. It failed to do justice to the major struggles most people experience in shifting material from one genre to another. Another failure was to focus exclusively on public genres.

(Swales, 1995: 16)

But I am very unsure whether I will ever use these particular materials again. As matters stand at the moment, these materials have been, I believe, an educational failure.

(Swales, 2001: 162)

Thus while normative and constraining, the rhetorical conventions of our communities are also the raw materials from which academics fashion their professional selves. Through recurring selection from a rhetorical repertoire, they create individual positions towards colleagues and arguments. These two experienced writers project *who they are* to readers in very different ways through the rhetorical choices they make to argue their ideas and engage with their readers. It also raises issues of power and authority, so that we might ask, with Bizzell (1989: 225), 'who gets to learn and use complex kinds of writing' and who has rights to manipulate or resist the conventions of a discipline rather than merely accommodate to them?

3.3 Participation as learning

Participation as a mode of learning can, of course, take different forms, but learning to write for publication necessarily involves writing: it is an active engagement *with* texts rather than learning *about* them. Many novices, in fact, learn to write by modelling their submissions on articles they have read, so that Li (2006), for example, found that her sample of Chinese graduate students often borrowed textual structures from published articles to help them plug the gaps in their knowledge of academic conventions. Casanave and Vandrick (2003): 5) point out, however, that this can mislead junior scholars into thinking that the polished prose they find in journals results from an expertise which is beyond them. But rather than seeing writing as something done by expert 'others', learning is often more productive through interactions with these others. Such interactions may include not only teachers and peers, but also members of tenure review committees, co-authors, and collaborators, and the journal editors, production editors, and reviewers of a journal. In this section I look at some of the main forms that participation takes in this process and also at some of the problems.

Supervisors and mentors

This situated learning through interactions with knowledgeable others is what Lave and Wenger (1991) refer to as *legitimate peripheral participation.* While somewhat vague about what constitutes sufficient 'participation', this draws attention to the fact that novice writers often participate in communities of practitioners in moving towards full competence as publishing authors. The most obvious site of such participation is at graduate school and, while experiences differ considerably, many novice authors benefit from an ability to rehearse authorial practices as students (see Berkenkotter and Huckin, 1995; Prior, 1998).

The guidance of supervisors and senior colleagues is often seen as valuable by junior scholars when learning to write for publication:

No, I don't seek advice about publishing but actually we just discuss with our senior because usually I partner with a professor. So, she has very good experiences in publication and then I usually I suggest sometimes that I think it is appropriate and then she will make the final decision if she supports it or not. Sometimes she does and sometimes she just counter proposes: 'Oh, forget that – try this'.

(Medical student)

This support is particularly valuable for EAL authors. While English was not mentioned as a major obstacle by participants in my interviews, they admitted to often relying on their supervisors to provide them with advice on drafting, using appropriate and accurate English:

I first write myself. Usually I start with the materials and the method so when I wrote myself, I send it to each supervisor and they make the corrections, make some comments and then we have the, you know, the ping pong process. Especially the corrections come for the English because I am French. I am not a native speaker and my supervisor is English. So he helps me very much with the English and also with the scientific comments and then we exchanging all the time until everybody is happy with the version and then we can submit it.

(Engineering student)

Yes, all the time co-author. I don't think I have the level to write it alone. So, my principal supervisor, he is in HK obviously, he's helping me very much and also my co-supervisor, he is in Chile. Because part of my project was done in Chile. So, these two supervisors helped me very much. I usually write first the paper and then they go through with their comments and they change. They help me very much with the English.

(Engineering student)

If some senior students have already gone through this process when they give me the first draft, I may work with them after that probably a few times, and then it may take about 3 to 4 months to submit it to a journal. If the draft is the first time a student has drafted, it may take up to a year. So, that's why, at the beginning, they need a lot of help.

(Biology professor)

Beyond immediate language issues, supervisors saw the most important development as assisting students to present a research story using appropriate conventions of argument. Several experienced writers referred to this development as 'maturity' or 'logical thinking':

They certainly get better in framing what we want to say, what the story is. At the beginning they lack maturity in writing and the biggest challenge for me is to point out why their writing doesn't make sense even if it presents the right information. To show them they need to think in the right way.

(Engineering professor)

The thing I have learnt most about writing for publication is logical thinking and setting things out to make what I say relevant for my colleagues. When you are writing you are looking at your work in a particular way. You try to criticize your work and look at it from other angles before you send it off. You are trying to second guess your reviewers; anticipating what they might say. You never know what's a problem until you organize it, then you can see the logical flaws.

(Economics professor)

For junior scholars in the sciences, interacting with senior colleagues and supervisors as a member of a research team is often central to gaining full participation:

Students need to be trained from the very beginning. It isn't just their writing skills but the content that should be in the manuscript that is generally not good enough. So, the first draft is usually still a long long way to publishable level and that involves a very long process for me to give them feedback on what to include or even the way of presenting the meaning or whatever. Pretty much all the time I have to rewrite the whole manuscript but all the elements start from their thoughts.

(Engineering professor)

Working with mentors not only provides science students with advice on drafting, but also a substantial leg up in publishing itself. Several students commented that these discussions helped them to differentiate between journals to target an appropriate publication, to observe its formatting demands, to address its preferred theoretical and methodological orientations, and to respond to reviewers' comments:

We have co-authored a couple of papers now and I was really pleased to be working with Professor X. She was able to see what journal was most likely to be interested in our work and the angle to take. I just thought you wrote the paper and sent it off but she knew who liked that kind of experimental approach. I had no idea.

(Junior Psychology scholar)

I remember when I started trying to publish. My supervisor really helped, you know. What would make a good paper from my thesis, how I might approach it, where I could send it, and such like. Vital stuff. Of course I knew the main journals, but didn't really have an idea about how they ranked or how to get started.

(Junior Business scholar)

Clearly these quotes do not represent the experiences of all novices, and many graduate students are simply left to learn for themselves in a hit and miss way. Others find participation in the projects of mentors to be discouraging, and perhaps even damaging, to their growth as academic authors.

Power and regulation

It would be wrong to assume that professors are experts engaged in a one-way initiation of passive novices into an academic community. Students usually welcome and trust the supervisor's corrections (see Li and Flowerdew, 2007), but the process of constructing disciplinary texts and an academic identity can be conflictual and communities themselves less than uniformly benign. Participation is a means of acquiring abstract knowledge and skills, but it is also a fuzzy set of relations among real people. A focus on learning through participation must acknowledge the constraints of such a process and recognize that even experts in a discipline do not always find themselves operating in a predictable site of shared values and conventions. Harris in fact reminds us that communities are:

polyglot, a sort of space in which competing beliefs and practices intersect with and confront one another. One does not need consensus to have community. Matters of accidents, necessity, and convenience hold groups together as well.

Harris (1989: 20)

A degree of change and struggle is normal within communities and within the relationships which comprise the contexts in which participation occurs.

Most obviously, supervisors can have a considerable impact on their students' publishing practices as well as on their attitudes to research and their self-conceptions as academics. This influence can be negative as well as positive, and the differences in power and expertise that underpin it can determine the publishing success of junior scholars. Blakeslee suggests that the relationship between supervisor and supervisee can hinder the students' acquisition of publication practices due to:

the location and distribution of authority in practitioner/newcomer relationships, which may inhibit newcomers as they struggle to acquire and establish their own authority by making original contributions to their fields.

(Blakeslee, 1997: 125)

Aguinis et al. (1996) similarly argue that students' perceptions of their advisers' power can lead to students' compliance with the requests of the supervisors. Academic work does not always live up to the cultural stereotype of a rational, asocial, and collaborative enterprise where individuals always act in the interests of science rather than themselves.

Such collaborations are always supported by power differences which can undermine an emerging confidence and academic identity:

> As a junior faculty member I was kind of expected to collaborate with a senior professor and to kind of run-errands for the project. I often wrote a first draft that would be corrected but we didn't really talk about it. He was an eminent professor with too many things to do. No, I didn't really learn much.
>
> (Applied Linguistics junior scholar)

Supervisors often have their own interests which may not coincide with those of their students, a situation which is most apparent where the supervisor is a grant-holder and provides the resources the student needs for their research:

> It is difficult to do independent research as we are part of a team and are funded by his grant. We can only publish what our advisor approves. It is impossible for you to submit a paper to journals on your own.
>
> (Chemistry student)

> My first supervisor was a German guy. He was very intrusive. He was always correcting, correcting and then at the end, even if I was writing, he wrote it. He was really leading me.
>
> (Biology student)

While profiting from a scholarship or stipend from a project grant which enables them to study, science students can find it frustrating that the cost of this benefit is a contribution to their supervisors' research. While students get their names on the list of authors on published papers, supervisors also benefit by gaining tenure, reputation, or more graduate students.

For second language authors the mentor's influence can extend to an appropriation of their texts, so that Li (2006), for example found that the Chinese graduate students in her study tended to blindly conform to the revisions made by their professors, so they became over-dependent on them rather than increasingly autonomous. This can be a critically disheartening experience and discourage students from efforts to improve their academic literacy abilities.

In her study of 11 PhD/post-doctoral Taiwanese science students, for example, Huang found that they were intimidated by the dominance exerted by their advisers in decision-making, so that their autonomy was gradually eroded until they lost ownership of their manuscripts. Without a sense of ownership, they consequently lost the motivation to improve writing skills. Huang summarizes matters like this:

> Because advisers hold the cultural and political privilege to decide the timeline of the students' publications and graduation, and control the channels for interaction with international scholars, students are reluctant to confront them and ask for more responsibility in writing for publication. As a result, students do not feel the need to make additional efforts to master writing skills for publication; they believe their 'bosses' would make every decision anyway.
>
> (Huang, 2010: 40)

Although this finding refers to a small number of participants in a single country, it shows that students' perception of a power imbalance between themselves and their supervisors can be extremely discouraging and undermine both their independence and emerging writing skills.

Networks and brokers

Connections built during PhD studies often continue to offer junior EAL scholars fruitful collaborative opportunities after graduation, but not all novices have mentors or formally structured and sustained institutional support. Instead, they participate less centrally in the organized practices of academic text creation, working alone, or constructing their own support networks. Lillis and Curry (2010) refer to the many and varied individuals who impact on an author's authoring practices as 'literacy brokers': the friends, peers, editors, translators, reviewers, and others not co-authoring the paper, who mediate academic text production. Interactions with others may be personal as much as intellectual and even fleeting and unrepeated. They may relate to something very specific, such as recommending a particular journal or advice on dealing with reviewers' criticisms, or they may involve sustained, and perhaps professional, interventions such as those with translators, proof-readers, and editing companies.

Lillis and Curry (2010) found that such brokers are a key resource for many EAL academics in Europe, and that even experienced researchers used English-language editing services and translators to cut the time involved in authoring texts for publication. Many of my Hong Kong informants confirmed the importance of language brokers to them:

> My friend always checks my drafts. She is a native speaker. Not in my field, but she reads my drafts to check the grammar.
> I know some PhD students here would send their articles to editing companies. They advertise online. Even though the editors sometimes don't understand the science.
> My last paper I paid an English teacher to read it. I know my grammar is not strong in English and I need to publish before I graduate. It is worth it.

Such professional editorial services are still used rarely in Hong Kong and in China more generally (Li and Flowerdew, 2007). Academics rely mainly on peers, although such services may increase if they become more affordable and reputable.

As mentioned above, supervisors are an important source of language support for EAL students, but for junior scholars struggling to publish papers early in their careers, co-authorship with native English speaking academics is a good way of tackling language issues:

> For me, I mean I am not a native writer, so, for me the major problem is really the English. The way to solve the problem is that I seek help from my co-author who is a British. So, yes, usually I come up with a draft of the paper and he will help me to look at it and correct it. The problem of my writing is that I write very long sentences and also the grammar mistakes and sometimes I don't write very scientific sentences. So, yes, it's really the English problem.
>
> (Biology student)

Such brokering not only takes the form of language support, but can also offer junior academics the professional expertise of experienced opinion, and the advice of colleagues and peers can be crucial forms of support here:

> I usually write the draft and then I send it to many people. I mean because we are all specialized in certain area but sometimes the work is not really just in touch with one area. It could be multiple areas and some we are not that familiar with, then we have to ask people's opinion from different expertise. So, usually I give the paper to different people, my colleagues and lab mates and other overseas collaborators and they give me comments and I try to improve it.
>
> (Junior Biology researcher)

Sometimes I have problem with statistics then I ask a particular person who is good at this to judge whether the analysis I did was right or not. Sometimes I just ask for general comments then I send it to people who are good at theories. They would comment on the whole thing. Some people are good at English, I would ask them for comment to improve the writing of the paper.

(Physics junior researcher)

Such brokers offer considerable time and material resources in supporting junior scholars (see Flowerdew, 2001), and developing relationships with a network of willing colleagues can be important to the success of junior scholars. Most, however, are from the Anglophone centre and so brokering can contribute to a system which privileges both English-centre rhetorical practices and the unequal power relations of the West in text production (Lillis and Curry, 2010).

This close connection to centre scholars certainly seems to give EAL doctoral students at universities in the mainstream centres of research more confidence than those in more peripheral countries or institutions (Cho, 2004; Tardy, 2004). For junior academics in non-Anglophone countries collaborating with others is often crucial. Li (2013), for example found her Chinese academics actively sought out international colleagues to work with, avoiding 'big names' and targeting hungry, under-pressure assistant professors like themselves who were keen to publish. My informants in Hong Kong were similarly aware of the importance of establishing research networks:

It's very important in terms of writing and research I mean to build up a network with people and yes the more people you meet, the better you have the more knowledge you have from different areas of expertise. So, it's very important I mean to make collaborations in doing research.

(Sociology PhD student)

I need people to criticize my papers because I really know everything in my manuscript and it's very difficult for me to be critical about it even when there are mistakes that have been there for weeks. I cannot see mistakes because I don't read sentence by sentence but I get the idea and I carry on. When someone reads it and makes a comment it makes me think about things I haven't seen. I really believe that people criticize your manuscript is good even sometimes it's really hard and involves more work but the point is to make something good.

(Physics student)

My Hong Kong scholars made use of conferences and international seminars to network and establish professional contacts, but finding potential collaborators has been simplified for many academics by the proliferation of free scholarly networking sites in recent years. *ResearchGate.net, Academia.edu and Mendeley.com* collectively claim over 20 million followers, offering users a way to organize their research, create personal profiles, and search for people with similar scholarly interests. By serving as research repositories and platforms for connecting researchers, they encourage research sharing and support networks for scholars across the globe (Mangan, 2012).

Editors and reviewers

A crucial aspect of participation in academic text production is negotiating the review process and this is also a key form of learning for novice authors. The painstaking reworking of a reviewed manuscript not only improves the text at hand, but also appears to contribute to the development of writing skills and an expertise in crafting successful papers. In a study of the 'text histories' of 24 papers by a team of Spanish researchers in finance written over five years, for example, Mur Dueñas (2012) discovered that these authors had progressed from having their drafts translated from Spanish into English to eventually writing the first version in English themselves. They had, she observes, 'become (un)consciously familiar with some of the rhetorical conventions of international publications in their field' (p. 145). Similarly, Moreno et al. (2012) found that 1700 Spanish post-doctoral researchers believed that their ability to write academic papers in English improved significantly with increasing experience, irrespective of their command of written English.

My respondents certainly felt they developed as academics from responding to editorial comments in the review process:

> The reviewer's comments are difficult because it takes time to take advice to answer them and go through the paper again but I am happy with that because their criticism and comment are on the point. Sometimes they were a strong one but it helps me a lot to improve my manuscript then at the end I believe I can send them something more strong. The next time I will know too.
>
> (Sociology student)

> My two reviewers were very very rigorous but I would say it's a very good point, so, I don't complain about this. It made me think hard about the argument but I am glad about it. I think it's the right way to do it because if you are not rigorous enough, then, the quality of your paper

will not be good. The reviews taught me to be more rigorous in thinking. So, that is good.

(Assistant professor-Education)

The process of peer review lays bare the highly contingent and tentative epistemological status of scientific knowledge claims, encouraging authors to reflect on their research and how they present it. For many junior scholars this is a constructive part of their acculturation into academic practices:

One thing I've learnt about writing articles is to take the reviewers seriously. You may not agree with them, but they give you another view and maybe you haven't thought things logically enough or said things clearly. You have to work with reviewers because you can learn from them.

(Assistant History professor)

I follow the reviewers' comments every time. I revise it point by point. Each point the reviewer is making. I have to choose to accept or reject and if I really disagree with this point then I have to appeal why. If I think this point is good comment, then, I will try to change it as much as I can. So, that was really point by point and trying to answer all the questions. Yes, it's difficult because it's always difficult but it helped me by writing.

(Engineering student)

Most of the respondents in my sample said they always did their best to respond to negative comments, often seeking help to address style-related comments and issues they were unsure of. Overall, they were trying to learn from each submission process and they felt that negotiations of this kind helped them to not only improve the overall quality and readability of their manuscripts, but also their expertise in constructing successful research papers. This quote from a marine biology scholar shows how one researcher matured as a scientist through this process:

The best advice I got was from reviewers and that was to read more papers. A lot of time the comments were that the scope of my papers is too narrow, that means I don't have a broad knowledge, I don't have a good knowledge in my field. So, the most important thing was to read more. For example, I am working on mating behaviour of marine snails but at the beginning of my study I focused only on the studies done on marine snails mating behaviour, but that's very specific. Later I shifted to read more papers on a broader scale. I looked at the behavioural ecology of marine snails or even behavioural ecology of terrestrial animals as well and also the theories in behavioural ecology. I found that my writing was actually improving because how broad the scope you could write depends on how broad the knowledge you have. So, really that reading is the major part of doing research and I learnt that from reviewers' criticisms.

It seems, then, that not only will authorial persistence eventually lead to successful publication (Flowerdew, 1999; Belcher, 2007), but that repeated passes through the process of submission leads to the improvement of authors and of academics.

3.4 Conclusions: A thoroughly social practice

Participation in academic text production is clearly a situated activity. How individuals pick up and use relevant jargon, take on and gradually manipulate argument forms, observe and imitate behaviour and arcane disciplinary practices, and how they shape a scholarly identity for themselves through publishing, are acts of everyday cognition (Lave and Wenger, 1991; Rogoff, 1990). Brown, Collins, and Duguid, for instance, observe that:

> Students can quickly get an implicit sense of what is suitable diction, what makes a relevant question, what is legitimate or illegitimate behaviour in a particular activity.
>
> (Brown, Collins, and Duguid, 1989: 34)

While expertise in academic text production is a more protracted process than this, both conceptual knowledge and the ability to apply it in relevant contexts is progressively developed through participation in particular communities.

This process of writing and publishing research papers also reveals the regulating mechanisms of a discourse community and the dynamism of its practices, at least partly because the norms of the discipline are constantly defined and changed through the participation of newcomers to the discipline. It also points to the exercise of power in relationships defined in terms of position, expertise and, because authors are typically trying to publish in international journals, of scholars from the Anglophone centre. Participation also shows something of the local, personal, and individual engagement of writers with the writing process and the ways graduate students and novice researchers manage this engagement as everyday activities. Various brokers, but particularly peers and supervisors, play important roles in this process and in the acquisition of disciplinary expertise while novices find their voice and gain the confidence to develop a scholarly identity and a position towards their work and their professional colleagues.

Clearly, then, academic writing is thoroughly social and, like other forms of writing, embedded in specific community and personal relationships and in chains of communication. Texts are jointly constructed by various collaborators, contributors, editors, reviewers, and literacy brokers. The concept of participation is a fruitful way of looking at this behaviour, at the acquisition of a discursive expertise, and at the ways this is shaped by context, but it has little to say about the processes of learning new practices (see Edwards, 2005). We can see, however, that academic knowledge production is mediated by texts, institutions, and people so that published texts cannot be said to emanate from a single author but from a matrix of relationships.

4

Community and individuality: Performing identity in applied linguistics

The relationship between language and identity has long been a major area of sociolinguistic investigation and has become particularly important over the last decade as identity has come to be seen as something that we actively and publicly accomplish in our interactions with each other (e.g., Benwell & Stokoe, 2006). Identity is a person's relationship to his or her social world, a joint, two-way production, and language allows us to create and present a coherent self to others because it ties us into webs of commonsense, interests, and shared meanings. *Who we are* and *who we might be* are built up through participation and linked to situations, to relationships, and to the rhetorical strategies and positions we adopt in engaging with others on a routine basis. This means that it is through our use of community discourses that we claim or resist membership of social groups to define who we are in relation to others. Identity therefore helps characterize both what makes us similar to and different from each other and, for academics, it is how they achieve credibility as insiders and reputations as individuals.

Negotiating a representation of self from the standardizing conventions of disciplinary discourses is clearly a skilled accomplishment for individuals involving both recognizing and exploiting community constraints. However, it is also a challenge for analysts. To take seriously the idea that identity is formed through discourse, we need a means of getting at the ways individuals routinely assemble markers of "who they are" through interaction. In this article I propose a novel method of exploring authorial identity as an aspect of discourse performance using corpus linguistic methods. By interrogating the published works of two leading figures in applied linguistics, John Swales and Deborah Cameron, and comparing these texts with mainstream work in the field, I seek to show how writers manage the tension between shared norms and individual traits. The research suggests how personal proclivities can contribute to an independent creativity shaped by shared practices and advances a methodology for uncovering this.

Identity: Individuality and disciplinarity in writing

Research on academic writing has long stressed the connection between writing and the creation of an author's identity (Hatch, Hill, & Hayes, 1993; Ivanic, 1998). Identity is said to be implicated in the texts we engage in and the linguistic choices we make, thus relocating it from the private to the public sphere and from hidden processes of cognition to its social and dynamic construction in discourse. Issues of agency and conformity, stability and change, remain controversial, however. Some writers question whether there is an absolute, unchanging self lurking behind such discourse and suggest that identity is a *performance* (e.g., Butler, 1990), while others see identity as the product of dominant discourses tied to institutional practices (Foucault, 1972). I want to suggest, however, that regular patterns of language choices help individuals to realize coherent and relatively consistent identities. Almost everything we say or write, in fact, says something about us and the kind of relationship we want to establish with our interactants. Our identities are only successful to the extent that they are recognized by others however, and this means employing, appropriating, and transforming existing discourses (Bakhtin, 1986).

For Bakhtin, all writing is produced in relation to previous texts and as writers draw on these discourses they textually construct social identities in the sense of representing themselves in alignment, or dissonance, with those discourses. In any context, one discourse is likely to be dominant and hence more visible, so that writers often consciously or unconsciously take up the identity options this privileged discourse makes available (Wertsch, 1991). This means that powerful discourses, such as those authorized by academic disciplines, act to restrict the rhetorical resources participants can bring from their past experiences and constrain what they might take from those made available by the context. Such discourses, of course, exhibit a certain stability and power; after all, they are the principal means by which disciplines produce, assess, and authorize knowledge, train neophytes, distinguish members, and legitimate their authority in the world.

Adopting a voice associated with a particular field of study thus involves aligning ourselves with its knowledge-making practices and these tend to exclude the performance of certain identities and favor identities that imply an autonomous, asocial, and impersonal observer (e.g., Ivanic & Simpson, 1992). Foucault (1972) is pessimistic in his emphasis on regulation and denial of agency in this regard, but while actors are positioned in terms of what disciplinary Discourses allow, they also position themselves in terms of personal stance and interpersonal alignments. Essentially, the requirements of academic conventions do not form a closed and determining system but can be seen as a pattern of options that allows writers to actively and publicly accomplish an identity through discourse choices, as Fairclough (1992) observes,

> Subjects are ideologically positioned, but they are also capable of acting creatively to make their own connections between the diverse practices and ideologies to which they are exposed, and to restructure positioning practices and structures. (p. 91)

There is, in other words, always potential for transformation as well as reproduction in academic discourses. Bakhtin (1986), for example, talks of a process of "becoming" as we develop an awareness of our tacit choices and habits of meaning making to gain control over our projections of self in writing. We draw on a repertoire of voices as we write, bringing to the task our own experiences, purposes, and conceptions of self to recombine the options offered by the genre we are writing in to perform a professional identity. Our diverse experiences and memberships, including those of class, ethnicity, and gender, influence how we understand our disciplinary participation and how we interact with our colleagues in the performance of this academic identity. So while the production of texts is always the production of self, individual agency is not eliminated by the cultural authority of convention and the editing of disciplinary gatekeepers. Precisely how we unpick this tangle of influences remains unclear, however, but corpus methods offer a way forward.

Bringing corpora to identity studies

Identity research is an area largely characterized by autobiographical methods, where discourse helps construct an identity through the ways that people explain and understand their lives (e.g., de Fina, Schiffrin, & Bamberg, 2006). But while this approach is profoundly social and emphasizes the continual interpretation and reinterpretation of experience through a cultural lens, narrative can only ever be a partial representation of who we are. It is a one-sided self-construction that underplays the fact that our interpretations must accord with the narratives of others and with the facts of actual events: we cannot claim to be whatever and whoever we want (Lawler, 2008). More directly, narrative is a self-conscious and reflective assembling of experience for the purpose of constructing an identity, usually for a researcher, in a relatively formal and contrived context. Most of the time, however, we are not performing identity work by narrating stories of ourselves but claiming identities while engaged in doing something else. If identity is really a *performance* and not simply an *interpretive recounting* then we need to find ways of capturing what people routinely do with language that is similar or different from what others do with it. Corpus studies help provide us with this.

The value of a corpus approach is that it provides evidence for how language is used by particular authors and so how individuals construct an identity through consistent patterns of rhetorical choices. Corpus analysis is a method based on the idea that the study of a collection of related texts can illuminate our understanding of the contexts and events those texts help create. In other words, it attempts to reveal interaction in a particular domain as a collection of rhetorical choices rather than as specific acts of writing. Its advantage for identity research is that it goes beyond claims made in interviews or decisions made on particular occasions of writing to explore the regularity and repetition of what is socially ratified and independently variant and therefore what represents preferred practices by both individuals and collectivities.

Corpus analysis is, however, a method more commonly used to describe genres, rhetorical practices, and form–function relationships (e.g., Hyland, 2004a) than to explore issues of identity. Increasingly though, corpora are opening new vistas of research in stylistics and authorship studies, where they have helped to discover what is distinctive about a particular author's work (e.g., Holmes, 1994; Semino & Short, 2004). Recently, corpora have also been used in forensic investigations to settle legal cases of disputed authorship (e.g., Olsson, 2004) or for profiling writers according to sets of sociolinguistic attributes such as gender (Koppel, Argamon, & Shimoni, 2002), language background (Vel, Corney, Anderson, & Mohay, 2002), and education level (Juola & Baayen, 2005).

This work recognizes the plausibility of identifying a "stylistic profile" or "linguistic fingerprint" (Hanlein, 1998) from the consistent patterns of choices authors make, so that if choice is constant then it is seen as an individual style marker. In other words, by "dematerializing texts" away from actual concrete instances a corpus approach has the potential to offer linguistic evidence of consistent rhetorical patterning rather than author impressions. Uncovering the regularities in frequency and patterning of words, senses and phraseology can therefore help identify authorial preferences and how writers seek to position themselves with their readers and so project a possible identity. The following sections explore how this works through the published research of two applied linguists: Deborah Cameron and John Swales.

The protagonists

I selected these two academics largely because their highly distinctive rhetorical styles offers a good starting point for this kind of analysis. They are also interesting because of their profiles and the fact they are likely to be known to readers of this journal. Both enjoy considerable disciplinary celebrity and are among the foremost researchers in their fields, holding professorships at leading research universities either side of the Atlantic, Cameron at Oxford and Swales at Michigan, with substantial research and writing careers. Deborah Cameron is a sociolinguist known most widely for her work on gender, globalization and language, and discourse in the workplace while John Swales is the doyen of the ESP movement and the most influential champion of *genre* in English language teaching and research. It should be mentioned that both authors encouraged this project and offered their comments on this article.

These writers are very different from each other in terms of experience and philosophy. Cameron is a committed left winger and active feminist from a working class background, while Swales had a conventional middle class upbringing. Both, however, are mavericks and didn't tread the traditional academic career path. Cameron left school at 17 doing low-paid, drudge jobs before going to university and starting her career relatively late, while Swales was a peripatetic English teacher wandering through Europe and Africa before taking a senior lectureship at Aston and finally settling in the United States. Neither went through conventional PhD training and both came to prominence on the basis of an early, highly influential, publication in a new field: *Feminism and Linguistic Theory* (Cameron, 1992) and *Genre Analysis* (Swales, 1990). Unusually in applied linguistics, both have continued to be more heavily cited for their books rather than articles and both experiment with nonacademic genres: Swales with textbooks and autobiography (e.g., Swales, 2009; Swales & Feak, 2004) and Cameron with polemics and popularizations (e.g., Cameron, 2006a) as a high-profile commentator in both print and broadcast media.

Perhaps most importantly, both are intensely self aware rhetoricians who have published on discourse analysis and grammar, and both are known for their accessibility. Cameron's books *Verbal Hygiene* (1995) and *Good To Talk?* (2000), both dealing in different ways with contemporary normative practices of regulating communication, and her collection of articles *On Language and Sexual Politics* (2006b), are very readable and have enjoyed success across a range of fields. Similarly, the popularity of Swales monographs *Other Floors, Other Voices* (1998) and *Research Genres* (2004) is partly due to their reader friendliness. Swales has edited a journal and both have edited important collections of research and regularly review others' writing. It is this rhetorical reflexivity, and the confidence to deploy it, which was decisive in selecting these writers for analysis. This apparent willingness to stray from disciplinary norms to locally manage a discoursal identity makes them ideal case studies for a fruitful analysis of variation.

Texts and method

The main investigative technique here is comparison. Comparing the features of target writers' texts with a much large reference corpus of work in the same discipline can help to determine what is general in the norms of a community and what represents more personal choices. Stubbs (2005) puts it like this:

> Individual texts can be explained only against a background of what is normal and expected in general language use, and this is precisely the comparative information that quantitative corpus data can provide. An understanding of the background of the usual and everyday—what happens millions of times—is necessary in order to understand the unique.

Extending this principle beyond individual texts, we can see that if a particular word, phrase, or usage is common in a corpus of a particular writer's work, then it might be said to be a consistent preference that reveals something of that individual's routine expression of self: of a relatively unreflective performance of identity.

The corpora used here represent a considerable proportion of each writer's single authored output over their careers (See appendix). My corpus of Cameron's published writing consists of 21 single authored articles made available by the author. It represents some 20 years of publishing and comprises 125,000 words. The Swales corpus was compiled at the Michigan ELI and provided with his approval. It consists of 14 single-authored articles together with the bulk of his three monographs, representing 18 years of output and comprising 342,000 words. These corpora were individually compared with a larger reference corpus representing a spectrum of current published work in applied linguistics and in the same genres as the target texts. It comprises 75 research articles from 20 leading international journals and 25 chapters from 12 books totaling 750,000 words.

I used Wordsmith Tools Version 4 (Scott, 2004) to generate word lists of the most frequent single words, and three- and four-word strings for each of the target authors and for the reference corpus. These strings, which Biber, Johansson, Leech, Conrad, and Finegan (1999) call *lexical bundles* and Scott (2004) *clusters,* are words that follow each other more frequently than expected by chance and so contribute to our sense of distinctiveness in a register. Particular bundles tend to differ according to discipline and genre (e.g., Hyland, 2008a), but it seemed worth exploring ideolect variation.

I then compared each author corpus with the reference corpus using the KeyWords tool. This program identifies words and phrases that occur significantly more frequently in the smaller corpus than the larger one using a log-likelihood statistic. This offers a better characterization of the differences between two corpora than a simple frequency comparison as it identifies items that are "key" differentiators across many files. That is, they are significantly more or less frequent in the author texts rather than simply being the most used. In this way, I could identify which words best distinguish the texts of these authors from those in applied linguistics more generally as represented in the reference corpus. After reviewing the keyword lists and identifying individual words and multiword clusters, I concordanced the more frequent items of group common devices into broad pragmatic categories to capture central aspects of their writing.

Personal interests and professional niches

I first examined the high frequency content words and keywords in these texts to discover how far the respective research fields of these authors threw up a specific lexis. Academics construct whatever status they manage to achieve by advancing knowledge, and in a field marked by considerable competition for space, this is an imperative that requires precise contributions. As a result, and over time, academics carve a niche of expertise from the mass of disciplinary subject matter, creating a specialization that forms the basis of their career and reputation. High frequency items therefore reflect the key themes of an individual's work and serve as motifs for their contribution to the field.

The most frequent content words in Cameron's writing are *women, language, gender, men, social, linguistic, talk, people, discourse,* and *work,* all of which occur more than 200 times and in 90% of the articles in her corpus. These high frequency items clearly identify the terrain marked out and occupied by Cameron as her own. They indicate her concern with the ways language functions to structure social relations in diverse settings, particularly in work contexts and in the ways gender-linked patterns of language use are made significant in social relations. Her studies of gender are acknowledged as pivotal in helping to undermine a binary model of gender to take account of intragender diversity, revealing both the ways gender is enacted locally and the institutional factors that operate to construct inequality. These observations are supported in her preferred multiword clusters, which are *men and women* (76 times), *language and gender* (74), and *women and men* (56).

The top eight content items from the larger Swales corpus are *research, genre(s), English, discourse, language, academic, writing,* and *students.* All these items occur over 500 times and, like *texts, community,* and *rhetorical,* which appear a little further down the list with over 300 occurrences and appear in 90% or more of the articles and chapters that make up the corpus. The most common multiword clusters are *nonnative speakers of English, the concept of discourse community, a genre-based approach,* and *English as a second language.* Again, these are the main areas by which we identify Swales as an individual academic, encompassing his work on genre, community, and his concern for international students using English as a foreign language.

More importantly than raw frequencies in this regard, however, is that of *keywords,* those that are most unusually frequent compared to a larger reference corpus. Table 4.1 summarizes the words and phrases that are far more typical of their work than those in the 725,000 word applied linguistics corpus (at $p < .1$ significance).

Keywords give a reasonably good idea about what a writer's work is about and which best distinguish it within the discourses of the community. The analysis thus returns the nouns and noun phrases which characterize the research interests of these two academics. Some odd forms in the Swales' list like *herbarium, species, specimen,* and *the North University Building* are attributable to his research into the lives and texts of those inhabiting a university building published as *Other Floors, Other Voices.* There are also some unexpected items in Cameron's list. *The gender genie,* a Web site that supposedly predicts the gender of an author of a supplied text, for instance, is critiqued in several articles. *The call centre* also appears in several papers as an example of the technicization of communication: how the commonplace social activity of talk has been transformed into a technical skill and what this means in the production line contexts of service calls. Unlike the data for Swales, the most frequent 4-grams in her corpus simply added another word to the most common 3-grams, suggesting that she uses a more restricted range of common phrases.

TABLE 4.1 Keywords in the Swales and Cameron Corpora

Cameron corpus		Swales corpus		
Singles	3-grams	Singles	3-grams	4-grams
women	language and gender	genre(s)	would seem to	The university of Michigan
gender	Men and women	Dissertation	In terms of	English for Specific Purposes
men	women and men	herbarium	various kinds of	as might be expected
female	top down talk	I	the research world	as far as I
gender is	the female voice	Michigan	the English language	I have tried to
male	in public contexts	have	the fact that	the North University Building
call	the gender genie	species	at this juncture	a genre-based approach
it	female verbal superiority	my	in the herbarium	turns out to be
genie	male female misunderstanding	specimens	The Testing Division	over the last decade
public	the call centre	ELI	the research article	have been able to

Beyond these items, noncontent words and phrases emerge in the keywords lists as consistent individual choices. Rhetorical conventions obviously reflect the epistemological assumptions of a discipline and applied linguistics tends to be seen as a "soft-applied" field: functional, oriented to the improvement of practice, and employing explicitly interpretive, data-informed methods (Hyland, 2004a). But within these broad institutional practices individuals have recourse to different "interpretive repertoires," or ways of constructing their versions of events, and in what follows I illustrate the potential of corpus methods to reveal what these can tell us about the creation of a disciplinary self.

Deborah Cameron—The radical linguist

Deborah Cameron has created, through her writing, a reputation as a radical linguist, challenging orthodox conceptions of workplace and gender discourse. Part of this impact is due to what is the most striking feature of her discourse: her willingness to engage in head-on debate with alternative positions, thus projecting a confident, combative identity. She does this, however, while simultaneously aligning herself with her disciplinary colleagues. In this section of the paper, I explore how she accomplishes this rhetorical identity through the ways she establishes claims, challenges others, and establishes solidarity.

Establishing truths

Classification and identification are commonplace in academic discourse, but in Cameron's writing they take on an assertive and confident quality. Wordsmith identified *is* as the fifth most Keyword in Cameron's corpus, representing a significantly above average use. This is, of course, one of the most common words in English (Sinclair, 1999, p. 176) and in academic prose usually specifies a logical relationship between referents, typically with full noun phrase subjects (Biber et al., 1999, pp. 448-450). In Cameron's work these are *gender* (62 times) and *language* (53), which are variously defined, described, and commented on, as here:

1. The term *gender is used* in this chapter primarily to refer to the social condition of being a man or a woman.

   . . . *gender is regulated* and policed by rather rigid social norms.

   . . . *language is actually* the symbolic arena in which some other ideological contest is being fought out.

More often, however, we find other collocational patterns with *is* in her writing. *It is* co-occurs most frequently (370 times) with a particularly high use of *it is + adj. + to infinitive* (161 times):

2. *It is reasonable to* suppose that a diner wouldn't enquire about the existence of a particular foodstuff out of idle curiosity. . .

It is important to distinguish between the ideological representations of gender found in texts like conduct books and the actual practice of real historical gendered subjects.

It is difficult to think of any human occupation whose performance does not depend on some kind of knowledge.

Thematic *it* introducing an embedded clause as subject helps to shift new or complex information towards the end of a sentence, to the rheme, where it is easier for readers to process. It also, however, functions to assert the writer's opinion and recruit the reader into it. But because it attempts explicitly to take control of readers' thinking, it is a potentially threatening strategy in rhetorical terms and, as a result, carries a high risk of rejection. To pull it off, Cameron has to recognize a diversity of viewpoints and be prepared to engage with these. She therefore creates a sense of solidarity by "writing the reader into the text" through adjective choices which encourage the addressee to share the conviction she has in her views. Essentially, however, she is willing to win them over to her position through the confident, unambiguous expression of her commitments.

This assertiveness in Cameron's authorial positioning is also realized through other uses of the verb *is*. It also, for instance, occurs frequently in the company of *that* (230 times) which is itself among the most highly listed keywords in Cameron's writing. A common use of this collocation in the corpus is to express what Hyland and Tse (2005) have called *"evaluative that,"* a grammatical structure in which a complement clause is embedded in a superordinate clause to project the writer's attitudes or ideas. These examples are typical:

3. *It is my own view that* generalization remains a legitimate goal for social science . . .

   What has not changed is my conviction that theoretical arguments about meaning are not just a side issue in debates on sexism in language.

   In this context *it is problematic that* unmarked or generic occupational terms are also often masculine.

This is a powerful construction for expressing evaluative meanings in academic discourse as it allows the writer to thematize the evaluation, making the attitudinal meaning the starting point of the message and the perspective from which the content of the *that* clause is interpreted. While rarely employing a first person subject, Cameron nevertheless leaves us in no doubt of her attitude in these examples, fronting her statements with a strong personal evaluation.

Challenging contrary positions

Another way in which Cameron deploys the linguistic resources of the discipline to construct a distinctive identity is through the use of rebuttal and counterargument, with *not* (904 times), *but* (572), and *though* (144), all in the top 20 keywords. Once again this is a forceful and dialogistic means of engaging with others' views, but instead of proclaiming a position it disputes alternatives.

Cameron employs negation far more than is common in applied linguistics, responding to possible viewpoints through direct challenge. This is a typical example:

4. The idea that access to higher education should be widened, that degree courses should be for the many and not just the few, has attained the status of received wisdom, and it is hard to dispute it without appearing snobbish, reactionary or simply out of date. What lies behind it is not, *however,* a desire to democratize the "life of the mind," *but* a set of ideas about the changing nature of work.

Negation is thus a resource for introducing an alternative position into the dialogue in order to reject it. Here Cameron appears to concur with the apparently reasonable policies promoting wider university access for nontraditional groups, agreeing with the implied reader that such policies are positive and democratic. She then steps back to question the assumptions that arise from it, presenting her own position that "knowledge work" is better characterized as a skill acquired for the benefit of employers. The reader is not bludgeoned by her argument but construed as potentially vulnerable to a pervasive ideology, which she then disputes. So once again, Cameron shows she is sensitive to the addressee's assumed beliefs and seeks to adjust these with her own decisive views.

This is also evident in cases where she counters a contrary position rather than negates it, mainly using the conjunctions *but, though,* and *however.* Like denials, these are dialogistic in that they acknowledge other voices only to dispute them. Often this is to contest a claim in the prior literature, as in this example where she discusses views on nonsexist language then offers a restrictive modification of this work:

5. Apart from their criticisms of it, Shortland and Fauvel seem curiously undecided as to whether nonsexist language makes any political difference. *But* once again, this entire discussion is locked into a framework dictated by false premises, for within the authors' problematic

the reformist's rationale can only be determinism (change language and you change the world) or else accurately (change language and you reflect reality better).

Alternatively, the view which is countered does not originate in the disciplinary literature, but is regarded as more widespread and projected onto readers themselves. In (6), for example, Cameron raises the widely held view that norms of verbal effectiveness are now seen to be increasingly influenced by female values and practices. Following the countering conjunction, she observes that "communication skills" is a cultural construct, not a natural phenomenon and that it is unwise to routinely attribute certain verbal skills to women while denying them to men:

6. Another argument that has sometimes been made is that the triumph of a "caring and sharing" interactional ethos reflects the growing feminization of British society. Certainly, new-style experts on communication tend to extol the virtues of women, while reserving their sternest warnings for stiff upper-lipped British men. *But* we should not be misled by the fact that therapeutic norms for interaction somewhat resemble the popular "Mars and Venus" stereotype of the way women interact.

Similarly, in (7) when commenting on an advertisement for hospital cleaners, she first voices the "accepted view" only to dispute it, overturning what she projects as normal expectations:

7. The specification just quoted attracted criticism in the mid-1990s as an instance of the "politically correct" impulse to dignify even the most menial positions by describing them in absurdly elevated terms. In my view, *however,* what it really illustrates is a more general discursive and rhetorical shift in the way experts think and talk about all kinds of work.

This kind of concession is a highly productive move in persuasive discourse (e.g., Azar, 1997), but while often labeled "adversative," it is both highly dialogistic, in that it invokes a contrary position, and reader sensitive. Cameron recognizes that persuasion requires the involvement of her readers and so seeks to acknowledge their value positions before leading them to her own. Interestingly, by marking the counter explicitly with "in my view" in (7), for example, Cameron both states her view unequivocally and presents it as just one possible opinion among others; the reader is invited to reserve judgment to follow her critique. I discuss this further below.

Establishing solidarity

By presenting her own position in the context of another, Cameron is not only able to situate her arguments and so better demonstrate their distinctiveness and superiority but also able to claim solidarity with her readers.

Claiming temporary agreement with a thesis before following up with a counter claim is common in the Cameron texts, a sensitivity to addressees' understandings which helps circumvent an early rejection of her argument. She first implies that it is not unreasonable to hold the countered position— after all, anyone might be deceived into doing so—and then adjusts their thinking to her own. This generally involves correcting rather than confronting readers' expectations and is typically prefaced with a stance adverbial, often *arguably* (which is proportionately 30 times more frequent in Cameron's texts):

8. It is true that both are most entrenched in the United States and are therefore easily seen as emanating from it. *But arguably* the diffusion of new norms is less a consequence of American cultural influence per se than a consequence of the spread of the same social conditions which have enabled certain practices to flourish in the United States.

   In most cases the styles of speech women are urged to adopt are presented as gender neutral; they are simply the most effective ways of using language in a particular domain, regardless of the speaker's sex. *Arguably however,* this is only a subtler form of androcentrism. *Undoubtedly,* the call centre industry is a hi-tech service industry which deals in symbols (words and bits); but as I will shortly seek to demonstrate by describing their work regime, the suggestion that operators have to deploy high levels of knowledge or skill in order to perform their functions is extremely misleading.

In other words, while she addresses issues head-on, she takes the trouble to avoid doing the same with her audience.

Forging an alignment with readers is also accomplished in Cameron's writing by the considerable use she makes of conditional arguments, which occur proportionately over 200 times more frequently in her corpus than in the reference corpus. By making one circumstance dependent on another, these raise the uncertainty of outcomes and are therefore often considered to be hedging devices, but they also bring the writer and reader closer to agreement. In Cameron's work, the specification of an "open condition" treats the possibility of the condition being fulfilled as dependent on the reader's agreement, as in these examples:

9. *If* we accept that women and men are internally diverse groups, the fact that some women do one thing while others do the opposite need not be considered a paradox at all.

If the hallmark of a mature academic field is its ability to set its own agenda for research and debate, should we not be addressing the questions we consider interesting rather than spending time debating other people's unquestioned assumptions?

So once again, she acknowledges the multivocal context of her argument, but addresses voices assumed to be shared by both the writer and the addressee.

The way Cameron aligns herself with her readers against an alternative viewpoint is nicely illustrated in the following extract. Here she employs a series of *if* clauses to patiently set out the arguments that support the ideological basis of education for the "knowledge society." Construing the reader as perhaps sharing this apparently reasonable paradigm, she then, using the stance marker *arguably* and the contrasting conjunction *though,* expresses her own view. The final conditional, combined with the writer–reader inclusive *we,* suggests that all readers need to do is consider the nature of "knowledge work" to arrive at the same conclusions she does:

10. There is a sense in which the trend to up skilling actually makes this assertion true. If even quite low-level employees are thought to require formal instruction in such matters as how to talk to customers/clients/patients, if this is considered to be a highly skilled form of behavior which needs to be supported by a body of codified knowledge, and if acquiring the knowledge and skills through training becomes an obligation imposed on the workers by their employer, then these employees do, in a sense, become "knowledge workers." Arguably, though, the sense in which they become knowledge workers is a very trivial and superficial one. And if we actually look at what is involved in many kinds of contemporary service work we will soon have cause to ask whether the rhetorical upskilling of these jobs masks a real deskilling of the workers who do them.

In Cameron's discourse then, we see a range of rhetorical features used to confidently and forcefully advocate particular realities, often arguing for a way of seeing the world in contradiction to others. Her preferred argument strategies actively construct a heteroglossic backdrop for the text by explicitly grounding propositions in her individual subjectivity, recognizing that her view is one among others and taking on alternatives through a combative and confident dialogue. One consequence of this is the emergence of a distinctive identity as a steadfast and committed academic, a disciplinary expert confident in her beliefs and determined in her assurance.

John Swales: The inquiring colleague

John Swales, while enjoying similar academic celebrity, projects a very different identity to Deborah Cameron. Here is an altogether more self-effacing and conciliatory writer, projecting the identity of a cautious and inquiring colleague exploring the mysteries of the ways people use language with the same curiosity and eye for classroom practice that his practitioner readers might. His rhetorical choices impart a clear personal attitude and a strong interpersonal connection to his readers, particularly through the use of self mention, hedges, and engagement.

Self mention and reflection

Frequent use of the first person is perhaps the most striking feature of Swales' discourse, with both *I* and *my* occurring in the top ten keywords. Self-referential *I, me,* and *my,* in fact, occur 9.1 times per 1,000 words in the Swales corpus compared with 5.2 in the applied linguistics reference corpus, imparting a clear authorial presence and a strong sense of personal investment to his writing. As these examples suggest, the reader finds a thoughtful and well-informed colleague in these texts: an impression of a real person thinking through issues:

11. But before *I attempt to* develop my main argument, it may first be helpful to place this aspect of applied linguistic research in a wider context lest *I am thought to* be even more obsessive-compulsive about the importance of genre analysis than is actually the case.

   I have on occasion proposed that students utilize models in their writing. *I have* done so only in those situations where *I feel* that research into the genre has reached a level of credibility to permit some generalization.

Here is a writer making decisions, weighing evidence and drawing conclusions, and engaging the reader in the discussion and investing his argument with personal experience. This self-reflexivity is apparent in this extract, where he comments on his changing teaching practices:

12. *My students* come from every conceivable department, but *I try* to make them a sociorhetorical community, a support group for each other. *I do a lot of* rhetorical consciousness raising and audience

> analysis . . . . *I take them* behind the scenes into the hidden world of recommendations, applications and evaluations . . . In actual fact, *I am much less sure* than I used to be that *I am* a language teacher. *I have* come to believe that *my classes* are, in the end, exercises in academic socialization.

This kind of writing conveys an openness and honesty that reaches out to readers as someone on the same wavelength and familiar with their own contexts and workplace challenges.

An interesting aspect of Swales' identity is the extent to which self-mention is used in a self-deprecatory way. Swales does not duck the fact that research involves uncertainties and failures, perhaps encouraging novice researchers by admitting that even the field's most illustrious figures have their setbacks:

13. Although Huddleston claims that it is comparatively easy to sort examples into the obligation and logical conclusion meanings, *I experienced greater difficulty* and *I have left 10% uncertainly classified.*

    But *I am very unsure* whether I will ever use these particular materials again. As matters stand at the moment, *these materials have been, I believe, an educational failure.*

    Indeed, despite some trying, *I have so far been unable to repeat my earlier success.* Perhaps in the same way that composers only seem able to write one violin concerto, discourse analysts can produce only one successful model.

More generally, a concordance of the first person in Swales' writing shows how far agency is explicitly associated with modality, or at least a deliberative attitude. The most frequent main verbs related to *I* are *think* (86), *believe* (71), *suspect* (35), *hope* (33), *tried* (31), and *guess* (29), all of which point to some degree of tentativeness and care in handling claims and readers. While *I* also appears frequently in Cameron's writing, suggesting that experienced writers may be less cowed by admonishments to author to evacuate their prose, it is the *extent* and the *use* of self-mention in Swales' writing that sets him apart and distinguishes his individual authorial identity.

Conveying, hedging, and attitude

A significant aspect of Swales' personal involvement in his writing is the extent to which he infuses his texts with commentary on the accuracy of claims and his stance toward them.

The use of language to express caution and commitment is a key feature of academic writing as it not only conveys the writer's assessments of reliability but also recognizes the heteroglossic character of statements (Hyland, 2004a; Martin & White, 2005). Swales employs hedges throughout his work, opening a discursive space that invites readers into a dialogue where they can consider and perhaps dispute his interpretations. This is, of course, if they are not beguiled by his candor. As these examples suggest, his arguments often accommodate any expectations that his reader have that their views will be acknowledged in the discourse:

14. The upshot of all these figures *would seem to suggest that* the anglophone grip on published research communications is both strong and tightening.

    I was, *I suspect, rather* too easily seduced by the concept of discourse community. *Perhaps* all too willingly I made common cause with all those who have their own agendas for viewing discourse communities as real, stable groups of consensus holders.

    I would suggest, therefore, that we need more HRD-type training for ESP instructors and practitioners, as an addition to advanced training in Applied Language Studies.

By marking statements as provisional in this way, Swales is able to both express his views and involve readers in their ratification, conveying respect for colleagues and their positions. This is because hedges help present statements as contingent and subjective, a product of the writer's reasoning and therefore open to challenge. But while offering space for dialogic alternatives suggests doubt and expands possibilities for debate, it is also disarming as it addresses views that are potentially in tension with his own. So in making room for alternatives, Swales presents an identity as a reasonable and open-minded seeker of truth, more interested in reaching a plausible interpretation for events than pushing his own.

The intent behind this readiness to concede and negotiate is perhaps demonstrated by a willingness to present claims with unambiguous robustness where necessary. The restrictive adverbs *indeed, doubtless, certainly,* and *especially,* for example, all occur proportionately 10 times as frequently in Swales' writing than in the applied linguistics corpus. Expressions that boost his claims and restrict alternatives are evident at key points of his arguments:

15. However, in other ways it is *definitely* nonstandard.

    Such pressures have *undoubtedly* contributed to the exponential growth of research journals and articles in the last few decades.

The key point I want to make here is that when matters do not go smoothly, we can find opportunities within encounters for conversation management.

But Swales never *demonstrates, proves, or establishes,* and only rarely *finds* or *shows.* Instead, his categorical assertions are more usually accompanied by an evaluative comment of some kind.

The expression of affect is relatively uncommon in academic articles and attitude usually concerns estimations of probability and value rather than ethical evaluations or emotions. Swales' writing, in contrast, is peppered with attitudinal lexis of various kinds, with *scholarly, important, best,* and *interesting* among a keywords list of around 30 items. These are almost always positive attributes that he largely employs to generously evaluate the research of others or underline strongly felt commitments to a particular viewpoint:

16. Certainly, I find it *remarkable* that even as *proficient* a nonnative user as Yao should have introduced such an *unexpected, subtle, and self-evaluative* question about her writing into the discussion.

    However, the *most interesting* feature of the above extract is the way in which the method is described.

    Some shift in the reading research area toward a genre perspective would seem *highly desirable.*

Through these acts of personal involvement and professional investment, we are invited to share his understandings and subscribe to his take on the ways that both people and language behave. By scattering expressions of attitude and mitigation through his texts, Swales creates for himself a distinctive discoursal style that allows him to convey ideas in a very personal way, engaging readers as a collegial guide, sharing their interests and creating a sense of participating in an unfolding exploration of issues.

Engaging with readers

In addition to this extremely personal authorial stance, Swales constructs a collegial identity by taking the trouble to recognize and respond to the potential objections, misunderstandings, and processing difficulties of his readers. As well as softening his arguments, he also draws readers into a collusive web of agreement by assembling a professional context in which they are construed as intelligent colleagues sensible enough to follow what he has to say.

One aspect of this, and extremely unusual in current practice, is a quaint and rather dated reference to *"the reader."* There are 16 mentions of *the reader* in the Swales data compared with just 1 in the reference corpus, and Swales uses it much like the 18th-century novelists to explicitly bring readers into the discourse at certain points, reminding them that they are linked by a common curiosity and engaged in the same fascinating endeavor. These cases are typical:

17. By now *the reader* may have recognized that all our encounters so far lack what Professor Erickson calls "leakage"—the leaking into the functional frame of social and interpersonal elements.

    Now, I can hear *the reader* thinking "Surely we can solve this problem by having the same teacher teach two matched groups of learners using two different methods."

This projects a sympathetic and almost avuncular, tone to the discourse while, at the same time, leading readers to the writer's view by putting thoughts, and even words, into their minds.

A more conventional way of engaging readers is the use of inclusive *we*. While binding the writer to the reader in this way is common in persuasive prose (Hyland, 2005b), it is particularly salient in the Swales' corpus where it is among the top 50 keywords. Unsurprisingly, most of these collocate with primary auxiliaries and modals, but we also see the considerable interactivity of this pronoun in Swales' writing by noting the most frequent main verbs it combines with. Table 4.2 lists these together with their frequencies for up to three words to the right of *we*.

The fact that cognition verbs (*see, find, know*) head the list suggest something of how Swales uses inclusive *we* to recruit the reader into the interpretation process by assigning them a researcher role, guiding them toward a preferred reading of the evidence. Examples, however, show how this shades into explicit positioning of the reader:

TABLE 4.2 Main Verbs (Lemmas) Collocated With "We" in the Swales Corpus

see	201	expect	24	go	14
need	61	note	22	want	12
find	60	use	22	seem	12
know	54	take	16	examine	10
recognize	25	look	15	learn	10

18. In retrospect, *we can see* that the great attractiveness of this approach lay in the fact that it seemed eminently manageable to early LSP practitioners.

I think *we know* in our hearts that the real issues are about how ESP operations are perceived in the wider administrative and operational environment.

Don't *we all* find that our scholarly drafting is slower than *we* had hoped, and don't *we* often feel that other scholars of comparable interests and experience must surely be writing faster than *we* do?

There is an attempt to build a relationship through an implicit claiming of solidarity with readers here, soliciting agreement by dialogue with equals. But there is also a more direct attempt to position readers and lead them along with the argument. The use of obligation modals with *we* signals a more assertive author seeking to focus readers' attention and navigate them through his exposition to a particular understanding:

19. *We can salvage* something of our hopes. First, *we need to* go back and review what we mean by discoursal competence. Here *we need to* recognize both the difference and the relationship between conversation management and oral genre skills.

I now believe that *we should see* our attempts to characterize genres as being essentially a metaphorical endeavor.

More usually, however, he dilutes the imperative force of such *directives* (Hyland, 2002a) by framing them with a modal to mitigate the imposition and transform an instruction into an invitation:

20. *We might conclude,* then, that the role of the subject specialist informant in RA genre analysis remains, given the current levels of evaluated experience, somewhat controversial.

However, *it could be noted that* in the research world there may be more occasions when we have (at least ostensibly) "a distinct communicative situation."

Once again, these linguistic resources allow him to present his arguments with consideration for the reader, while not compromising the strength of his convictions.

Finally, in addition to the devices Swales uses to impart a particular interpersonal tenor, he engages readers through an array of *interactive* metadiscourse options: resources that set out an argument for readers (Hyland, 2005a). There are numerous expressions among the keywords that indicate the attention Swales gives to monitoring his evolving text to make it coherent for readers, and particularly assessing what needs to be made explicit by frequently comparing and summarizing material as he goes along. An interesting, and quirky, variation on this regular gisting of material is his use of introductory prefaces like *it turns out that* (14 occurrences) and *as it happens* (20), which cataphorically alert the reader to findings that might be considered somehow unexpected or counterintuitive:

21. Thus *it turns out that* certain legal, academic and literary texts all point to another kind of contract that can exist between writer and reader.

    The Advanced Learners Dictionary (ALD) lists 21 meanings for *point*, the last of these consisting of a large grab bag of widely different idiomatic uses. *As it happens,* not all of these are represented in the current MICASE database, including the very first use given in the ALD.

These expressions help readers to navigate the discussion, but they do so by lending a strong interpersonal element to it, injecting an attitude of conviviality as Swales shares a certain surprise with readers at the unfailingly interesting nature of rhetorical and human behavior.

Conclusions: Thoughts on the discursive production of identity

In this article I have presented the view that identity is, at least in part, constituted through our consistent language choices. The ways we talk and write are not simply a mimicry of community patterns but complicated means of constructing who we are, or rather, how we would like others to see us. They are the most obvious and unselfconscious displays of our routine engagement with the world, highlighting the ways identity is embedded in interaction and sociality. I have also argued that corpus linguistics is not only a viable means of uncovering such routinized choices but perhaps an indispensible way of operationalizing the concept of *identity formation.*

Corpus analysis can help illuminate the ways individuals construct fairly consistent authorial orientations by using the disciplinary resources available to them. I hope to have shown that while normative and constraining, the rhetorical conventions of our communities are also the raw materials from which we fashion our professional selves, creating, through recurring selection of a rhetorical repertoire, the people we want to be. Clearly this identity work does not preclude other identity choices in the writing of these authors, and on particular occasions they may well adopt different subject positions. But the analyses suggest that these two experienced writers project *who they are* to readers over time in very different ways through the rhetorical choices they make to argue their ideas and engage with their readers.

It has to be said that I have taken a relatively easy route into the corpus analysis of identity by selecting two of the most rhetorically aware individuals writing in applied linguistics today. Both writers are professional discourse analysts and so are highly attentive to the effects of their choices. In his recently published memoir, for example, Swales (2009) observes,

> Tim Johns used to say in our Birmingham days in the 1980s: "A good writer is one who makes a friend of his or her reader" and that, as much as anything, that is what I am still trying to do. (p. 206)

Similarly, Cameron's own response to this article underlines the importance she attaches to the "aesthetics" of language and to "avoiding verbal clutter." She points out,

> I would say I am a pretty deliberate and self-aware writer of prose, I think about what I'm doing and am conscious of at least some of the recurrent features that make my style what it is. I'd rather come across as crude or even arrogant than leave the reader struggling to parse my sentences, or wondering at the end of them "what the hell is she actually saying?"

Interestingly, these orientations do not stand outside broad social models but draw on recognizable cultural traditions. Cameron's energetically and intelligently combative style, which explicitly pits her ideas against others, seems to be informed by British traditions of open debate. The fact that she positions herself as a very public intellectual, at home in the media and in popular genres, and with a variety of high profile issues, brings a wider significance to her writing. Swales' style, on the other hand, seems to represent a different kind of intellectual in public discourse—quieter, more urbane, and gently self-deprecating. This is not to say of course that they have not given these styles an individual stamp influenced by their own backgrounds and experiences, or perhaps added innovations to the repertoire, but simply that the identity options provided by academic disciplines do not exist in isolation from wider social and cultural practices.

Like all corpus work, the method is informed by numbers, largely frequency counts of keywords and collocations but constructed on interpretation. While these repeated uses represent each writer's more or less conscious choices to project themselves and their work in particular ways, my take on them is necessarily subjective. I have, however, tried to work as closely as possible with the data and this clearly points to consistently distinctive rhetorical choices within broad disciplinary boundaries. The analysis suggests that the performance of an identity is always shaped by our goals and by the demands of the context as we walk a tightrope between projecting an individual persona and taking on social roles and qualities valued by community members. As Gee (1999, p. 23) points out, discourses are "ways of being in the world," so that language choices are always made from available resources and involve interactions between the conventions of the literacy event and the values and prior cultural experiences of the participants.

In summary, this article has sought to reveal something of how authorial positioning is consistently accomplished through repeated rhetorical acts. At the same time, I have also tried to make a small contribution to the methodologies we might use to explore identity construction and to shed some light on the social processes at work in academic discourse communities. It might be argued, however, that this kind of discourse analysis fails to provide sufficient context to understand identity performance. After all, I've conducted no interviews, explored no detailed biographies, and unpacked no narratives. Instead I have looked at texts and what, over and over again, is on the page. What this shows, above all else perhaps, is that academic communities are human institutions where actions and understandings are influenced by the personal and biographical as well as the institutional and sociocultural. They are sites where differences in worldview or language usage intersect as a result of the myriad backgrounds and overlapping memberships of participants.

This methodology therefore points to new ways of understanding and exploring identity that takes us beyond what individuals say about themselves to what they do in interaction on repeated occasions and how they build a consistent persona through discourse. In this view, identity can only be understood through close analysis of the ways writers routinely draw on the rhetorical repertoires of their communities to position themselves in recognizable ways as both individuals and as members of collectivities. It is a methodology, moreover, that offers a way of exploring other unanswered questions about disciplinary constraints. Do all academic writers have a relatively consistent stylistic "signature," for example, or is this something that only develops over time? Are novice writers more tightly constrained by conventions? What changes in their repertoire with greater experience and confidence? What variations exist across disciplines and between individuals in other fields? Not least it makes sense to address the wider political operation of discourse communities and to ask, with Bizzell (1989), "who gets to learn and use complex kinds of writing" (p. 225) and who has rights to manipulate or resist the conventions of a discipline rather than merely accommodate to them?

Appendix: Corpus texts

Deborah Cameron corpus

Cameron, D. (1984). Sexism and semantics. *Radical Philosophy*, *36*, 14-16.

Cameron, D. (1992). Naming of parts gender, culture and terms for the penis among American college students. *American Speech*, *67*, 364-379.

Cameron, D. (1994). Verbal hygiene for women: Linguistics misapplied? *Applied Linguistics*, *15*, 382-398.

Cameron, D. (1997). Performing gender identity: Young men's talk and the construction of heterosexual masculinity. In S. Johnson & U. Meinhof (Eds.), *Language and masculinity* (pp. 86-107). Oxford, UK: Blackwell.

Cameron, D. (1998). Is there any ketchup, Vera? Gender, power and pragmatics. *Discourse & Society*, *9*, 437-455.

Cameron, D. (2000). Styling the worker. *Journal of Sociolinguistics*, *4*, 323-347.

Cameron, D. (2003). Gender as an issue in language change. *Annual Review of Applied Linguistics*, *23*, 187-201

Cameron, D. (2005). Relativity and its discontents: Language, gender and pragmatics. *Intercultural Pragmatics*, *2*, 321-334.

Cameron, D. (2006a). Unanswered questions and unquestioned assumptions in the study of language and gender. Female verbal superiority. *Gender and Language*, *1*(1), 15-25.

Cameron, D. (2006b). Theorizing the female voice in public contexts. In J. Baxter (Ed.), *Speaking out: The female voice in public contexts* (pp 3-20). NewYork: Palgrave.

Cameron, D. (2006c). Language and gender. In B. Aarts & A. MacMahon (Eds.), *Handbook of English linguistics*. Oxford, UK: Blackwell.

Cameron, D. (2007a). Dreaming of Genie. In S. Johnson & S. A. Ensslin (Eds.), *Language and the media* (pp. 234-249). London: Continuum.

Cameron, D. (2007b). Ideology and language. *Journal of Political Ideologies*, *11*, 141-152.

Cameron, D. (2007c). Redefining rudeness. In M. Gorji (Ed.), *Rude Britannia* (pp. 61-73). London: Routledge.

Cameron, D. (2007d). Language endangerment and verbal hygiene. In A. Duchene & M. Heller (Eds.), *Discourses of endangerment* (pp. 268-285). London: Continuum.

John Swales corpus

Swales, J. (1993). More discourse analysis in communicative language teaching: more felicity and less breakdown? Proceedings of the XIIIth MATE Annual Conference, Ouarzazate, March 29th-April 2nd, 1993. (pp 35-46). Proceedings of the MATE Conference, Morocco.

Swales, J. (1994). ESP in and for human resource development. *ESP Malaysia*, *2*, 1-18.

Swales, J. (1995). Field guides in strange tongues: A workshop for Henry Widdowson. In G. Cook & B. Seidelhofer (Eds.), *Principles and practice in the study of language: Studies in honour of H. G. Widdowson* (pp. 215-228). Oxford, UK: Oxford University Press.

Swales, J. (1990). *Genre analysis: English in academic and research settings.* Cambridge, UK: cambridge University Press.

Swales, J. (1995). The role of the textbook in EAP writing research. *English for Specific Purposes, 14,* 3-18.

Swales, J. (1996a). Occluded genres in the academy: The case of the submission letter. In E. Ventola & A. Mauranen (Eds.), *Academic writing: Intercultural and textual issues* (pp. 45-58). Amsterdam: Benjamins.

Swales, J. (1996b). Teaching the conference abstract. In E. Ventola & A. Mauranen (Eds.), *Academic writing today and tomorrow* (pp. 45-59). Helsinki, Finland: University of Helsinki Press.

Swales, J. (1998a). Language, Science and Scholarship. *Asian Journal of English Language Teaching, 8,* 1-18.

Swales, J. (1998b). *Other floors, other voices: A textography of a small university building.* Mahwah, NJ: Lawrence Erlbaum.

Swales, J. (2000). Languages for specific purposes. *Annual Review of Applied Linguistics, 20,* 59-76.

Swales, J. (2001a). Integrated and fragmented worlds: EAP materials and corpus linguistics. In J. Flowerdew (Ed.), *Academic discourse.* (pp.153-167). London: Longman.

Swales, J. (2001b). Metatalk in American academic talk: The cases of "point" and "thing." *Journal of English Linguistics, 29,* 34-54.

Swales, J. (2002a). On models in applied discourse analysis. In C. C. Candlin (Ed.), *Research on discourse and the professions.* (pp. 61-77). Hong Kong, People's Republic of China: City University Press.

Swales, J. (2002b). Issues of genre: Purposes, parodies, pedagogies. In A. Moreno & V. Colwell (Eds.), *Recent perspectives on discourse* (pp. 11-26). Leon, Spain: University of Leon Press.

Swales, J. (2004). Evolution in the discourse of art criticism: The case of Thomas Eakins. In H. Backlund, U. Melander Marttala, U. Börestam, & H. Näslund (Eds.), *Text i arbete/Text at work* (pp. 358-365). Uppsala, Sweden: Institute for Nordic Languages.

Swales, J. (2004). *Research genres: Explorations and applications.* Cambridge, UK: Cambridge University Press.

Swales, J. (2004). Then and now: A reconsideration of the first corpus of scientific English. *Iberica, 8,* 5-22.

Commentary on Part I

Chuck Bazerman

Ken Hyland, steeped in the methods of applied linguists, has recognized that writing is a broad and complex activity that encompasses many dimensions beyond language, but which are brought together and crystallized in the language of the text. Written language is what writers produce and what readers read. It is what goes between reader and writer and is the concrete symbolic manifestation, interpersonal action and material residue of the communicative process. The language provides the tools and external signs for writers to realize, shape and transmit meaning by serving as the stimulus for the readers' meaning-making. Yet the meaning and intent are formed in the minds of readers and writers, as are the perception of the situation (including power, economic, social and material relations) and relevant knowledge brought to bear on interpretation, including histories and intertexts evoked but not mentioned in the texts. Additionally, the world represented is known through material interaction and experience even though the processes of representation foster certain appearances and memories. Finally the social relations enacted through the texts exist in complexes that are not visible in single texts and cannot be fully understood even by full collections of related texts without evidence about the social relations in the world beyond the texts.

Hyland, along with other applied linguistic researchers engaged in language education, particularly written language education, exposes the nuances of the linguistic medium, so that the language learners can increase their interpretive, communicative and expressive potential. Hyland has made it his business to understand through many corpus studies what the resources of language are and how they are differentially deployed in different academic and scientific domains. As indicated in the chapters in the later sections of this book, his accomplishments in exposing the variations of different dimensions of linguistic practices in different academic domains through corpora studies are exemplary and set him apart from other applied linguists in their comprehensiveness, nuance and subtlety of understanding, particularly in the elusive areas of identity, affiliation and positioning. Nonetheless, Hyland understands that ultimately the choices of language, selection of what to represent in the text and the construction of the texts depend on understanding much beyond the subtleties of language and require methods of research that go beyond the examination of texts as autonomous objects, even if examined in large collections or corpora. In order to put his corpus linguistic studies in the context of cultures, social organization, institutions, roles, statuses, processes and activities, Hyland draws on readings from a number of the social sciences which explicate these social issues within the academic and research worlds.

Ultimately, however, the study of writing cannot just rely secondhand on the work of other social sciences, which may not be attuned to the important mediating role of language and language production. The study of writing needs to engage in social research specifically in relation to language issues but that goes beyond the analysis of the linguistic realization. To initiate that investigation, Hyland has added interviewing to his methodological repertoire. His use of interviews goes beyond the common applied linguistic method of gaining expert confirmation or explication of practices and resources. Hyland's extended use of interviewing is an important first step further getting into the mind and goals of the writer, but is only a first step into a much more thorough examination of these social issues, drawing on the methods of the other social sciences. This use of interviews, in particular, takes him a step closer to rhetorical analysis, which is concerned with the purposive reasoning and goals of the individual language producer. But the traditional rhetorical analysis is also limited as it considers social, political and material situation and consequences projected from the author's point of view or interpreted from produced text. Context is also constructed within specifics of situations, rarely considering larger structural or structurational dynamics, nor the actual uptake, unfolding or consequences of rhetorical events in sequence.

Hyland also has focused his studies on an important, but limited, domain of writing – namely academic writing at the turn of the twenty-first century, particularly as expressed in the published scientific and scholarly literature, and by extension student work aimed at building competence in contributing to such publications. His investigations, as well, are limited to English language writing, bolstered by the recognition of the spread of English as the global language of science and other scholarly specialties. His studies further are largely attentive to the situation of non-native speakers of English needing to publish within the English language scholarly system. This focus is well justified as it serves the urgent priorities of many scholars seeking publication, students preparing for academic careers and universities seeking global research recognition and excellence. More specifically many of his studies make explicit nuance and tools of language that would not be evident or subtly used by writers who are not native speakers of English. Explication of these tools can aid in language instruction and support.

Hyland's focuses mean that scholarly publications in Chinese, Spanish, French, Portuguese, Farsi or other languages with substantial academic literatures escape his scrutiny, as does the work of students in universities where these languages are the dominant medium of instruction and evaluation. Since he is particularly focused on language features and not underlying cognitive or social processes, the consequences of his monolingual orientation is that the findings may be language-bound. Even if the educational target is academic publication in English, to support English language instruction it would be useful to understand how such instruction intersects with the first language educational writing processes, practices, textual forms and goals experienced by these non-native English students. Even the struggles of native English students with academic writing are only at the periphery of his vision. Assignments and uses of writing that are not directly framed within publication standards and purposes, but serve other educational functions, are not part of his inquiry, even if they are in English. Accordingly all the writing texts, practices and development in primary and secondary education are also beyond the scope of his inquiry. University and academic writing practices and forms, as well, change over time, and thus examining the particulars and dynamics of change would be of value. Writing practices in careers, businesses, institutions and other sites are also worth understanding both because they may engage with scientific or academic knowledge and because they may be important in students' lives once they leave the university.

Hyland's focus is a feature of his studies, not a fault. Other applied linguists have indeed studied some of the issues mentioned in the previous paragraph, and I believe Hyland himself is looking into histories of change. No one can study everything all at once, and he has chosen an important, consequential and practically useful domain to which to apply his linguistics. However, expanding his discoveries, realizations and theories to a broader understanding of writing requires caution. From my own work I have found the study of scientific and academic writing generative of much understanding of writing that is potentially more broadly applicable to writing, but the danger of overgeneralization is ever present. Making extensions of conclusions based on academic writing needs to examine and compare the nature of domains, their activities, their timing and locations, and the differing participants and actors with other commitments, interests, knowledge, social positions and social engagements.

In the chapters in this section, Hyland presents his broader view of writing and offers samples of the evidence corpus linguistics is able to contribute, carefully limiting his domain to the research-oriented university. In Chapter 1, he lays out the case why writing is central within the work of the university and, in turn, is sensitive to many key dimensions of academic life. He further argues that academic writing is of central concern for the societies within which the universities reside in light of the democratization of university entrance and the role of academic publication in knowledge societies. He notes that research-oriented higher education writing presents challenging novelty to students despite prior education, embedded as the writing is within the varieties of disciplinary and national research cultures. He illustrates these themes through corpus linguistic evidence of some textual features indicating certainty and assertiveness of claims as well as of lexical variation. These indications that textual variation does exist begin to suggest how writing is produced within and for local social and institutional contexts, and is indeed part of the practices that are carried out within, maintain, and even construct, those contexts. Understanding these processes more fully can contribute to education and aiding researchers to build careers and reputations as researchers, particularly in the world of English language research publication, as Hyland discusses. Understanding these processes also has broader social benefits in improving communicative practices and creating more reliable knowledge. Identifying lexical variation is a first step indicating social complexity, but a fuller inquiry requires other kinds of data that go beyond the text.

In Chapter 2, Hyland considers some of the textual means by which the author can adjust relationships to readers and identity within a discipline. These include drawing on disciplinary literatures, using disciplinary ways of using language, positioning the claims within the ideologies of the field and asserting the novelty of claims. He offers examples of how various research authors display and manage textual resources to carry out these functions. Such enumeration of issues and exemplification can guide writers to attend to relationships and identity with a field and can direct them to the linguistic resources that will aid their representation. Deploying textual resources can help researchers make the most of what they have, but first they must know what they have to work with and the playing field they are working on. They must also know how to create the knowledge goods that will make a difference in the market, before they craft a final representation in the article. Elucidating such issues, as Hyland knows, requires other kinds of inquiries and data. I should note that Hyland does preface his discussion of textual resources with a discussion over the nature of disciplinarity, coming up with uncertainty about the definition of disciplines. Inquiring, however, into disciplines as varieties of social configurations having at times institutional status and certain forms of institutionalized organization often conditioning publication outlets, rather than seeking a fixed definition of disciplines, may allow a deeper understanding of the differing sites which publishing researchers contribute to and act within.

In Chapter 3, Hyland does go beyond the examination of textual corpora in order to understand participation, a concept that can be viewed from both psychological and sociological perspectives. Here he uses interviews with novices and experts to see how they perceive writing and reviewing situations, the roles of power and brokers, and how they construct relations with others. The inquiry includes how the interviewees see learning as related to participation and mentoring relations they have experienced. Their perceptions of such thoroughly social practices are important because these perceptions guide engagement, learning, knowledge and action choices. As Hyland notes in his conclusions, the response to situation is deeper and broader than adopting the linguistic coloration of a chosen specialty. He further notes that students readily recognize and adopt recognizable discursive forms, but effective expert writing in a discipline requires extensive processes of engagement in the knowledge, social relations, discussions and inquiries of a field, shaping the meaning and substance of the representational choices. The actors' perceptions revealed in interviews provide important phenomenological windows onto the communicative choices and practices of the disciplines as well as how they are learned; other windows can give us further insight into the unfolding of communal discussions, processes and patterns that emerge across multiple participants, and the differential consequences of different practices, as well as the relation of these practices to the nature of the knowledge produced, evaluated and disseminated through these practices.

In Chapter 4, Hyland moves beyond generalized participation within disciplinary spaces to consider the formation of distinct individualized personalities. These personalities embody stances towards academic tasks, a field's knowledge and ideas, relations with one's colleagues and assertiveness of one's own claims. Using a corpus formed from the publications of two prominent applied linguists of widely recognized distinctive personalities, Deborah Cameron and John Swales, Hyland examines how the authors manage resources to represent coherent and consistent individualized personalities within disciplinary boundaries. There is, however, no separate analysis to confirm that patterns found in early publication match later; the assumption of continuity is based only on the aggregate weight of the evidence. Hyland does, nonetheless, find distinctive, patterned differences between the two and which vary from disciplinary norms in particular word choice and word strings that contribute to authorial stances. Cameron's challenging, confident stance is managed through the use of *is*, boosting and evaluative infinitives, negations and adversative and concessive conjunctions among other features. Swales projects a reflexive modesty through use of the first person (*I* limiting assertions to personal views and the inclusive *we* showing affiliation with readers), statements of uncertainty, hedges, boosters used for evaluative rather than assertive purposes and praise of others' work.

While the positions and stances one can credibly take and maintain depend on such things as the substance of the knowledge and concepts of the field, the nature of each individual's inquiry, the relation to the group inquiry, one's history of contributions and institutional location, and the concrete contributions one has to offer, these linguistic tools of personality management can present an authorial personality consistent with these underlying factors to establish professional presence. While Hyland documents that the largest number of distinctive words used by each of the subjects refers to the terrain they study and make claims about, he does not analyse them. Such an analysis would require going beyond the immediate texts in the corpus to examine the conceptual, epistemic and social structure of the field as well as of the individuals relation to them, which would require different kinds of data as well as analytic tools.

Hyland opens a door for applied linguists to look beyond the texts, to see that competence in producing effective texts requires knowing more than language resources in order to make effective textual choices within specific situations. As he recognizes, walking through this door requires both awareness of literatures of the fields that have explored the domains where communication occurs and investigative tools to understand these situations more fully. In so doing he disrupts in substantial and consequential ways the *langue/parole* distinction that facilitated the study of language as an autonomous stable object. Hyland's vision recognizes that *langue* is only a codification of what is deployed within *parole* and comprehensive knowledge of *langue* alone does not make effective users of language. Language education to make effective language users must go beyond *langue*, and not just peer out from world of *langue*. Language education needs to gather evidence of that great world of language use beyond and develop systematic methods of inquiry and analysis.

PART TWO

Interaction, stance and metadiscourse

Introduction

This section focuses on the area which informs almost all my research and which represents what may be my main contribution to the field of academic writing, that of interaction. Interestingly, in Western cultures at least, academic writing is often held to represent a privileged form of argument precisely for its objectivity and absence of interaction, offering a model of rationality and detached reasoning in stark contrast to the partisan discourses of commerce and politics. The natural sciences, of course, imply all that is most objective and empirically verifiable about academic knowledge. But scientific methods offer probabilities rather than proof, and interpretation always depends on the assumptions scientists bring to the problem. All reporting occurs in a context and in relation to a theory which fits observation and data into patterns meaningful to a community, and the fact that there is always at least one possible interpretation for research findings shifts attention to the ways that academics argue their claims. Readers always have the option of refuting interpretations, which means that at the heart of academic persuasion is writers' attempts to anticipate possible negative reactions to their statements. We have to look for proof in the textual practices for producing agreement and in the language used to acknowledge, construct and negotiate social relations.

These chapters illustrate the importance of interaction in academic writing and focus on the approaches I have developed to understand it: metadiscourse and stance. I don't claim originality for these ideas or models and in fact they exemplify my own interactions with the work of others. Most notably the work of Carol Berkenkotter and Tom Huckin on genre knowledge and situated cognition, of Charles Bazerman on community and sociohistorical context, of Greg Myers on the ways textual choices produce authoritative knowledge, of John Swales on genre, purpose and function, and of Avon Crismore on metadiscourse (see reference list for details). Those familiar with this work will see the influence of these authors as they read these chapters.

Chapter 5 is the first chapter of my most successful and, if Google Scholar citations are anything to go by, most influential book. The book was originally published (with a striking image by Malevich on its cover) in 2000 in Chris Candlin's *Applied Linguistics and Language Study* series but was almost immediately withdrawn when Longman scrapped the series after twenty years. Kelly Sippel at the University of Michigan Press saw some merit in the book and it has been available on that list since 2004. Essentially the book sets out to explore the cultures of academic communities and their discoursal practices, tracing through a series of studies of eight disciplines and various genres, the ways that academics represent themselves, their work and their readers in their texts, to establish their credibility as disciplinary insiders and the value of their claims.

The view that it is through texts that individuals collaborate to create both knowledge and their disciplines was probably unfamiliar to many at the time the book was published but has become commonplace today. Research has firmly shifted to the interpersonal dimension of language and the ways academics select their words to engage with others and present their ideas in ways which will most effectively persuade them. The book also contributed to the move away from studying interactions as the individual creation of particular texts to the repeated patterns of particular features in corpora. The chapter selected here seeks to offer arguments for both these, tying interactions, disciplines and cultures into a matrix within which people make sense to each other and conduct their professional lives. The chapter is perhaps more theoretical than much of what I have published and I may have aimed for greater reader friendliness if I wrote it today. I include it here, however, as it contextualizes my work and has strong practical implications for the understanding of academic writing and classroom practice.

Chapter 6 was written to consolidate some of the ideas discussed above into a single model which would allow teachers to identify particular features of interaction and researchers to extend our understanding of how it works in particular genres, disciplines and registers. The framework suggests that there are two aspects to interaction. The first is *stance.* This is the community-recognized persona of writers as they choose to intervene in the text to mark their commitments, doubts, attitudes and presence or to remain anonymous and project a more objective, impersonal authorial self. The second aspect of interaction is the writer's attempts to bring the reader into the text through *engagement* features, recognizing the role that readers play in validating the writer's arguments. Writers do not merely project the most commanding self possible nor make the strongest claims they can while readers do not expect to be passive recipients of a monologue. Writers and readers act out a dialogic relationship where the writer anticipates the reader's prior knowledge, likely attitude to the argument, processing needs and rhetorical expectations while readers assume they will be part of the discourse and their views on the subject will be addressed.

Chapter 7 takes a wider perspective on interaction by including those features which contribute to textual organization, signalling the direction, purpose and internal structure of a text. The chapter is subtitled 'a reappraisal' to mark its engagement with, and development of, the earlier work of Crismore and others on metadiscourse. The chapter was the outcome of a funded research project which allowed me to hire a research assistant, Polly Tse, whose substantial contribution in collecting graduate dissertations for the corpus and insightful conversations on their analysis we recognized with co-authorship. Polly's early death through cancer in 2016 at the age of thirty-five was a tragic loss to the academic community and a bitter pill to swallow personally.

Metadiscourse refers to how we use language out of consideration for our readers or hearers based on our estimation of how best we can help them process and comprehend what we are saying. It is a recipient design filter which helps to spell out how we intend a message to be understood by offering a running commentary on it. While sometimes restricted to text-organizing features only, the 'interpersonal model' represents a coherent set of options which includes text-organizing material together with the ways speakers and writers project themselves into their discourse to signal their understandings of the material and their audience. In so doing, they are able to shape their messages for particular listeners or readers and this, in turn, allows discourse analysts to say something about the preferences of language-using communities, such as disciplines, in what ways these preferences differ and what they might say about the communities themselves.

This model has been extremely successful and is now one of the most commonly employed methods for analysing specialist-written texts, with Google Scholar returning over 185,000 documents containing the term and the Web of Science encompassing nearly 300 papers on the topic. It has, in brief, become a monster, and as a result is in danger of devouring itself. While recent work has seen developments in methodology and the features which are understood to realize metadiscourse, much research in the field follows well-trodden paths and risks simply repeating work has preceded it.

Chapter 8 is a more recent paper and was co-authored with my student Kevin Jiang. This paper takes the features of evidentiality, affect and presence discussed as stance markers in Chapter 6 and explores how they have changed over the last fifty years. Diachronic studies are relatively uncommon in academic discourse research. Although several studies look back to the origins of research writing in the seventeenth century and study the evolution of argument and changes in conventional genre forms, surprisingly little work has explored how things have changed in more recent times. However, the last thirty years have witnessed major changes in how research is conducted and written. There has been, for example, an explosion of writing with the globalization of research and the encroaching demands of publishing metrics on scholars across the planet; the growth of collaboration and multiple authorship; the expansion of access to a massive online literature; the fragmentation and specialization of research and the imperatives to reach new audiences and funders. It would be surprising if these factors did not have an impact on rhetorical practices and interaction in academic writing, and this is what this chapter sets out to determine.

To conclude the section, Brian Paltridge, professor of TESOL at the University of Sydney and a leading figure in academic writing research, offers a commentary on these chapters and some of the influence they have had.

5

Disciplinary cultures, texts and interactions

This book ("Disciplinary Identities", CUP), is a study of social interactions in published academic writing, looking at why members of specific disciplines use language in the ways they do. It focuses on texts as the outcome of interactions and explores the view that what academics do with words is to engage in a web of professional and social associations. The book, therefore, is essentially about relationships between people, and between people and ideas. I seek to show that in research articles, abstracts, book reviews, textbooks, and scientific letters, the ways writers present their topics, signal their allegiances, and stake their claims represent careful negotiations with, and considerations of, their colleagues. Their writing therefore displays a professional competence in discipline-approved practices. It is these practices, I suggest, and not abstract and disengaged beliefs and theories, that principally define what disciplines are.

Successful academic writing depends on the individual writer's projection of a shared professional context. That is, in pursuing their personal and professional goals, writers seek to embed their writing in a particular social world which they reflect and conjure up through particular approved discourses. Here then, I follow Faigley's (1986: 535) claim that writing 'can be understood only from the perspective of a society rather than a single individual' and Geertz's (1983) view that knowledge and writing depend on the actions of members of local communities. Looking at writing in this way evokes a social milieu which influences the writer and activates specific responses to recurring tasks.

Rather than regarding linguistic features as regularities of academic style, or the result of some mental processes of representing meaning, I examine them for traces of social interactions with others engaged in a common pursuit. To do this we need to see academic writing as collective social practices, and to focus on published texts as the most concrete, public and accessible realisation of these practices. These texts are the lifeblood of the academy as it is through the public discourses of their members that disciplines authenticate knowledge, establish their hierarchies and reward systems, and maintain their cultural authority. To study the social interactions expressed through academic writing is not only to see how writers in different disciplines go about producing knowledge, it is also to reveal something of the sanctioned social behaviours, epistemic beliefs, and institutional structures of academic communities.

This view is not new, of course, and strong foundations for it are provided in the work of Berkenkotter and Huckin (1995), Bazerman (1988), Myers (1990) and others. In fact, it is perhaps a truism that writing involves interactions. Writers and readers clearly consider each other, try to imagine each other's purposes and strategies, and write or interpret a text in terms of these imaginations. While it might be obvious that writing is interaction, it is not at all evident just what a particular text tells us about that interaction or about those who participate in it. What motivates interactions in academic writing? What linguistic features realise these interactions? What strategies are involved and what principles are employed? What do these tell us about the beliefs and practices of the disciplines?

These are by no means trivial questions for the analysis of discourse, the understanding of disciplinary communities, or the teaching of academic writing. We need to understand these transient regularities and why particular features seem to be so useful to writers that they become regular practices, often institutionalised as approved disciplinary literacies. An improved awareness of such interactions is, then, the key to understanding how academic discourse works in English, whether seen as professional training or as published emblems of scholarship. Such an understanding, in turn, allows users to question both prevailing discursive practices and the relations they express, offering teachers, novices and expert writers greater alternatives in their choice of discourse forms and in their ability to negotiate and establish a plurality of cultural norms in disciplines.

This, then, provides the starting point, a theoretical and pedagogical imperative which urges us to research texts, their contexts and their ideologies – to see how genres are written, used and responded to as part of the wider social and intellectual culture of a particular group and historical period. First, I want to provide some theoretical background as a context for what follows in subsequent chapters. I shall do this by highlighting four central themes of the book: the notions of academic writing, knowledge-construction, disciplinary cultures, and social interactions.

The importance of academic writing

The written genres of the academy have attracted increasing attention from fields as diverse as philosophy, sociology of science, history, rhetoric, and applied linguistics. There is, however, a clear consensus on the importance of written texts in academic life – a recognition that understanding the disciplines involves understanding their discourses. There are two main reasons for this.

The first reason is that disciplinary discourse is considered to be a rich source of information about the social practices of academics. Kress (1989: 7), for example, argues that discourses are 'systematically-organised sets of statements which give expression to the meanings and values of an institution'. Texts, that is, are socially produced in particular communities and depend on them for their sense. As Bazerman (1993: vii) observes:

> . . . everything that bears on the professions bears on professional writing. Indeed, within the professions, writing draws on all the professional resources, wends its way among the many constraints, structures, and dynamics that define the professional realm and instantiates professional work.

In academic fields this means that texts embody the social negotiations of disciplinary inquiry, revealing how knowledge is constructed, negotiated and made persuasive. Rather than simply examining nature, writing is actually seen as helping to create a view of the world. This is because texts are influenced by the problems, social practices and ways of thinking of particular social groups (Kuhn, 1970; MacDonald, 1994). In other words, discourse is socially constitutive rather than simply socially shaped; writing is not just another aspect of what goes on in the disciplines, it is seen as *producing* them.

The second reason for the attention given to academic writing is the fact that what academics principally do is write: they publish articles, books, reviews, conference papers and research notes; they communicate with colleagues by e-mail, reprint requests, and referee evaluations; they communicate with students by handouts, study guides and textbooks; they contribute to electronic lists and to university reports; and they submit applications for grants and equipment. Latour and Woolgar (1979), for example, have suggested that the modern research lab devotes more energy to producing papers than to making discoveries, and that scientists' time is largely spent in discussing and preparing articles for publication in competition with other labs. The popular view of the 'academic' as a solitary individual experimenting in the laboratory, collecting data in the field or wrestling with ideas in the library, and then retiring to write up the results, is a modern myth. Research is essentially a *social* enterprise, both in the sense that it is an immediate engagement with colleagues and that it is mediated by the social institutions within which it occurs. It is hardly surprising, therefore, that the written communications of academics express this social imperative.

But while disciplines are defined by their writing, it is *how* they write rather than simply *what* they write that makes the crucial difference between them. An article may discuss garlic proteins, motherese or the existence of truth without people, but we see more than differences of content when we start to read them carefully. Among the things we see are different appeals to background knowledge, different means of establishing truth, and different ways of engaging with readers. Scholarly discourse is not uniform and monolithic, differentiated merely by specialist topics and vocabularies. It is an outcome of a multitude of practices and strategies, where what counts as convincing argument and appropriate tone is carefully managed for a particular audience. These differences are a product then of institutional and interactional forces, the result of diverse social practices of writers within their fields.

One reason for these differences in disciplinary discourses is that texts reveal generic activity (e.g. Reynolds and Dudley-Evans, 1999; Swales, 1990). They build on the writer's knowledge of prior texts and therefore exhibit repeated rhetorical responses to similar situations with each generic act involving some degree of innovation and judgement. This kind of typification not only offers the individual writer the resources to manage the complexities of disciplinary engagement, but also contributes to the stabilisation of reproduction of disciplines. Our attention is therefore directed towards textual variation, not only in the content of the texts we examine in a particular discipline, but in the structure of those texts and the kinds of rhetorical strategies they allow. By focusing on the stereotypical and the commonplace we catch a glimpse of what is largely unattended to by writers themselves, the pragmatic expectations and beliefs which are taken for granted in their naturalness. Thus, because these textual regularities apparently reflect the unremarked and automatic practices of disciplinary situated writers, they offer insights into the routine understandings which guide social interactions in those disciplines.

Until recently these disciplinary variations were often obscured by the practicalities of preparing learners for academic studies in English – an enterprise that has tended to emphasise genre rather than discipline and similarity rather than difference. A purely formal view of academic writing tended to dominate early practice in English for Academic Purposes (EAP). This was a view which largely took for granted the academy's perception of its texts as objective, rational and impersonal, and set out to provide students with the generic skills they needed to reproduce them. By ignoring context it was possible to ignore variation and to marginalise language itself as simply a set of skills for clearly communicating ideas from one person to another. Moving away from a process approach to writing, and with little research to help them, teachers were principally concerned to offer second language students survival training in universities where English was the medium of instruction. Textbooks and materials thus emphasised 'common core skills' such as describing, summarising, expressing causality, and so on as general principles of a universal academic literacy (e.g. Murray, 1989; Spack, 1988).

Only in the last ten years has the importance of a more socially situated analysis of genres been fully understood. This social conception of academic writing has been illuminated from a number of directions. The most important of these have been the theory of language developed by Halliday (1978, 1994) which emphasises the mutually constituting relationship of language and context; Miller's (1984) notion of genre as typified rhetorical action; and Toulmin's (1958) conception of disciplinary-specific argument forms. Together these perspectives have led to a view of genre as a means of routinely representing information in ways that reflect the social contexts of their construction and the beliefs of their users, providing insights into the norms, epistemologies, values and ideologies of particular fields of knowledge (e.g. Candlin and Hyland, 1999a). However, while genres develop from repeated situations, thereby helping to stabilise participants' experiences and give them a sense of solidarity, we must not see them as frozen artefacts. Genres are also in a state of constant evolution as members respond to professional and private exigencies in new and innovative ways (e.g. Bhatia, 1999).

The importance devoted to academic writing has not been entirely restricted to English. In recent years greater attention has been given to the rhetorical conventions of academic exposition in a number of different languages and genres. Rhetorical patterns of academic writing in the humanities in Czech (Čmejrková, 1996), the sciences in Malay (Ahmad, 1995), linguistics in Bulgarian (Todeva, 1999) and sociology in Russian (Namsaraev, 1997), for instance, differ considerably from those described for English. There has been a strong tradition of research into Languages for Specific Purposes in Scandinavia and Central Europe over the past 25 years, exemplified by the collection in Lundquist et al. (1998) and in the reviews by Schröder (1991) and Gunnarsson (1995). Academics writing in German (e.g. Schröder, 1995; Ylonen et al., 1993), Finnish (e.g. Luukka, 1995), and French (e.g. Eurin Balmet and Henao de Legge, 1992; Gambier, 1998; Lerat, 1995) have also paid considerable attention to professional and academic discourse, although much of the European work has tended to focus on terminological issues rather than generic description.

Writing, therefore, is not simply marginal to disciplines, merely an epiphenomenon on the boundaries of academic practice. On the contrary, it helps to create those disciplines by influencing how members relate to one another, and by determining who will be regarded as members, who will gain success and what will count as knowledge. Texts therefore contain traces of disciplinary activities in their pages; a typical clustering of conventions – developed over time in response to what writers perceive as similar problems – which point beyond words to the social circumstances of their construction. They offer a window on the practices and beliefs of the communities for whom they have meaning.

The social creation of knowledge

To a large extent disciplinary discourse has evolved as a means of funding, constructing, evaluating, displaying and negotiating knowledge. Merton's view that the goal of science is to add to a body of certified knowledge has also been adopted as a dominant practice in the humanities and social sciences. It is the display of this ideology which distinguishes academic discourse from other kinds of writing, and allows us to examine variations between academic disciplines. Examining texts as disciplinary practices moves us from the individual to the collective, from the boundaries of the page to the activities of social beings. Because of this, such writers as Geertz (1983) and Bruffee (1986) reject a representational view of knowledge and instead argue that knowledge emerges from a disciplinary matrix. In Rorty's (1979: 170) words, it is 'the social justification of belief'.

Social constructionism has undermined an earlier, objectivist, model which regarded writing as a means of simply dressing the thoughts that one sent into the world, and which saw texts as channels of communicating independently existing truths. The basic premise of constructionism is that academics work within communities in a particular time and place, and that the intellectual climate in which they live and work determine the problems they investigate, the methods they employ, the results they see, and the ways they write them up. In the words of Knorr-Cetina (1981), all they do is 'theory impregnated'.

In this view, academic knowledge is both situated and indexical, 'inextricably a product of the activity and situations in which it [is] produced' (Brown et al., 1989: 33); that is, it is embedded in the wider processes of argument, affiliation and consensus-making of members of the discipline. As Weimer (1977: 5) observes, all reporting occurs within a pragmatic context and in relation to a theory which fits 'observation and data in meaningful patterns. . . . Theories argue for a particular pattern or way of seeing reality'. Similarly, Toulmin (1972: 246) states that 'nature has no language in which she can speak to us on her own behalf, while the eminent theoretical physicist Stephen Hawking (1993: 44) notes that all our notions of the universe are based on models:

> A theory is a good theory if it is an elegant model, if it describes a wide class of observations, and if it predicts the results of new observations. Beyond that it makes no sense to ask if it corresponds to reality, because we do not know what reality is independent of a theory.

Thus it is naive to regard texts as accurate representations of what the world is like because this representation is always filtered through acts of selection, foregrounding and symbolisation; reality is seen as constructed through processes that are essentially social, involving authority, credibility and disciplinary appeals.

So from the perspective of the social constructionist, academic texts do more than report research that plausibly represent an external reality, they work to transform research findings or armchair reflections into academic knowledge. This knowledge is not a privileged representation of non-human reality, but a conversation between individuals and between individuals and their beliefs. This is not to fall into a world of idealism divorced from the physical world. Scientists and sociologists need their sensory experience of the world, but this experienced reality *under-determines* what they can know and say about it. They must therefore draw on principles and orientations from their cultural resources to organise it. We cannot step outside the beliefs or discourses of our social groups to find a justification for our ideas that is somehow 'objective'.

The acceptance of theories is located in the discourse community and the constraints on justifiable belief are socially constructed among individuals. The model is of 'independent creativity disciplined by accountability to shared experience' (Richards, 1987: 200). 'Objectivity' thus becomes 'consensual intersubjectivity' (Ziman, 1984:107), as methods and findings are coordinated and approved through public appraisal and peer review. We are then concerned with knowledge and knowing as social institutions, with something collectively created through the interactions of individuals.

The importance of social factors in transforming research activities into academic knowledge is perhaps most clearly illustrated by the socio-historical variability of rhetorical practices. The conventional linguistic means for securing support for scientific knowledge are not defined by a timeless idea, but developed in response to particular rhetorical situations. The work of Robert Boyle and his colleagues in the 1650s, for example, was crucial in establishing rules of discourse to generate and validate facts and create a 'public' for experimental research (Shapin, 1984). Boyle laid down literary and social means of verifying facts by a multiplication of the witnessing experience, stating how writers should reproduce the phenomena in their texts and establish themselves as providers of reliable testimony. An appropriate moral and interactional posture were therefore essential to the process of gaining assent to one's results from the beginning.

Atkinson (1996) similarly observes that research writing in the seventeenth and eighteenth centuries was substantially influenced by author-centred norms of genteel conduct, and that these were gradually transformed by changing social conditions and the growing professionalism of research communities. Using a cluster analysis of features, Atkinson found that papers steadily became less affectively and 'narratively' focused and more 'informational' and abstract over this period, shifting to an 'object-centred' rhetoric organised around specific community-generated research problems rather than around the experiencing gentleman-scientist. Halliday (1988, 1998) has also shown how the use of nominalisation and strings of nominal groups has increased in science to allow writers to package complex phenomena into single semiotic entities. Forms such as *'the rate of glass crack growth depends on the magnitude of the applied stress'* have become a powerful resource for constructing and communicating increasingly complex concepts. This practice serves to 'semiotically reconstruct' experience and bring into being a new construction of knowledge and new ways of seeing the world.

In other words, today's rhetorical situation has emerged from the *political* establishment of a scientific community and increasingly refined over the centuries to reflect a changing audience and material conditions. As publication became essential to research, a network of scientists evolved which required institutionalised standards of public debate. Changes in argument, referencing and length of research articles in physics, for example, appear to reflect increasing knowledge and changes in audience requirements (Bazerman, 1988). More recent changes have been tracked by Berkenkotter and Huckin (1995) who see the emergence of a news-oriented text schema in biological research articles since 1944. They argue that this increasing promotion of results in research papers has developed to accommodate the increasingly selective reading patterns of researchers swamped by the explosion of information in the sciences.

What is regarded as 'truth' is thus a 'best guess', relative to a particular time and community, and what may appear as self-evident in the practices we find today simply contributes to the illusion that these conventions are somehow natural. Persuading readers to accept a particular observation as a fact, or at least as a worthwhile contribution to disciplinary knowledge, involves relying on situated assumptions concerning what issues to address and how best to contextualise results. Knowledge claims are the outcome of socially agreed-upon ways of discussing academic problems that are always subject to dialectical revision (Prelli, 1989; Bizzell, 1992). The persuasiveness of academic discourse, then, does not depend upon the demonstration of absolute fact, empirical evidence or impeccable logic, it is the result of effective rhetorical practices, accepted by community members. Texts are the actions of socially situated writers and are persuasive only when they employ social and linguistic conventions that colleagues find convincing.

Considerable evidence shows that academic papers are written to provide an account that reformulates research activity in terms of an appropriate, but often contested, disciplinary ideology (e.g. Gilbert and Mulkay, 1984; Myers, 1990). It has been argued, for example, that in the sciences research articles are constructed to conceal contingent factors, downplaying the role of social allegiance, self-interest, power and editorial bias, to depict a disinterested, inductive, democratic and goal-directed activity. This is, moreover, a perspective that occasionally receives support from within the disciplines themselves.[1] In sum, the discursive practices which certify knowledge rely more on subjective decisions of plausibility than universal principles of rationality. Rational argument is a social matter, governed by disciplinary norms and oriented to achieving an intersubjective consensus through persuasive means. Notions of what counts as convincing argument, appropriate theory, sound methodology, impressive logic and compelling evidence are community-specific.

Disciplinary cultures

The view that knowledge is constructed within social communities draws attention to the homogeneity of disciplinary groups and practices. Each discipline might be seen as an academic tribe (Becher, 1989) with its particular norms, nomenclature, bodies of knowledge, sets of conventions and modes of inquiry constituting a separate culture (Bartholomae, 1986; Swales, 1990). Within each culture individuals acquire specialised discourse competencies that allow them to participate as group members. These cultures differ along social and cognitive dimensions, offering contrasts not only in their fields of knowledge, but in their aims, social behaviours, power relations, political interests, ways of talking and structures of argument (Toulmin, 1972; Whitley, 1984). Through the code of their specialised languages, these tribes consecrate their cultural privilege (Bourdieu and Passeron, 1996).

Academics talk to each other within the frameworks of their disciplines and generally have little difficulty in identifying the most central journals, main grant-awarding agencies, essential conferences, leading figures and most prestigious departments in their fields. The notion of *discourse community* has therefore proved useful here as it seeks to locate writers in particular contexts to identify how their rhetorical strategies are dependent on the purposes, setting and audience of writing (e.g. Bruffee, 1986). Bizzell (1982: 217), for example, has discussed them in terms of 'traditional, shared ways of understanding experience' including shared patterns of interaction, and Doheny-Farina (1992: 296) refers to the 'rhetorical conventions and stylistic practices that are tacit and routine for the members'. Killingsworth and Gilbertson (1992: 7) make the interesting point that communities are actually a kind of communication media in that they affect the manner and meaning of any message delivered through it.

The concept, however, has not found universal favour. Harris (1989), for example, argues that we should restrict the term to specific local groups, and labels other uses as 'discursive utopias' which fail to state either their rules or boundaries. Chin (1994), Cooper (1989) and Prior (1998) more pointedly view the term as altogether too structuralist, static and deterministic, giving too much emphasis to a stable underlying core of shared values which removes writing from the actual situations where individuals make meanings. Clearly there is something in this. If we see communities as real, stable groups conforming to rules and values and upholding a consensus we are clearly obscuring the potentially tremendous diversity and variation of members' roles, allegiance and participation in their disciplines. We are also neglecting the innovation and momentum that is possible in disciplines – which I have discussed briefly in the last section.

The fact is, of course, that discourse communities are not monolithic and unitary. They are composed of individuals with diverse experiences, expertise, commitments and influence. There are considerable variations in the extent to which members identify with their myriad goals, methods and beliefs, participate in their diverse activities, and identify themselves with their conventions, histories or values. In addition to committed researchers, influential gatekeepers and high profile proselytisers, communities comprise competing groups and discourses, marginalised ideas, contested theories, peripheral contributors and occasional members. The student neophyte, the laboratory research assistant, the professorial theorist and the industrial applied scientist interact with and use the same texts and genres for different purposes, with different questions and different degrees of engagement. Disciplines are, in short, human institutions where actions and understandings are influenced by the personal and interpersonal, as well as the institutional and sociocultural.

Questioning the construct of community, however, has sharpened some of its meaning. Killingsworth (1992) thus distinguishes between local groups of readers and writers who habitually interact, and global communities defined exclusively by a commitment to particular actions and discourses. Porter (1992) understands a community in terms of its *forums* or approved channels of discourse such as publications, meetings, and conferences which carry traces of its orientations and practices. The view that academic groups might be constituted by their characteristic genres of interaction, of how they get things done, rather than existing through physical membership has also attracted Swales (1993). In other words, an individual's engagement in its discourses can comprise his or her membership of that discipline – an idea Swales (1998) elaborates as a 'textography of communities'.

So while it remains a contested concept, the notion of community does foreground what is an important influence on social interaction. It draws attention to the fact that discourse is socially situated and helps to illuminate something of what writers and readers bring to a text, emphasising that composition and interpretation both depend on assumptions about the other.

The equally powerful, and equally inexact, metaphor 'communities of practice' has also been employed to avoid a strong reliance on a shared core of rather abstract knowledge and language. Here the central aspect of disciplinary groups is the emphasis on situated activity and 'a set of relations among persons, activity, and world, over time and in relation with other tangential and overlapping communities of practice' (Lave and Wenger, 1991: 98). This shifts the focus from language or social structure to the situated practices of aggregations of individuals; to communities strongly shaped by a collective history of pursuing particular goals within particular forms of social interaction. Irrespective of whether we choose to label disciplines as tribes, cultures, discourse communities or communities of practice, these concepts move us from a concern with the abstract logicality and substance of ideas of academic writing to the world of concrete practices and social beliefs. They put community decision-making and engagement at centre-stage and underline the fact that disciplinary discourse involves language users in constructing and displaying their roles and identities as members of social groups.

The idea of disciplinary cultures therefore implies a certain degree of interdisciplinary diversity and a degree of intradisciplinary homogeneity. Writing as a member of a disciplinary group involves textualising one's work as biology or applied linguistics and oneself as a biologist or applied linguist. It requires one to give a tangible and public demonstration that one has legitimacy. There are then disciplinary constraints on discourse which are both restrictive and authorising (Foucault, 1972), allowing one to create successful texts which display one's disciplinarity, or tacit knowledge of its expectations, for the practical purposes of communicating with peers. This points to the power relations hidden in text, the unspoken assumptions of a largely undiscussed world which is the basis for cooperative action (Bourdieu, 1980: 269).

Sullivan (1996) argues that there are four central elements of such disciplinary constraints: an ideological perspective of the discipline and the world; assumptions about the nature of things and methodologies; a system of hierarchical power relationships; and a body of doctrinal knowledge of external reality. Nevertheless, while such factors help draw the boundaries of cooperative action among disciplinary members, we might be cautious in emphasising the degree to which a consensus exists and how far the authority of a single overarching disciplinary paradigm determines behaviour. Instead, as I noted earlier, disciplines might be seen as systems in which multiple beliefs and practices overlap and intersect.

Most disciplines are characterised by several competing perspectives and embody often bitterly contested beliefs and values. Rauch (1992), for instance, highlights the very different, and largely incompatible, values and assumptions underlying a debate between conservation biologists on the species status of the red wolf. In other fields empiricists contest the same ground with phenomenologists, cognitivists with behaviourists, existentialists with Freudians, and relativists with realists. Communities are frequently pluralities of practices and beliefs which accommodate disagreement and allow subgroups and individuals to innovate within the margins of its practices in ways that do not weaken its ability to engage in common actions. Seeing disciplines as cultures helps to account for what and how issues can be discussed and for the understandings which are the basis for cooperative action and knowledge-creation. It is not important that everyone agrees but members should be able to engage with each others' ideas and analyses in agreed ways. Disciplines are the contexts in which disagreement can be deliberated.

While all academic discourse is distinguished by certain common practices, such as acknowledging sources, rigorous testing, intellectual honesty, and so on, there are differences which are likely to be more significant than such broad similarities. The ways that writers chose to represent themselves, their readers and their world, how they seek to advance knowledge, how they maintain the authority of their discipline and the processes whereby they establish what is to be accepted as substantiated truth, a useful contribution and a valid argument are all culturally-influenced practical actions and matters for community agreement.

These practices are not simply a matter of personal stylistic preference, but community-recognised ways of adopting a position and expressing a stance. 'Doing good research' means employing certain post-hoc justifications sanctioned by institutional arrangements. As a result, the rhetorical conventions of each text will reflect something of the epistemological and social assumptions of the author's disciplinary culture. In the sciences, for example, this often requires a public commitment to experimental demonstration, replicability and falsification of results. In philosophy it may involve narratives containing 'twin-Earth fantasies', 'imaginary conversations' and argumentative point scoring (Bloor, 1996).

In sum, disciplinary communicative practices involves a system of appropriate social engagement with one's material and one's colleagues. The writing that disciplines produce, support and authorise can therefore be seen as linked to forms of power in those organisations. They are representations of legitimate discourses which help to define and maintain particular epistemologies and academic boundaries. Because texts are written to be understood within certain cultural contexts, the analysis of key genres can provide insights into what is implicit in these academic cultures, their routine rhetorical operations revealing individual writer's perceptions of group values and beliefs. Genres are not therefore only text types but imply particular institutional practices of those that produce, distribute and consume them (Fairclough, 1992: 126). Individual and social purposes interact with discourse features at every point of choice and in every genre, and to analyse these is to learn something of how each discipline views knowledge and defines itself.

Texts as social interaction

This discussion of academic texts, disciplinary cultures, and the social construction of knowledge, highlights the interactive and rhetorical character of academic writing. It leads us to see writing as an engagement in a social process, where the production of texts reflects methodologies, arguments and rhetorical strategies designed to frame disciplinary submissions appropriately. Creating a convincing reader environment thus involves deploying disciplinary and genre-specific conventions such that 'the published paper is a multilayered hybrid *co-produced* by the authors *and* by members of the audience to which it is directed' (Knorr-Cetina, 1981: 106). Textual meanings, in other words, are socially mediated, influenced by the communities to which writers and readers belong.

If texts are a means of studying social negotiations between academics, how is this achieved in particular texts? For what reasons? What are we looking for? From the discussion so far it is clear that academic writing is broadly concerned with knowledge-making and that this is achieved by negotiating agreement among colleagues. In most academic genres then, a writer's principal purpose will be persuasive: convincing peers to assent to a knowledge claim in a research paper, to fund a project in a grant-submission, to accept an evaluation in a review, to acknowledge a disciplinary schema in a textbook, and so on. These shared purposes both help us to identify what is similar in these genres and what is disciplinarily distinctive, and also suggest some areas where common practices will involve social interactions.

In knowledge-creating genres such as the research article, for example, these practices will include (at least) the need to:

- establish the novelty of one's position

- make a suitable level of claim

- acknowledge prior work and situate claims in a disciplinary context

- offer warrants for one's view based on community-specific arguments and procedures

- demonstrate an appropriate disciplinary ethos and willingness to negotiate with peers.

The means by which academics present knowledge claims and account for their actions thus involves not only cognitive factors, but also social and affective elements, and to study these necessarily moves us beyond the ideational dimension of texts to the ways they function at the interpersonal level.

Both are involved in the interactions needed to secure peer agreement because the writer's ability to influence the reader's response is severely restricted. If 'truth' does not lie exclusively in the external world, there is always going to be more than one credible interpretation of a piece of data and more than one way of looking at a certain problem. This plurality of competing interpretations, with no objective means of absolutely distinguishing the actual from the plausible, means that while readers may be persuaded to judge a claim acceptable, they always have the option of rejecting it. All statements require community ratification, and because readers are guarantors of the negatability of claims this gives them an active and constitutive role in how writers construct them.

In other words, the social interactions in academic writing stem from the writer's attempts to anticipate possible negative reactions to his or her persuasive goals. The writer will choose to respond to the potential negatability of his or her claims through a series of rhetorical choices to galvanise support, express collegiality, resolve difficulties and avoid disagreement in ways which most closely correspond to the community's assumptions, theories, methods and bodies of knowledge.

Opposition to statements can come from two principal sources (Hyland, 1996a, 1998a). First, readers may reject a statement on the grounds that it fails to correspond to what the world is thought to be like, i.e. it fails to meet *adequacy conditions.* Claims have to display a plausible relationship with 'reality' (the discipline's epistemological framework), and writers must take care to demonstrate this satisfactorily by using the specialised vocabularies and argument forms of the discourse community. Here writers have to encode ideational material, establish relationships, employ warrants, and frame arguments in ways that the potential audience will find most appropriate and convincing. Thus what writers choose to emphasise from among the array of physical and conceptual phenomena in their fields, their decision to adopt an attitude of systematic doubt or personal conviction, and their choice of reasoning, are part of their strategic skill in establishing authority and credibility.

Second, statements have to incorporate an awareness of interpersonal factors, addressing *acceptability conditions,* with the writer attending to the affective expectations of participants in the interaction. Rhetorical strategies for social interaction are employed here to help the writer to create a professionally acceptable persona and an appropriate attitude, both to readers and the information being discussed. This means representing one's self in a text in a way that demonstrates one's flawless disciplinary credentials; showing yourself to be a reasonable, intelligent, co-player in the community's efforts to construct knowledge and well versed in its tribal lore. Critical here is the ability to display proper respect for colleagues and give due regard for their views and reputations. In this sense linguistic choices seek to establish an appropriate, discipline-defined balance between the researcher's authority as an expert-knower and his or her humility as a disciplinary servant (Myers, 1989). This is principally accomplished through a judicious balance of tentativeness and assertion, and the expression of a suitable relationship to one's data, arguments and audience.

In sum, the interactions of academic writing indicate the writer's acknowledgement of the community's epistemological and interpersonal conventions and connect texts with disciplinary cultures. Academic knowledge is not simply a databank of general, and generally agreed upon, facts, but networks of values, beliefs and routines that guide practice and define disciplines. The academic writer must make assumptions about the nature of the world and about how it will best be received by a particular audience, the question of adequacy corresponding to the objective negatability of a proposition and acceptability to its subjective negatability. Texts thus reveal how writers attempt to negotiate knowledge in ways that are locally meaningful, employing rhetorical skills which establish their credibility though an orientation towards arguments, topics and readers.

Approaches to academic interactions

Theorists have sought to account for the social interactions in texts in two main ways, focusing on either models of actors or models of social structures. The former, notably Grice's (1975) Cooperative Principle and Nystrand's (1989)

Reciprocity Principle, posit general principles of interaction based on what people are believed to be like, examining linguistic features for evidence of the operation of these principles. More structural approaches, influenced in particular by Foucault and taken up by Kress and Hodge (1993) and Fairclough (1992, 1995), seek to establish links between discourse and society by focusing on aspects of context such as power and ideology. Both views assume then that discourse is socially embedded and has social consequences, although they look at this from different directions.

Grice's (1975) well-known Cooperative Principle has been very fruitful for those seeking a theoretical model of interaction. This is built on the idea that readers understand texts by drawing on both their knowledge of the context and their belief that writers will consider their interpretive needs. Thus, when approaching a text we assume communication will be shaped by various universal maxims of cooperation, most generally that what we find will be consistent with the accepted purpose of the discourse. Anticipating that such maxims are being observed, we interpret apparent violations of expected relevance, informativeness, honesty and clarity as meaningful, i.e. as conveying pragmatic 'implicatures'. Particular importance has been given to the reader's ability to recover contextual relevance from a text (e.g. Sperber and Wilson, 1986). Once again this is principally a philosophical theory which does not systematically elaborate the idea of context, it is clear, however, that the reader's ability to supply assumptions about the relevant social institutions and relationships are likely to be a crucial dimension of interpretation.

Nystrand (1989) similarly develops an approach to interaction based on the writer and readers' 'mutual co-awareness' and the social group's taken-for-granted rules of conduct. For him, an effective text is one which 'balances the reciprocal needs of the writer for expression and the reader for comprehension' (1989: 81) and where participants draw on certain 'elaborations' in texts to overcome interpretive difficulties.

Importantly, then, linguistic choices are seen to work as a result of the way participants attribute to each other the intention to communicate, and the principles these writers propose are seen as culturally and contextually variable means of systematising the process of inferencing. Both Nystrand and Grice (indirectly) posit an interactive connection between writers and readers that makes communication possible, linking them together via principles that allow texts to be interpreted in relatively stable and predictable ways. Participants are seen as interactants who engage in strategic reasoning to determine a consistent purpose in rhetorical choices. Rationality is therefore attributed to exchanges based on a collective commitment to conduct discourse in a certain way, drawing on the community as a kind of normative framework for interpretation.

This approach tends to emphasise the ways that writers imagine readers' expectations, knowledge and interests in constructing their texts, deciding what to include, what to highlight and how to show the relationships between elements. It is 'information-oriented' (Thompson and Thetela, 1995) in the sense that linguistic choices are based on the clarification of meaning. This, however, omits a great deal of what writers do when they directly engage with others to influence their response by conveying an attitude to the text or to the readers themselves. In other words, for a fuller picture we need to fill out this view of interaction with an interpersonal dimension.

Such a dual perspective has formed the cornerstone of many genre accounts of writing and has perhaps received its most explicit realisation in Brown and Levinson's (1987) politeness model of interaction. Brown and Levinson seek to explain variations in linguistic features as politeness strategies rationally selected to accomplish communicative goals. Modifying Goffman's (1967) concept of face as the desire to be approved of (positive face) and to act without being impeded (negative face), Brown and Levinson argue that writers are motivated to protect both aspects of their own face and that of their readers. In this theory the need to balance face needs is a consequence of the fact that interaction is seen as inherently imposing, involving numerous Face Threatening Acts (FTAs). In academic settings, Myers (1989) has argued that knowledge claims, criticisms and denials of claims constitute FTAs, both against readers engaged in the same research area and a wider disciplinary audience, and that we can reconstruct the reasoning behind linguistic choices such as hedges and solidarity pronouns as strategies to mitigate threats to face.

Politeness is thus used to explain why academic writers might deviate from the strictly utilitarian principles of communication suggested by Grice and Nystrand. The approach has been extremely useful in repositioning discourses which are apparently autonomous, impersonal and informational as socially grounded. It highlights, in other words, competitive and cooperative interactions which are conducted through features often seen as merely conventions of academic style.

The view that certain assumptions of conversational behaviour serve to motivate strategy choices in knowledge-creating discourse may be suggestive, but it is not entirely satisfactory. Conversation, generally, is not argument, and the model neglects significant differences in purposes and consequences between these registers, generalising the avoidance of imposition in phatic contexts to the regularities of academic discourse. Clear distinctions can be drawn between these discourses, however, not only because these communicative sites have distinct purposes, but also because they generally have very different consequences for participants. There is more at stake in most academic encounters as, for most writers, their careers ride on successful participation in them.

The impositions of conversation are not those entailed in presenting a claim or an evaluation in a research article because these writing practices not only seek to legitimate the views of their holders, but also their rights to employment, prestige and material support. Communication is the key cognitive and social impetus in academia and as such reflects central aspects of disciplinary cultures, representing the point where knowledge is advanced, reputations are gained and merit is bestowed. Engagement in disciplinary forums therefore involves norms of interpersonal behaviour which are underpinned by the sanctions inherent in a system of academic recognition and rewards that hinges on publication (Hyland, 1997). Although writers may well calculate the impositional impact of their statements, politeness cannot explain the full significance of their rhetorical choices.

Brown and Levinson's model, while an interesting theory of how linguistic features might represent strategic choices, is insufficiently grounded in social structures to account for the motives, roles and conventions of disciplinary interaction. It is undoubtedly correct that academic writers engage in conflict avoidance and that they weigh up the effects of their statements when communicating with their peers, but this is not to say that interaction is based on individual judgements of imposition or entitlements to deference. If we seek to reconstruct interaction from the features of texts based on assumptions about the strategic accomplishments of model actors, then we also need to locate these actors in particular physical, social and economic realities. We need to recognise their behaviour as influenced by their pursuit of goals in given institutional contexts, and to see that these behaviours have real consequences. In other words, an adequate framework for the study of facework must be based on a more complex understanding of social life, and in particular for disciplinary discourse, on the ambitions, constraints and rewards of academic engagement.

Accounts of discourse as interaction which focus more explicitly on the social structures of language use, attempt to provide just such an understanding of social life, seeking to fill in the causal relations between language and society that social actor models leave implicit. Approaches associated with the work of writers such as Kress and Fairclough problematise the cooperative model of discourse sketched above by highlighting the often asymmetrical distribution of discoursal and social rights. This view criticises such models for exaggerating the extent to which strategic actions are cooperatively negotiated and failing to identify the power relations in texts, leaving unexamined *who* actually defines interactional concepts like 'truth', relevance', 'informativity' and so on (Fairclough, 1995: 47). These writers therefore adopt what is widely termed a 'critical approach' and set themselves the task of revealing how social structures impinge on the ways that discourse is created and interpreted. These structures are seen as being principally ideological, and so to study discourse is to study the ways in which meaning ideologically serves to sustain relations of power (Thompson, 1984).

Ideologies serve to link individuals and groups, providing ways to maintain communities and structure common purposes and beliefs; they help, in other words, to constitute what is real. All knowledge, including disciplinary knowledge of the world and members' knowledge of the social practices of their communities, is contingent upon social structures; neither the content of natural knowledge nor the practical knowledge of investigation and reporting can be separate from the social processes that produce it (Dant, 1991; Thompson, 1984). Importantly, then, both language and the ways it is used have meanings only in virtue of the discursive formations in which they occur. Disciplinary discourses assist solidarity, facilitate coordinated activities, offer frameworks to achieve personal goals, and contribute to the successful reproduction of communities. In using language, however, we engage in reinforcing our relations with others and in sustaining the asymmetrical relations of power of those contexts.

At the community level, academics write as group members. They adopt discoursal practices that represent an authorised understanding of the world (and how it can be perceived and reported) which acts to reinforce the theoretical convictions of the discipline and its right to validate knowledge. In the hard sciences, for example, discourse functions to convey ideological representations which help to impart an authority to a discipline-defined epistemology based on a detached attitude to an external reality of objective facts. This discourse has been spectacularly successful in gaining credibility, influence and prestige for its practitioners, providing a standard of rhetorical inquiry which other disciplines have not always imitated sufficiently successfully to gain similar respectability. Ideologies thus serve to define disciplines and to establish their status and power in a wider social structure, offering members an identity based on a set of typical practices, norms and objectives, and building the authority of the group, both in relation to society and to other groups. The fact that ideologies often become naturalised and unnoticed disguises their inherently social functions.

At a more individual level, disciplinary ideologies commit members to engage cooperatively in approved discoursal practices as a way of securing personal goals. Success often involves a display of disciplinary competence and members must present a narrative that is perceived by the community as persuasive both in terms of the propositions that the writer sets out and the credibility of the *persona* he or she seeks to convey. Engagement in a rhetorical forum that is discursively adversarial, and where there are always multiple plausible interpretations for any given phenomenon, encourages writers to accommodate their textual practices to those that readers, particularly journal editors and referees, are likely to find most convincing. Studies focusing on the process of peer review, for example, have shown the influence of disciplinary gatekeepers on revision practices in order to secure publication (Berkenkotter and Huckin, 1995; Myers, 1990). In these ways, the conventions of the research article are shaped by the ideological assumptions of the discipline and reinforced by both routine use and the customary procedures of academic quality control.

Participation in the social interactions of academic discourse, therefore, involve writers in discoursal practices that are largely argumentative and competitive. In Brown and Levinson's terms, the imposition of strategies is going to be high, but attributing them to politeness fails to give sufficient importance to the ideological/institutional constraints which distinguish the allowable from the doubtful. The social actor approaches remind us that people and actions are at work within texts: that writers and readers are aware of each other's aims and have common assumptions about how texts function. However, they fail to make explicit the connections between social norms and linguistic conventions, and how wider social practices and beliefs are embedded in the rhetorical features of texts.

While we can learn about texts by studying social action, we cannot just read off social action from texts. Participants may not always act strategically and they always have the option of adopting a personal or idiosyncratic relation to the text. So while we might point to possible norms and conventions as reflecting ideological behaviour, we must always recognise that the social world is not always a stable and predictable place. Moreover, as we have seen, this view of writer–reader interaction is largely a theoretical construct, devised to explain how individuals make decisions while writing, rather than an account of the various purposes and uses that the completed text may actually fulfil in the hands of the consumers. The important point is, however, that texts reflect writers' expectations of how they will be read, and therefore provide clues to the wider understandings underlying their creation.

I am, then, interested less in specific occasions of writing than in what this writing tells us of how writers shape their ideas by directing them beyond the immediate situation to the expectations of a disciplinary audience. Specific individual experiences or goals may impinge on the settings of composition, but it is the writer's ability to project from these immediate contexts to a perceived social reality that is important in determining disciplinary structures and meanings. From intensive study of large numbers of texts it is possible to see how much academic writing is the result of situated choices, and how writers typically select the forms and patterns that are most likely to help them to negotiate their purposes with an anticipated audience.

In sum, the notion of writer–reader interaction provides a framework for studying texts in terms of how knowledge comes to be socially constructed by writers acting as members of social groups. It offers an explanation for the ways writers frame their understandings of the world and how they attempt to persuade others of these understandings. But while the norms and ideologies that underpin these interactions provide a framework for writing, they are, essentially, a repertoire of choices rather than sets of binding and immutable constraints. As Mulkay (1979: 72) observes, norms should be seen

> not as defining clear social obligations to which scientists generally conform, but as flexible vocabularies employed by participants in their attempts to negotiate suitable meanings for their own and others' acts in various social contexts.

The resulting texts constitute a social and institutional code, both rhetorical and interpersonal, for achieving agreement.

These issues form the background and the motivation for what follows, and many are revisited there. In the ensuing chapters I shall try to give concrete expression to these ideas through a series of linked studies, each examining particular discoursal features and informants' accounts of their practices in selected genres. Because my goal is to explore the social interactions expressed in the features of texts, I give priority to published writing as the principal sites of academic engagement. It is here that writers most significantly interact with their peers, their students and their disciplines.

Note

1 An interesting insider's narrative of how chance, competition and self-interest are often disguised in the published account is offered in James Watson's (1968) discussion of the events leading to the discovery of the structure of DNA.

6

Stance and engagement: A model of interaction in academic discourse

Over the past decade or so, academic writing has gradually lost its traditional tag as an objective, faceless and impersonal form of discourse and come to be seen as a persuasive endeavour involving interaction between writers and readers. This view sees academics as not simply producing texts that plausibly represent an external reality, but also as using language to acknowledge, construct and negotiate social relations. Writers seek to offer a credible representation of themselves and their work by claiming solidarity with readers, evaluating their material and acknowledging alternative views, so that controlling the level of personality in a text becomes central to building a convincing argument. Put succinctly, every successful academic text displays the writer's awareness of both its readers and its consequences.

As this view gains greater currency, more researchers have turned their attention to the concept of evaluation and how it is realized in academic texts. Indeed, much of my own work over the past decade or so has been devoted to this. Consequently, a variety of linguistic resources such as hedges, reporting verbs, *that*-constructions, questions, personal pronouns, and directives have been examined for the role they play in this persuasive endeavour (e.g. Hyland, 2000; Hyland and Tse, 2004; Swales, 1990; Thompson, 2001). Despite this plethora of research, however, we do not yet have a model of interpersonal discourse that unites and integrates these features and that emerges from the study of academic writing itself. How *do* academic writers use language to express a stance and relate to their readers? This is the question addressed in this article, that brings together a diverse array of features by drawing on interviews and a corpus of 240 research articles to offer a framework for understanding the linguistic resources of academic interaction. My aim, then, is to consolidate my previous work using this corpus to offer a model of stance and engagement in academic texts.

Interaction and evaluation

Evaluation, as Bondi and Mauranen (2003) have recently observed, 'is an elusive concept'. For while we recognize interaction and evaluation in academic texts, it is not always clear how they are achieved. The ways that writers and speakers express their opinions have long been recognized as an important feature of language, however, and research has attempted to account for these meanings in a number of ways. Hunston and Thompson (2000) use the term 'evaluation' to refer to the writer's judgements, feelings, or viewpoint about something, and others have described these varied linguistic resources as *attitude* (Halliday, 1994), *epistemic modality* (Hyland, 1998a), *appraisal* (Martin, 2000; White, 2003), *stance* (Biber and Finegan, 1989; Hyland, 1999), and *metadiscourse* (Crismore, 1989; Hyland and Tse, 2004).

Interest in the interpersonal dimension of writing has, in fact, always been central to both systemic functional and social constructionist frameworks, which share the view that all language use is related to specific social, cultural and institutional contexts. These approaches have sought to elaborate the ways linguistic features create this relationship as writers comment on their propositions and shape their texts to the expectations of their audiences. Perhaps the most systematic approach to these issues to date has been the work on appraisal which offers a typology of evaluative resources available in English (Martin, 2000). For Martin, appraisal largely concerns the speaker's attitudinal positions, distinguishing three sub-categories of affect, judgement, and appreciation, roughly glossed as construing emotion, moral assessments, and aesthetic values respectively, and the ways these are graded for intensity. While this broad characterization is interesting, however, it is unclear how far these resources are actually employed in particular registers and to what extent they can be seen as comprising core semantic features in given contexts of use.

Because the work on evaluation and stance is relatively new, much of it has tended to concentrate on mass audience texts, such as journalism, politics, and media discourses, which are likely to yield the richest crop of explicitly evaluative examples. Yet these public genres tend to offer writers far more freedom to position themselves interpersonally than academic genres. Because we do not yet have a model of evaluative discourse that emerges from the study of academic writing itself, we cannot say which features are *typical* in scholarly writing, rather than which are *possible*. It seems, for example, that 'attitude' in academic texts more often concerns writers' judgements of epistemic probability and estimations of value, with affective meanings less prominent (Hyland, 1999, 2000a). The role of hedging and boosting, for instance, is well documented in academic prose as communicative strategies for conveying reliability and strategically manipulating the strength of commitment to claims to achieve interpersonal goals.

An important consideration here is that evaluation is always carried out in relation to some standard. Personal judgements are only convincing, or even meaningful, when they contribute to and connect with a communal ideology or value system concerning what is taken to be normal, interesting, relevant, novel, useful, good, bad, and so on. Academic writers' use of evaluative resources is influenced by different epistemological assumptions and permissible criteria of justification, and this points to and reinforces specific cultural and institutional contexts. Writers' evaluative choices, in other words, are not made from all the alternatives the language makes available, but from a more restricted sub-set of options which reveal how they understand their communities through the assumptions these encode. Meanings are ultimately produced in the interaction between writers and readers in specific social circumstances, which means that a general categorization of interactional features is unable to show how academic writers, through their disciplinary practices, construct and maintain relationships with their readers and thus with their communities.

To be persuasive, writers need to connect with this value system, making rhetorical choices which evaluate both their propositions, and their audience. In sum, to understand what counts as effective persuasion in academic writing, every instance of evaluation has to be seen as an act socially situated in a disciplinary or institutional context.

Stance and engagement

Interaction in academic writing essentially involves 'positioning', or adopting a point of view in relation to both the issues discussed in the text and to others who hold points of view on those issues. In claiming a right to be heard, and to have their work taken seriously, writers must display a competence as disciplinary insiders. This competence is, at least in part, achieved through a writer–reader dialogue which situates both their research and themselves, establishing relationships between people, and between people and ideas. Successful academic writing thus depends on the individual writer's projection of a shared professional context. That is, in pursuing their personal and disciplinary goals, writers seek to create a recognizable social world through rhetorical choices which allow them to conduct interpersonal negotiations and balance claims for the significance, originality and plausibility of their work against the convictions and expectations of their readers.

The motivation for these writer–reader interactions lies in the fact that readers can always refute claims and this gives them an active and constitutive role in how writers construct their arguments. Any successfully published research paper anticipates a reader's response and itself responds to a larger discourse already in progress. This locates the writer intertextually within a larger web of opinions (Bakhtin, 1986), and within a community whose members are likely to recognize only certain forms of argument as valid and effective. Results and interpretations need to be presented in ways that readers are likely to find persuasive, and so writers must draw on these to express their positions, represent themselves, and engage their audiences.

Evaluation is therefore critical to academic writing as effective argument represents careful considerations of one's colleagues as writers situate themselves and their work to reflect and shape a valued disciplinary ethos. These interactions are managed by writers in two main ways.

1 *Stance.* They express a textual 'voice' or community recognized personality which, following others, I shall call *stance.* This can be seen as an attitudinal dimension and includes features which refer to the ways writers present themselves and convey their judgements, opinions, and commitments. It is the ways that writers intrude to stamp their personal authority onto their arguments or step back and disguise their involvement.

2 *Engagement.* Writers relate to their readers with respect to the positions advanced in the text, which I call *engagement* (Hyland, 2001a). This is an alignment dimension where writers acknowledge and connect to others, recognizing the presence of their readers, pulling them along with their argument, focusing their attention, acknowledging their uncertainties, including them as discourse participants, and guiding them to interpretations.

The key resources by which these interactional macro-functions are realized are summarized in Figure 6.1 and discussed in more detail below.

Together these resources have a dialogic purpose in that they refer to, anticipate, or otherwise take up the actual or anticipated voices and positions of potential readers (Bakhtin, 1986). Stance and engagement are two sides of the same coin and, because they both contribute to the interpersonal dimension of discourse, there are overlaps between them. Discrete categories inevitably conceal the fact that forms often perform more than one function at once because, in developing their arguments, writers are simultaneously trying to set out a claim, comment on its truth, establish solidarity and represent their credibility. But it is generally possible to identify predominant meanings to compare the rhetorical patterns in different discourse communities.

It should also be borne in mind that evaluation is expressed in a wide range of ways which makes a fine-grained typology problematic. While writers can mark their perspectives explicitly through lexical items (such as *unfortunately, possible, interesting*, etc.), they can also code them less obviously through conjunction, subordination, repetition, contrast, etc. (e.g. Hunston, 1994). Moreover, because the marking of stance and engagement is a highly contextual matter, members can employ evaluations through a shared attitude towards particular methods or theoretical orientations which may be opaque to the analyst. Nor is it always marked by words at all: a writer's decision not to draw an obvious conclusion from an argument, for example, may be read by peers as a significant absence. It may not always be possible therefore to recover the community understandings and references embedded in more implicit realizations.

Distinguishing between these two dimensions is a useful starting point from which to explore how interaction and persuasion are achieved in academic discourse and what these can tell us of the assumptions and practices of different disciplines. This is what I set out to do below. Following a description of the corpus, I sketch out some of the key resources of stance

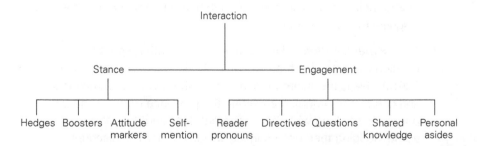

FIGURE 6.1 *Key resources of academic interaction.*

and engagement and discuss what these differences in functionality tell us about the epistemological and social beliefs of disciplinary cultures.

Corpus and methods

My view of stance and engagement is based on a series of studies which draw on both qualitative and quantitative approaches, comprising the analysis of a corpus of published articles and interviews with academics. The text corpus consists of 240 research articles comprising three papers from each of ten leading journals in eight disciplines selected to represent a broad cross-section of academic practice and facilitate access to informants. The fields were mechanical engineering (ME), electrical engineering (EE), marketing (Mk), philosophy (Phil), sociology (Soc), applied linguistics (AL), physics (Phy) and microbiology (Bio). The value of exploring such a large corpus is that it makes available many instances of the target features in a naturally occurring discourse, replicating the language-using experience of community members.

The texts were converted to an electronic corpus of 1.4 million words and searched for specific features seen as initiating writer–reader interactions using *WordPilot 2000,* a text analysis and concordance programme. A list of 320 potentially productive search items was compiled based on previous research into interactive features (e.g. Biber and Finegan, 1989; Bondi, 1999; Hyland, 1999, 2000a), from grammars (Biber et al., 1999; Halliday, 1994), and from the most frequently occurring items in the articles themselves. All cases were examined to ensure they functioned as interactional markers and a sample was double-checked by a colleague working independently.

The interviews were conducted with experienced researcher/writers from the target disciplines using a semi-structured format. These employed open-ended interview prompts which focused on subjects' own and others' writing, but allowed them to raise other relevant issues. Subjects could therefore respond to texts with insider community understandings of rhetorical effectiveness, while also discussing their own discoursal preferences and practices.

Stance and features of writer positioning

Stance concerns *writer-oriented features* of interaction and refers to the ways academics annotate their texts to comment on the possible accuracy or credibility of a claim, the extent they want to commit themselves to it, or the attitude they want to convey to an entity, a proposition, or the reader. I take it to have three main components: *evidentiality, affect* and *presence.* Evidentiality refers to the writer's expressed commitment to the reliability of the propositions he or she presents and their potential impact on the reader; affect involves a broad range of personal and professional attitudes towards what is said, including emotions, perspectives and beliefs; and presence simply concerns the extent to which the writer chooses to project him or herself into the text. It is comprised of four main elements:

1　Hedges.

2　Boosters.

3　Attitude markers.

4　Self-mentions.

Hedges are devices like *possible, might* and *perhaps,* that indicate the writer's decision to withhold complete commitment to a proposition, allowing information to be presented as an opinion rather than accredited fact. Because all statements are evaluated and interpreted through a prism of disciplinary assumptions, writers must calculate what weight to give to an assertion, attesting to the degree of precision or reliability that they want it to carry and perhaps claiming protection in the event of its eventual overthrow (Hyland, 1998a). Hedges, therefore, imply that a statement is based on plausible reasoning rather than certain knowledge, indicating the degree of confidence it is prudent to attribute to it (Example 1):

> (1) <u>Our results suggest</u> that rapid freeze and thaw rates during artificial experiments in the laboratory <u>may</u> cause artifactual formation of embolism. Such experiments <u>may</u> not quantitatively represent the amount of embolism that is formed during winter freezing in nature. In the chaparral <u>at least</u>, low temperature episodes <u>usually</u> result in gradual freeze-thaw events.
>
> (Bio)

Equally importantly, hedges also allow writers to open a discursive space where readers can dispute their interpretations. Claim-making is risky because it can contradict existing literature or challenge the research of one's readers, which means that arguments must accommodate readers' expectations that they will be allowed to participate in a dialogue and that their own views will be acknowledged in the discourse. By marking statements as provisional, hedges seek to involve readers as participants in their ratification, conveying deference, modesty, or respect for colleagues views (Hyland, 1998a). Two of my informants noted this:

> Of course, I make decisions about the findings I have, but it is more convincing to tie them closely to the results.
>
> (Phy interview)

> You have to relate what you say to your colleagues and we don't encourage people to go out and nail their colours to the mast as maybe they don't get it published.
>
> (Bio interview)

Boosters, on the other hand, are words like *clearly, obviously* and *demonstrate,* which allow writers to express their certainty in what they say and to mark involvement with the topic and solidarity with their audience. They function to stress shared information, group membership, and engagement with readers (Hyland, 1999). Like hedges, they often occur in clusters, underlining the writer's conviction in his or her argument (Example 2):

> (2) This brings us into conflict with Currie's account, for static images <u>surely</u> cannot trigger our capacity to recognize movement. If that were so, we would see the image as itself moving. With a few interesting exceptions we <u>obviously</u> do not see a static image as moving. Suppose, then, that we say that static images only depict instants. This too creates problems, for it suggests that we have a recognitional capacity for instants, and this seems <u>highly</u> dubious.
>
> (Phil)

Boosters can therefore help writers to present their work with assurance while effecting interpersonal solidarity, setting the caution and self-effacement suggested by hedges against assertion and involvement.

Both boosters and hedges represent a writer's response to the potential viewpoints of readers and an acknowledgement of disciplinary norms of appropriate argument. They balance objective information, subjective evaluation and interpersonal negotiation, and this can be a powerful factor in gaining acceptance for claims. Both strategies emphasize that statements not only communicate ideas, but also the writer's attitude to them and to readers. Writers must weigh up the commitment they want to invest in their arguments based on its epistemic status and the effect this commitment might have on readers' responses. These comments from my interview data suggest the importance of getting this balance right:

> I'm very much aware that I'm building a façade of authority when I write, I really like to get behind my work and get it out there. Strong. Committed.

That's the voice I'm trying to promote, even when I'm uncertain I want to be behind what I say.

(Soc interview)

You have to be seen to believe what you say. That they are *your* arguments. It's what gives you credibility. It's the whole point.

(Phil interview)

I like tough minded verbs like 'think'. It's important to show where you stand. The people who are best known have staked out the extreme positions. The people who sit in the middle and use words like 'suggest', no one knows their work.

(Soc interview)

Attitude markers indicate the writer's affective, rather than epistemic, attitude to propositions, conveying surprise, agreement, importance, frustration, and so on, rather than commitment. While attitude is expressed throughout a text by the use of subordination, comparatives, progressive particles, punctuation, text location, and so on, it is most explicitly signalled by attitude verbs (e.g. *agree, prefer*), sentence adverbs (*unfortunately, hopefully*), and adjectives (*appropriate, logical, remarkable*). By signalling an assumption of shared attitudes, values and reactions to material, writers both express a position and pull readers into a conspiracy of agreement so that it can often be difficult to dispute these judgements (Example 3):

(3) these learner variables should prove to be promising areas for further research.

(Bio)

. . . . two quantities are rather important and, for this reason, the way they were measured is re-explained here.

(ME)

The first clue of this emerged when we noticed a quite extraordinary result.

(Phil)

Student A2 presented another fascinating case study in that he had serious difficulties expressing himself in written English.

(AL)

Self-mention refers to the use of first person pronouns and possessive adjectives to present propositional, affective and interpersonal information (Hyland, 2001). Presenting a discoursal self is central to the writing process (Ivanic, 1998), and writers cannot avoid projecting an impression of themselves and how they stand in relation to their arguments, their discipline, and their readers. The presence or absence of explicit author reference is generally a conscious choice by writers to adopt a particular stance and disciplinary-situated authorial identity. In the sciences it is common for writers to downplay their personal role to highlight the phenomena under study, the replicability of research activities, and the generality of the findings, subordinating their own voice to that of unmediated nature. Such a strategy subtly conveys an empiricist ideology that suggests research outcomes would be the same irrespective of the individual conducting it. One of my respondents expressed this view clearly:

> I feel a paper is stronger if we are allowed to see what was done without 'we did this' and 'we think that'. Of course we know there are researchers there, making interpretations and so on, but this is just assumed. It's part of the background. I'm looking for something interesting in the study and it shouldn't really matter who did what in any case.
>
> (Bio interview)

In the humanities and social sciences, in contrast, the use of the first person is closely related to the desire to both strongly identify oneself with a particular argument and to gain credit for an individual perspective. Personal reference is a clear indication of the perspective from which a statement should be interpreted, enabling writers to emphasize their own contribution to the field and to seek agreement for it (Example 4):

> (4) I argue that their treatment is superficial because, despite appearances, it relies solely on a sociological, as opposed to an ethical, orientation to develop a response.
>
> (Soc)

> I bring to bear on the problem my own experience. This experience contains ideas derived from reading I have done which might be relevant to my puzzlement as well as my personal contacts with teaching contexts.
>
> (AL)

In these more discursive domains, then, self-mention clearly demarcates the writer's role in the research:

> Using 'I' emphasizes what you have done. What is yours in any piece of research. I notice it in papers and use it a lot myself.
>
> (Soc interview)

The personal pronoun 'I' is very important in philosophy. It not only tells people that it is your own unique point of view, but that you believe what you are saying. It shows your colleagues where you stand in relation to the issues and in relation to where they stand on them. It marks out the differences.

(Phil interview)

Engagement and features of reader positioning

In comparison with stance, the ways writers bring readers into the discourse to anticipate their possible objections and engage them in appropriate ways have been relatively neglected in the literature. Based on their previous experiences with texts, writers make predictions about how readers are likely to react to their arguments. They know what they are likely to find persuasive, where they will need help in interpreting the argument, what objections they are likely to raise, and so on. This process of audience evaluation therefore assists writers in constructing an effective line of reasoning and, like stance options, also points to the ways language is related to specific cultural and institutional contexts (Hyland, 2001a). There are two main purposes to writers' uses of engagement strategies:

1 Acknowledgement of the need to adequately meet readers' expectations of inclusion and disciplinary solidarity. Here we find readers addressed as participants in an argument with reader pronouns and interjections.

2 To rhetorically position the audience. Here the writer pulls readers into the discourse at critical points, predicting possible objections and guiding them to particular interpretations with questions, directives and references to shared knowledge.

Again, these two functions are not always clearly distinguishable, as writers invariably use language to solicit reader collusion on more than one front simultaneously. They do, however, help us to see some of the ways writers project readers into texts and how this is done in different disciplines. There are five main elements to engagement:

1 Reader pronouns.

2 Personal asides.

3 Appeals to shared knowledge.

4 Directives.

5 Questions.

Reader pronouns are perhaps the most explicit way that readers are brought into a discourse. *You* and *your* are actually the clearest way a writer can acknowledge the reader's presence, but these forms are rare outside of philosophy, probably because they imply a lack of involvement between participants. Instead, there is enormous emphasis on binding writer and reader together through inclusive *we,* which is the most frequent engagement device in academic writing. It sends a clear signal of membership by textually constructing both the writer and the reader as participants with similar understanding and goals. This was recognized by my informants:

> Part of what you are doing in writing a paper is getting your readers onside, not just getting down a list of facts, but showing that you have similar interests and concerns. That you are looking at issues in much the same way they would, not spelling everything out, but following the same procedures and asking the questions they might have.
>
> (Bio interview)

> I often use 'we' to include readers. I suppose it brings out something of the collective endeavour, what we all know and want to accomplish. I've never thought of it as a strategy, but I suppose I am trying to lead readers along with me.
>
> (ME interview)

In addition to claiming solidarity, these devices also set up a dialogue by weaving the potential point of view of readers into the discourse, thereby anticipating their objections, voicing their concerns, and expressing their views. Thus, we helps guide readers through an argument and towards a preferred interpretation, often shading into explicit positioning of the reader (Example 5):

> (5) Now that we have a plausible theory of depiction, we should be able to answer the question of what static images depict. But this turns out to be not at all a straightforward matter. We seem, in fact, to be faced with a dilemma. Suppose we say that static images can depict movement. This brings us into conflict with Currie's account, ...
>
> (Phil)

Although we lack knowledge about a definitive biological function for the transcripts from the 93D locus, their sequences provide us with an ideal system to identify a specific transcriptionally active site in embryonic nuclei.

(Bio)

Personal asides allow writers to address readers directly by briefly interrupting the argument to offer a comment on what has been said. While asides express something of the writer's personality and willingness to explicitly intervene to offer a view, they can also be seen as a key reader-oriented strategy. By turning to the reader in mid-flow, the writer acknowledges and responds to an active audience, often to initiate a brief dialogue that is largely interpersonal. As we can see, such comments often add more to the writer–reader relationship than to the propositional development of the discourse (Example 6):

(6) And – as I believe many TESOL professionals will readily acknowledge – critical thinking has now begun to make its mark, particularly in the area of L2 composition.

(AL)

He above all provoked the mistrust of academics, both because of his trenchant opinions (often, it is true, insufficiently thought out) and his political opinions.

(Soc)

What sort of rigidity a designator is endowed with seems to be determined by convention (this, by the way, is exactly the target of Wittgensteinian critiques of Kripke's essentialism).

(Phil)

This kind of engagement builds a relationship between participants which is not dependent on an assessment of what needs to be made explicit to elaborate a position or ease processing constraints. It is an intervention simply to connect: to show that both writer and readers are engaged in the same game and are in a position to draw on shared understandings. While all writing needs to solicit reader collusion, this kind of engagement is far more common in the soft fields. Because they deal with greater contextual vagaries, less predictable variables, and more diverse research outcomes, readers must be drawn in and be involved as participants in a dialogue to a greater extent than in the sciences.

Appeals to shared knowledge seek to position readers within apparently naturalized boundaries of disciplinary understandings. The notion of 'sharedness' is often invoked by writers to smuggle contested ideas into their argument, but here I am simply referring to the presence of explicit markers where readers are asked to recognize something as familiar or accepted. Obviously readers can only be brought to agree with the writer by building on some kind of implicit contract concerning what can be accepted, but often these constructions of solidarity involve explicit calls asking readers to identify with particular views. In doing so, writers are actually constructing readers by presupposing that they hold such beliefs, assigning to them a role in creating the argument, acknowledging their contribution while moving the focus of the discourse away from the writer to shape the role of the reader (Example 7):

(7) Of course, we know that the indigenous communities of today have been reorganized by the catholic church in colonial times and after,........

(Soc)

This tendency obviously reflects the preponderance of brand-image advertising in fashion merchandising.

(Mk)

Chesterton was of course wrong to suppose that Islam denied 'even souls to women'.

(Phil)

This measurement is distinctly different from the more familiar NMR pulsed field gradient measurement of solvent self-diffusion.

(Phy)

Over three-quarters of such explicit appeals to collective understandings in the corpus were in the soft papers. Writers of scientific papers expect their readers to have considerable domain knowledge and to be able to decode references to specialized methods, instruments, materials, and models, but these understandings are signalled less explicitly.

Directives instruct the reader to perform an action or to see things in a way determined by the writer. They are signalled mainly by the presence of an imperative (like *consider, note,* and *imagine*); by a modal of obligation addressed to the reader (such as *must, should,* and *ought*); and by a predicative adjective expressing the writer's judgement of necessity/importance (*It is important to understand* . . .). Directives can be seen as directing readers to engage in three main kinds of activity (Hyland, 2002a):

1 Textual acts.

2 Physical acts.

3 Cognitive acts.

First, *textual acts* are used to metadiscoursally guide readers through the discussion, steering them to another part of the text or to another text (Example 8):

(8) See Lambert and Jones (1997) for a full discussion of this point.

(Soc)

Look at Table 6.2 again for examples of behavioristic variables.

(Mk)

Consult Cormier and Gunn 1992 for a recent survey

(EE)

Second, *physical acts* instruct readers how to carry out research processes or to perform some action in the real world (Example 9):

(9) Before attempting to measure the density of the interface states, one should freeze the motion of charges in the insulator.

(EE)

Mount the specimen on the lower grip of the machine first, . . .

(Bio)

Set the sliding amplitude at 30mm traveling distance.

(ME)

Finally, *cognitive acts* guide readers through a line of reasoning, or get them to understand a point in a certain way and are therefore potentially the most threatening type of directives. They accounted for almost half of all directives in the corpus, explicitly positioning readers by leading them through an argument to the writer's claims (Example 10) or emphasizing what they should attend to in the argument (Example 11):

(10) Consider a sequence of batches in an optimal schedule.

(EE)

Think about it. What if we eventually learn how to communicate with aliens.

(Soc)

(11) It is important to note that these results do indeed warrant the view that..

(AL)

What has to be recognised is that these issues. . . . .

(ME)

Questions are the strategy of dialogic involvement *par excellence,* inviting engagement and bringing the interlocutor into an arena where they can be led to the writer's viewpoint (Hyland, 2002d). They arouse interest and encourage the reader to explore an unresolved issue with the writer as an equal, a conversational partner, sharing his or her curiosity and following where the argument leads. Over 80 percent of questions in the corpus, however, were rhetorical, presenting an opinion as an interrogative so the reader appears to be the judge, but actually expecting no response. This kind of rhetorical positioning of readers is perhaps most obvious when the writer poses a question only to reply immediately, simultaneously initiating and closing the dialogue (Example 12):

(12) Is it, in fact, necessary to choose between nurture and nature? My contention is that it is not.

(Soc)

What do these two have in common, one might ask? The answer is that they share the same politics.

(AL)

Why does the capacitance behave this way? To understand we first notice that at large B there are regular and nearly equal-spaced peaks in both C3,(B) and C31(−B).

(Phy)

Stance and engagement practices: corpus findings

Analysis of the research article corpus shows that the expression of stance and engagement is an important feature of academic writing, with 200 occurrences in each paper, about one every 28 words. Table 6.1 shows that stance markers were about five times more common than engagement features and that hedges were by far the most frequent feature of writer perspective in the corpus, reflecting the critical importance of distinguishing fact from opinion and the need for writers to present their claims with appropriate caution and regard to colleagues' views.

The significance of these frequencies can be more clearly understood in comparison to other common features of published academic writing. Biber et al. (1999), for instance, give figures of 18.5 cases per 1000 words for passive voice constructions and 20 per 1000 words for past tense verbs. These overt interaction markers can therefore be seen as an important element of academic

TABLE 6.1 Stance and engagement features in the research articles

Stance	Items per 1000 words	% of total	Engagement	Items per 1000 words	% of total
Hedges	14.5	46.6	Reader pronouns	2.9	49.1
Attitude markers	6.4	20.5	Directives	1.9	32.3
Boosters	5.8	19.2	Questions	0.5	8.5
Self-mention	4.2	13.7	Knowledge ref	0.5	8.2
			Asides	0.1	1.9
Totals	30.9	100		5.9	100

TABLE 6.2 Stance and engagement features by discipline (per 1000 words)

Feature	Phil	Soc	AL	Mk	Phy	Bio	ME	EE	Total
Stance	**42.8**	**31.1**	**37.2**	**39.5**	**25.0**	**23.8**	**19.8**	**21.6**	**30.9**
Hedges	18.5	14.7	18.0	20.0	9.6	13.6	8.2	9.6	14.5
Attitude mkrs	8.9	7.0	8.6	6.9	3.9	2.9	5.6	5.5	6.4
Boosters	9.7	5.1	6.2	7.1	6.0	3.9	5.0	3.2	5.8
Self-mention	5.7	4.3	4.4	5.5	5.5	3.4	1.0	3.3	4.2
Engagement	**16.3**	**5.1**	**5.0**	**3.2**	**4.9**	**1.6**	**2.8**	**4.3**	**5.9**
Reader ref	11.0	2.3	1.9	1.1	2.1	0.1	0.5	1.0	2.9
Directives	2.6	1.6	2.0	1.3	2.1	1.3	2.0	2.9	1.9
Questions	1.4	0.7	0.5	0.3	0.1	0.1	0.1	0.0	0.5
Shared knowledge	1.0	0.4	0.6	0.4	0.5	0.1	0.3	0.4	0.5
Asides	0.2	0.2	0.1	0.1	0.0	0.0	0.0	0.0	0.1
Total	**59.1**	**36.2**	**42.2**	**42.7**	**29.9**	**25.4**	**22.6**	**25.9**	**36.8**

prose. Perhaps more interesting, however, are the disciplinary distributions. Table 6.2 shows the density of features in each discipline normalized to a text length of 1000 words. As can be seen, the more discursive 'soft' fields of philosophy, marketing, sociology and applied linguistics, contained the highest proportion of interactional markers with some 75 percent more items than the engineering and science papers.

Stance, engagement and disciplinarity

It is clear that writers in different disciplines represent themselves, their work and their readers in different ways, with those in the humanities and social sciences taking far more explicitly involved and personal positions than those in the science and engineering fields. As I noted at the beginning of this article, the reason for this is that the resources of language mediate their contexts, working to construe the characteristic structures of knowledge domains and argument forms of the disciplines that create them.

In broad terms, rhetorical practices are inextricably related to the purposes of the disciplines. Natural scientists tend to see their goal as producing public knowledge able to withstand the rigours of falsifiability and developed through relatively steady cumulative growth (Becher, 1989). The fact that this research often occupies considerable investments in money, training, equipment, and expertise means it is frequently concentrated at a few sites and commits scientists to involvement in specific research areas for many years. Problems therefore emerge in an established context so that readers are often familiar with prior texts and research, and that the novelty and significance of contributions can easily be recognized. The soft knowledge domains, in contrast, are more interpretative and less abstract, producing discourses which often recast knowledge as sympathetic understanding, promoting tolerance in readers through an ethical rather than cognitive progression (Dillon, 1991; Hyland, 2000a). There is, moreover, less control of variables and greater possibilities for diverse outcomes, so writers must spell out their evaluations and work harder to establish an understanding with readers.

While there are clear dangers in reifying the ideologies of practitioners, these broad ontological representations have real rhetorical effects. They allow, for instance, the sciences to emphasize demonstrable generalizations rather than interpreting individuals, so greater burden is placed on research practices and the methods, procedures and equipment used. New knowledge is accepted on the basis of empirical demonstration, and science writing reinforces this by highlighting a gap in knowledge, presenting a hypothesis related to this gap, and then conducting experiments and presenting findings to support the hypothesis. In soft areas, however, the context often has to be elaborated anew, its more diverse components reconstructed for a less cohesive readership. Writers are far less able to rely on general understandings and on the acceptance of proven quantitative methods to establish their claims and this increases the need for more explicit evaluation and engagement. Personal credibility, and explicitly getting behind arguments, play a far greater part in creating a convincing discourse for these writers.

The suggestion that 'hard' knowledge is cumulative and tightly structured not only allows for succinct communication, but also contributes to the apparently 'strong' claims of the sciences. The degree to which the background to a problem and the appropriate methods for its investigation can be taken for granted means there are relatively clear criteria for establishing or refuting claims and this is reflected in writers' deployment of evidential markers. While writers in all disciplines used hedges in the evaluation of their statements, they were considerably more frequent in the soft disciplines, perhaps indicating less assurance about what colleagues could be safely assumed to accept. The use of a highly formalized reporting system also allows writers in the hard disciplines to minimize their presence in their texts. In the soft disciplines where what counts as adequate explanation is less assured, interpretative variation increases and writers must rely to a greater extent on a personal projection into the text, through self-mention and attitude markers to invoke an intelligent reader and a credible, collegial writer.

In addition to creating an impression of authority, integrity and credibility through choices from the stance options, writers are able to either highlight or downplay the presence of their readers in the text. As we have seen, the most frequent engagement devices in the corpus were reader pronouns and over 80 percent of these occurred in the soft discipline papers where they functioned to appeal to scholarly solidarity, presupposing a set of mutual, discipline-identifying understandings linking writer and reader. They also claim authority as well as communality, however, addressing the reader from a position of confidence as several of my informants noted:

> I suppose we help to finesse a positive response – we are all in this together kind of thing. I use it to signal that I am on the same wavelength, drawing on the same assumptions and asking the same questions.
>
> (Mk interview)

> It helps to locate you in a network. It shows that you are just doing and thinking what they might do and think. Or what you would like them to, anyway.
>
> (Soc interview)

Similarly, questions were largely confined to the soft fields. The fact they reach out to readers was seen as a distraction by my science informants:

> Questions are quite rare in my field I think. You might find them in textbooks I suppose, but generally we don't use them. They seem rather intrusive, don't they? Too personal. We generally prefer not to be too intrusive.
>
> (ME interview)

> I am looking for the results in a paper, and to see if the method was sound. I am looking for relevance and that kind of dressing is irrelevant. People don't ask questions as it would be seen as irrelevant. And condescending probably.
>
> (EE interview)

In contrast, the soft knowledge writers saw them as an important way of relating to readers:

> In my field that's all there are, questions. Putting the main issues in the form of questions is a way of presenting my argument clearly and showing them I am on the same wavelength as them.
>
> (Phil interview)

> Often I structure the argument by putting the problems that they might ask.
>
> (Mk interview)

Finally, directives were the only interactive feature which occurred more frequently in the science and engineering papers. Generally, explicit engagement is a feature of the soft disciplines, where writers are less able to rely on the explanatory value of accepted procedures, but directives are a potentially risky tactic and, as a result, most directives in the soft fields were textual, directing readers to a reference rather than informing them how they should interpret an argument. Two of my respondents noted this in their interviews:

> I am very conscious of using words like 'must' and 'consider' and so on and use them for a purpose. I want to say 'Right, stop here. This is important and I want you to take notice of it'. So I suppose I am trying to take control of the reader and getting them to see things my way.
>
> (Soc interview)

> I am aware of the effect that an imperative can have so I tend to use the more gentle ones. I don't want to bang them over the head with an argument I want them to reflect on what I'm saying. I use 'consider' and 'let's look at this' rather than something stronger.
>
> (AL interview)

The more linear and problem-oriented approach to knowledge construction in the hard knowledge fields, on the other hand, allows arguments to be formulated in a highly standardized code. Articles in the sciences also tend to be much shorter, probably due to editorial efforts to accommodate the rapid growth of knowledge and high submission rates in many sciences. These factors place a premium on succinctness, and directives provide an economy of expression highly valued by space-conscious editors and information-saturated scientists, as several informants noted:

> I rarely give a lot of attention to the dressing, I look for the meat – the findings – and if the argument is sound. If someone wants to save me time in getting there then that is fine. No, I'm not worried about imperatives leading me through it.
>
> (EE interview)

> I'm very conscious of how I write and I am happy to use an imperative if it puts my idea over clearly. Often we are trying to work to word limits anyway, squeezing fairly complex arguments into a tight space.
>
> (ME interview)

In sum, these different features, taken together, are important ways of situating academic arguments in the interactions of members of disciplinary communities. They represent relatively conventional ways of making meaning and so elucidate a context for interpretation, showing how writers and readers make connections, through texts, to their disciplinary cultures.

Conclusion

My claim has been that effective academic writing depends on rhetorical decisions about interpersonal intrusion and I have suggested a model which attempts to show how writers select and deploy community-sensitive linguistic resources to represent themselves, their positions and their readers. The account I have provided, however, is necessarily a partial one, representing only the broadest categories of rhetorical function. There are certainly more fine-grained distinctions to be made among these resources which are likely to offer further insights into the rhetorical options available to writers and the patterns of effective persuasion employed by different communities.

There are also obvious limitations with the kind of corpus approach I have adopted. Unlike the detailed studies of part genres, such as Swales' (1990) work on introductions, Brenton's (1996) study of conference abstracts, or Brett's (1994) analysis of results sections, for example, a corpus study is unable to provide information about where these features are likely to cluster. Several studies suggest that greater writer intrusion is a characteristic of Introduction and Discussion sections, where argument is emphasized and decisions, claims and justifications are usually found (e.g. Gosden, 1993; Hanania and Akhtar, 1985). While it seems to be an intuitively reasonable assumption that stance and engagement work is most likely to be done here, the division of research papers into rhetorically simple and detached Methods and Results, and complex, subjective and author-centred Introductions and Discussions might be unwise. Even the most rhetorically innocent sections reveal writers' efforts to persuade their audience of their claims, so that stance and engagement are likely to figure, in different ways, across the research paper. Indeed, as Knorr-Cetina (1981) pointed out many years ago, the IMRD structure is itself a rhetorical artefact.

It should also be noted that this creation of an authorial persona is an act of personal choice, and the influence of individual personality, confidence, experience, and ideological preference are clearly important. We are not the instruments of our disciplines and variables such as individuality and ideolecticity are important limitations on the kind of analysis presented here. It may even be the case, as John Swales (pers. comm.) has observed, that a few famous writers (perhaps Heidegger, Wittgenstein, Sartre, and Halliday) do not play this interactive game with their audiences. However, writers do not act in a social vacuum, and knowledge is not constructed outside particular communities of practice. Such communities exist in virtue of a shared set of assumptions and routines about how to collectively deal with and represent their experiences. The ways language is used on particular occasions are not wholly determined by these assumptions, but a disciplinary voice can only be achieved through a process of participating in such communities and connecting with these socially determined and approved beliefs and value positions. In this way, independent creativity is shaped by accountability to shared practices.

I hope to have shown, then, that stance and engagement are important elements both of a writer's argument and of a disciplinary context as they seek to bring writer and readers into a text as participants in an unfolding dialogue. The model presented here offers a plausible description of academic interaction and suggests how writers anticipate and understand their readers' background knowledge, interests, and interpersonal expectations to control how they respond to a text and to manage the impression they gain of the writer.

7

Metadiscourse in academic writing: A reappraisal

Ken Hyland and Polly Tse

Introduction

Metadiscourse, often wrongly characterized as 'discourse about discourse', is a concept familiar to many engaged in research and instruction in composition, reading, and text structure. Based on a view of writing as a social and communicative engagement between writer and reader, metadiscourse focuses our attention on the ways writers project themselves into their discourse to signal their attitude towards both the content and the audience of the text. As a result, it has been taken up by researchers of both social constructionist and functional orientations to discourse and by corpus analysts attracted by the possibility of tracing patterns of interaction and cohesion across texts.

Metadiscourse is an intuitively attractive concept as it seems to offer a motivated way of collecting under one heading the range of devices writers use to explicitly organize their texts, engage readers, and signal their attitudes to both their material and their audience. This promise, however, has never been fully realized because metadiscourse remains under-theorized and empirically vague. The failure to pin the term down precisely has meant that analysts have been unable to confidently operationalize the concept in real texts, making analysis an elusive and frustrating experience.

The view of metadiscourse and the descriptive framework discussed in this paper emerges from a corpus analysis of 240 dissertations written by

L2 postgraduate students from five Hong Kong universities totalling 4 million words. The corpus consists of 20 masters and 20 doctoral dissertations from each of six academic disciplines: Electronic Engineering (EE), Computer Science (CS), Business Studies (BS), Biology (Bio), Applied Linguistics (AL), and Public Administration (PA). Our purpose is to offer a reassessment of metadiscourse, present some key principles, and propose a more robust model of the concept based on our study of this corpus.

Conceptions of metadiscourse

Metadiscourse is defined here as the linguistic resources used to organize a discourse or the writer's stance towards either its content or the reader (Hyland 2000a: 109). It is typically used as an umbrella term to include a heterogeneous array of cohesive and interpersonal features which help relate a text to its context by assisting readers to connect, organize, and interpret material in a way preferred by the writer and with regard to the understandings and values of a particular discourse community (Hyland 1998d). While some analysts have narrowed the focus of metadiscourse to features of textual organization (Bunton 1999; Mauranen 1993a, b; Valero-Garces 1996) or explicit illocutionary predicates (Beauvais 1989), metadiscourse is more generally seen as the author's linguistic and rhetorical manifestation in the text in order to 'bracket the discourse organisation and the expressive implications of what is being said' (Schiffrin 1980: 231).

With the judicious addition of metadiscourse, a writer is able to not only transform a dry, difficult text into coherent, reader-friendly prose, but also relate it to a given context and convey his or her personality, credibility, audience-sensitivity, and relationship to the message (Hyland 2000a). Metadiscourse is, therefore, a functional category and, as shown in these extracts from our L2 postgraduate corpus, can be realized through a range of linguistic units, from exclamatory punctuation and scare quotes (1), to whole clauses (2), and even sequences of several sentences (3):

(1) I admit that the term 'error' may be an undesirable label to some teachers.   (AL PhD)

The geography curriculum teaches about representative fractions, scales and ratios in Form 1 (age 12+) whilst mathematics study does not deal with this topic until Form 2!   (Bio MSc)

(2) First, let us consider an oversaturated cross cut.   (CS PhD)

The rest of this chapter will be divided into four sections.   (PA MA)

(3) In this section, we will discuss what classifications scholars have made in the past. Based on their work, a multiple classification system will be developed to group puns under different categories. With the help of this classification system, puns (particularly the data collected for this study) can be analysed more easily.   (AL MA)

The organization of this paper will be as follows. Chapter 2 is a review of Hong Kong air cargo industry. Chapter 3 is a literature review. Chapter 4 is a model on measuring the multiplier effects brought by air cargo industry to Hong Kong labour market. Drivers and constraints for future growth of Hong Kong air cargo industry follow in Chapter 5. And the last Chapter is conclusions and recommendations.   (BS PhD)

These varied realizations mean there are no simple linguistic criteria for identifying metadiscourse. Not only is it an open category to which new items can be added to fit the writer's needs, but the same items can function as metadiscourse in some parts of the text and not in others. Consequently, metadiscourse studies begin with functional classifications and analyses of texts.

These classifications embrace those ways which allow the author to intrude into the evolving text to direct readers' reception of it. Generally, metadiscoursal comments are argued to have two main functions: textual and interpersonal. The first kind helps to organize the discourse by pointing out topic shifts, signalling sequences, cross-referencing, connecting ideas, previewing material, and so on. The second kind modifies and highlights aspects of the text and gives the writer's attitude to it with hedges, boosters, self-reference, and features generally labelled as *evaluation* (Hunston and Thompson 2001) or *appraisal* (Martin 2001). Broad functions are thus subdivided into more specific functions through which the writer regulates ongoing interaction and helps make the text comprehensible to a particular readership.

While the term is not always used in the same way (for example, Swales 1990: 188), metadiscourse has been a concern in a range of recent work in text analysis. It has informed studies into the properties of texts, participant interactions, historical linguistics, cross-cultural variations, and writing pedagogy. Studies have suggested the importance of metadiscourse in casual conversation (Schiffrin 1980), school textbooks (Crismore 1989), science popularizations (Crismore and Farnsworth 1990), undergraduate textbooks (Hyland 2000a), postgraduate dissertations (Bunton 1999; Swales 1990), Darwin's *Origin of the Species* (Crismore and Farnsworth 1989) and company annual reports (Hyland 1998b). It appears to be a characteristic of a range of languages and genres and has been used to investigate rhetorical differences in the texts written by different cultural groups (Mauranen 1993b; Crismore, Markkanen, and Steffensen 1993; Valero-Garces 1996). It has also been shown to be present in medieval medical writing (Taavitsainen 1999), to be a quality of scientific discourse from the late seventeenth century (Atkinson 1999), a feature of good ESL and native speaker student writing (Intraprawat and Steffensen 1995; Cheng and Steffensen 1996) and an essential element of persuasive and argumentative discourse (Crismore and Farnsworth 1990; Hyland 1998b).

In summary, metadiscourse is recognized as an important means of facilitating communication, supporting a writer's position and building a relationship with an audience. Yet despite this research interest, metadiscourse has never become a major analytical focus in the study of written discourse, nor has it produced the insights into language registers that were originally hoped for. Even in the area of academic writing, where most research is concentrated, metadiscourse studies have been suggestive rather than definitive, and analysts have turned to other concepts such as evaluation (Hunston and Thompson 2001) and engagement (Hyland 2001a) as potentially more productive ways of exploring interpersonal features of discourse. Essentially, its origins in pedagogic style guides (Williams 1981) and intuitive reflection (Vande Kopple 1985), provide an insufficiently solid theoretical foundation on which to analyse real texts or to understand how writers communicate effectively.

Key principles of metadiscourse

We wish to suggest a new model for metadiscourse in academic writing which builds on three key principles of metadiscourse. These are:

1 that metadiscourse is distinct from propositional aspects of discourse;

2 that the term 'metadiscourse' refers to those aspects of the text that embody writer–reader interactions;

3 that metadiscourse distinguishes relations which are external to the text from those that are internal.

In this section we will briefly discuss these principles, and then go on to suggest a robust framework which sees metadiscourse as a means of conceptualizing interpersonal relations in academic writing.

Propositional vs. non-propositional discourse

Definitions of metadiscourse usually make a clear distinction between metadiscourse and propositional content, often regarding the latter as 'primary'. Thus Vande Kopple (1985) defines metadiscourse as 'the linguistic material which does not add propositional information but which signals the presence of an author' and Williams (1981: 226) says it is 'whatever does not refer to the subject matter being addressed'. Similarly, Crismore, Markkanen, and Steffensen (1993) state that metadiscourse is:

> Linguistic material in texts, written or spoken, which does not add anything to the propositional content but that is intended to help the listener or reader organize, interpret and evaluate the information given. (Crismore *et al*. 1993: 40)

What is understood by the term 'proposition' is often left vague, but it is generally used to refer to all that which concerns thoughts, actors, or states of affairs in the world outside the text. Halliday (1994: 70), for example, states that propositional material is something that can be argued about, affirmed, denied, doubted, insisted upon, qualified, tempered, regretted, and so on.

Unfortunately however, this idea of propositional content does not rule out much of what is typically considered as metadiscourse. In fact, it is sometimes difficult to distinguish what is content from what is not and the traditional philosophical test of falsifiability is often of little assistance. In formal semantics, the term 'proposition' refers to the logico-semantic unit capable of being assigned a truth value, but both propositional and metadiscoursal aspects of texts are subject to similar infelicities or misfires. Mao (1993: 267) points out, for example, the explicit act of hypothesizing fails if what is hypothesized is a well-acknowledged fact. The picture is further clouded by inconsistencies in the metadiscourse literature itself. Crismore (1989; Crismore and Farnsworth 1990), for instance, includes 'referential, informational metadiscourse' in her classification, apparently referring to Halliday's ideational function of language or the ways writers express their ideas and experiences, and thus reintroduces propositional material back into metadiscourse.

In contrast, other writers have drawn the line between metadiscourse and propositional matter more firmly. Vande Kopple (2002), for instance, talks of different *levels of meaning*:

> On one level we expand ideational material. On the levels of metadiscourse, we do not expand ideational material but help our readers connect,

organise, interpret, evaluate, and develop attitudes towards that material. (Vande Kopple 2002: 93)

It is difficult to see, however, how metadiscourse can constitute a different level of meaning. It is certainly possible, even commonplace, to distinguish the propositional content of a text from the particular way it is expressed, for even the most idiosyncratic readings are constrained by the text and the conventions of a community of readers. Such content can be rewritten, summarized, paraphrased, and reformulated in different ways and, indeed, academic texts often undergo transformations of this kind, from their original appearance in research articles to new forms in popularizations, textbooks, grant proposals, abstracts, and undergraduate essays (for example Myers 1990). However, it is axiomatic that the meaning of a text depends on the integration of its component elements, and these cannot be separated into independent 'meanings'. Such retextualizations for different genres, purposes, and audiences will have different meanings, but a recognizable identity of content.

The point that we are making here is that a propositional content–metadiscourse distinction is required as a starting point for exploring metadiscourse in academic writing, but it is unwise to push this distinction too far. It is true that academic texts are usually concerned with issues other than themselves. They seek to inform readers of activities, objects, or people in the world, to persuade them to some action or thought, or seek to promote the writer's scholarly claims and credentials. Equally though, a large proportion of every text is not concerned with the world, but with its internal argument and its readers. Further, this is not somehow 'secondary' to the meaning of the text, simply supporting propositional content, but the means by which propositional content is made coherent, intelligible, and persuasive to a particular audience.

Both propositional and metadiscoursal elements occur together in texts, often in the same sentences, and we should not be surprised that a stretch of discourse may have both functions. Such integration is common, with each element expressing its own content: one concerned with the world and the other with the text and its reception. Like propositional discourse, metadiscourse is able to convey the writer's intended meaning in a given situation; it is part of the message, not an entirely different one.

A rigid conceptual separation between proposition and metadiscourse relegates the latter to a commentary on the main informational purpose of the text rather than seeing it as an integral process of communicating meaning. Metadiscourse is not simply the 'glue' that holds the more important parts of the text together, but is itself a crucial element of its meaning —that which helps relate a text to its context, taking readers' needs, understandings, existing knowledge, prior experiences with texts, and relative status into account. In other words, we blur the unhelpful distinction between 'primary' propositional discourse and 'secondary' metadiscourse and seek to recover the link between the ways writers intrude into their texts to organize and comment on it so that it is appropriate for a particular rhetorical context.

Writer–reader interactions

A second principle of our model sees metadiscourse as embodying the interactions necessary for successful communication. As such, it rejects the strict duality of textual and interpersonal functions found in much of the metadiscourse literature (for example Crismore and Farnsworth 1990; Crismore *et al.* 1993; Hyland 1998a, 2000a; Vande Kopple 1985). We suggest instead that all metadiscourse is interpersonal in that it takes account of the reader's knowledge, textual experiences, and processing needs and that it provides writers with an armoury of rhetorical appeals to achieve this.

The textual–interpersonal categorization ostensibly draws on Halliday's (1994) tripartite conception of metafunctions which distinguishes between the ideational elements of a text, the ways we encode our experiences of the world, and its textual and interpersonal functions. But while Halliday's terminology lends a certain theoretical respectability to the idea of metadiscourse, the concept plays no part in his thinking, and metadiscourse researchers do not necessarily subscribe to a functional grammar or to Halliday's assertion that all three functions are realized simultaneously. Instead, they separate those aspects which help to organize material as coherent discourse and those which convey the writer's attitudes to the text. Thus, Vande Kopple (1985: 87) believes that *textual metadiscourse* 'shows how we link and relate individual propositions so that they form a cohesive and coherent text and how individual elements of those propositions make sense in conjunction with other elements of the text'. Interpersonal metadiscourse, on the other hand 'can help us express our personalities and our reactions to the propositional content of our texts and characterise the interaction we would like to have with our readers about that content'.

In practice there are serious difficulties with this attempt to identify two single, discrete functions of metadiscourse. Most importantly this is because textual resources do not constitute a neatly separable set which can be clearly distinguished from either propositional or interpersonal aspects. Most textual metadiscourse signals are realized by conjuncts (*so, because, and*) and adverbials (*subsequently, first, therefore*), together with their respective metaphorical or paraphrasing expressions (*as a result, on the other hand, needless to say*), but these do not *only* create textual links. Unlike propositional and interpersonal meanings, which orient to extra-linguistic phenomena, the textual function is intrinsic to language and exists to construe both propositional and interpersonal aspects into a linear and coherent whole. Textual elements thus have an *enabling* role (Halliday 1994), facilitating the creation of discourse by allowing writers to generate texts which make sense within their context. Their role is crucial to expressing propositional and interpersonal functions, not something they do independently of them.

For Halliday, and those working in a systemic linguistics tradition, the textual function is principally realized by cohesive devices and by the choices a writer makes in giving prominence to information as 'given' or 'new' by locating it at either the beginning or the end of the clause. Theme choices help illustrate the simultaneity of functions as they not only provide for the development of a text, but also what the writer sees as key elements. The theme helps to signpost what writers have in mind as a starting point, the frame they have chosen for their message, and so also highlights the particular ideational or interpersonal information that best reflects their intentions and assessments of reader needs in developing the message. In other words, we should see text as a process in which writers are simultaneously creating propositional content, interpersonal engagement, and the flow of text as they write, which means that their linguistic choices often perform more than one function.

Two clear examples of this overlap are the roles of conjunctions and modal adjuncts in thematic position. Conjunctions, for instance, function textually to relate a clause to the preceding text, but they also function ideationally to signal the writer's understanding of the logical relationships between ideas. They therefore not only glue the text together, but extend, elaborate, or enhance propositional meanings (4). Similarly, by exercising the option to thematize modal or comment adjuncts, writers both signal a textual relationship to preceding discourse and indicate an interpersonal relationship to the reader or the position being taken (5):

(4) The author accepted the shortcomings of the study due to the fact that it was a non-random sample. *Nevertheless,* the study did highlight that ageism is not confined to Western countries alone.   (SA PhD)

A parametric estimation technique using global optimization is introduced for the output space partition. *But we first discuss* the optimization technology in the next section.   (EE PhD)

(5) *Probably* the most interesting and significant category of lexical errors is 'word class' since it is the major type of error made by the subjects.
(AL PhD)

Undoubtedly, there are limitations to the findings of this thesis. (Bio MSc)

I believe the following aspects should be seriously considered and reviewed by the SAR government if they want to maintain the prospect of this industry. (Bus MA)

Distinguishing a purely textual role for metadiscourse is therefore rather more problematic than many metadiscourse writers acknowledge, and this is also the case when considering cohesive markers. For those working in metadiscourse, conjunctive relations (called 'text connectives' by Vande Kopple (1985) and 'logical connectives' by Crismore *et al.* (1993)), are treated as 'straightforward and unproblematic' textual markers (Crismore *et al.* 1993: 48). But like other features of 'textual metadiscourse', the transitions that conjunctions mark between clauses can be oriented either towards the experiential or the interactional, referring to either propositional or interpersonal meanings. Our tendency to see conjunctions as expressing connections between ideas is perhaps a result of our primarily ideational orientation to the world. But while we expect academic texts to favour ideational meanings, we can also see conjunctions as interactionally motivated, contributing to the creation and maintenance of shifting interpersonal orientations.

The interpersonal orientation of conjunctions is perhaps most apparent in the use of concessive forms, as these both mark what the writer anticipates will be unexpected and also monitor the reader's response to the discourse (for example Martin and Rose 2003). In academic writing, tracking readers' expectations in this way is a vital interpersonal strategy. Concessives rhetorically acknowledge voices other than the writer's own by demonstrating a sensitivity to audience understandings and explicitly attempting to engage with these. In the following examples, for instance, writers are clearly doing more than creating a textually cohesive text; they are manoeuvring themselves into line with community expectations and shaping the reader's role to gain a more sympathetic hearing for their own views:

(6) *Even if* we assume that interlanguage is a viable research resource, exactly what constitutes input and output in relation to oral task performance is a definitional question which has no easy answers. (AL PhD)

Admittedly, the data collection of the present study may be classified as 'opportunistic', rendering the representativeness of the research findings very limited. (PA PhD)

The use of contrastive connectives is particularly important when writers seek to respond to potentially detracting information or competing interpretations.

This is why they are often used to mitigate counterclaims (Barton 1995), introducing an alternative statement in a two part structure:

(7) The multi-database language approach bypasses the schema integration problem. It extends the standard query facilities in a database model to cover the functions that are available in the query language for the other database model. One such example is MSQL. This approach, *however,* requires end-users to learn new data manipulation language, and new standard features.   (CS PhD)

Of course, these survey findings provided a more objective and independent perspective on police performance, *but* the findings are relevant to the service as a whole and cannot be reduced to individual and team performance.   (PA MA)

The markers in (7) are doing interpersonal work here, allowing the writer to display disciplinary membership and familiarity with community knowledge by expressing what he or she hopes will be a shared response to a claim. Concessive connectives are also commonly used to foreground a shared emphasis when making claims in support of the main thesis.

(8) We should, *however,* identify and assess the high risk factors first so that they become predictable.   (CS MSc)

In contrast, our sub-problem at the lower level is to minimize query cost with maintenance cost under different controls.   (BS MA)

Marking a contrast with prior knowledge in such cases as those in (8) helps to appeal to academic ideologies which value contrast in creating knowledge, and so direct the reader to a positive response.

In sum, because it overlooks the ways that meanings can overlap and contribute to academic arguments in different ways, the distinction between textual and interpersonal metadiscourse is unhelpful and misleading. Rather, textuality is a general property of the realization of discourse, perhaps analogous to syntax. The explicit signalling of connections and relationships between elements in an argument is related to the writer's awareness of self and of the reader when writing. By making reference to the text, the audience, or the message, the writer indicates his or her sensitivity to the context of the discourse and makes predictions about what the audience is likely to know and how they are likely to respond. What is commonly referred to as *textual metadiscourse* is therefore actually the result of decisions by the writer to highlight certain relationships and aspects of organization to accommodate readers' understandings, guide their reading, and make them aware of the writer's preferred interpretations. It therefore contributes to the interpersonal features of a text.

Internal vs. external relations

If we accept that many so-called textual items can realize either interpersonal or propositional functions depending on their context, then we need a means of distinguishing their primary function in the discourse. This brings us to the third key feature of metadiscourse: the distinction between 'internal' and 'external' reference.

Connective items offer the clearest example of this division as they can function to either connect steps in an exposition (internal), organizing the *discourse* as an argument, or connect activities in the world outside the text (external), representing *experiences* as a series of events (Martin 1992). An internal relation thus connects the situations described by the propositions and is solely communicative, while an external relation refers to those situations themselves. Halliday (1994) provides an unambiguous statement of this difference when discussing temporal connectors:

> Many temporal conjunctives have an 'internal' as well as an 'external' interpretation; that is, the time they refer to is the temporal unfolding of the discourse itself, not the temporal sequence of the processes referred to. In terms of the functional components of semantics, it is interpersonal not experiential time. (Halliday 1994: 325)

An example of the distinction is shown below. The connectors in (9) set up relations between propositions and express metadiscoursal functions, while those in (10) express a relation between processes and so are experientially oriented

(9) 93 questionnaires were received with 84 valid responses. *Therefore* the response rate for the questionnaire is 37 per cent.   (CS MSc)

In contrast, these findings were not found among the low collectivists.
(PA PhD)

In this paper, we investigate the effort allocation to construction under the BOT trend to check whether this kind of approach can improve the misallocation of the effort, *and furthermore*, we compare the allocation of the effort level under different ownership structures.   (BS MA)

(10) Though there are three psychogeriatric wards in Kwai Chung Hospital, the bed occupancy is only up to 41 for both long-stay and acute elderly

patients. *Therefore,* both Lai Chi Kwok and Kowloon Hospital Psychiatric Units have been used to provide additional beds for the long-stay and elderly patients over the last ten years, so as to ease the burden faced by the two main mental hospitals. (PA PhD)

However, in contrast to Western culture, Asian societies put emphasis on interdependent view of self and collectivism (SA PhD)

Initially, r(O) is set to 1.0 *so that* the normalized autocorrelation lags {r(i), for 1 < I < 10} can be computed by applying Equation 5.3.3, 5.3.4 and 5.3.2 recursively. (EE PhD)

This relationship can also be seen in the use of sequencing devices, which can be used to refer to either the linguistic interaction which is unfolding in the discourse itself (11), or to the steps involved in the particular research process being described (12):

(11) Crops accounted for a significant proportion of heavy metals dietary intake. The reasons are two folds. *Firstly,* crops are being the bottom positions of many food chains and food webs. *Secondly,* vegetables are one of the major dietary components of Hong Kong people. (Bio MSc)

Firstly, the importance of complete images in compression is described in section one. *Secondly,* predictors used for lossless image coding is introduced. *Thirdly,* the results and analysis are used to show the performance of the proposed compression. (EE PhD)

(12) For the boric acid indicator, *firstly,* 5g of boric acid crystals was dissolved in 200ml of warm distilled water, *then,* 40ml of methyl red indicator [0.02 per cent (w/v) in 60 per cent ethanol] and 15ml of bromocresol green indicator [0.1 per cent (w/v) in 60 per cent ethanol] were added to the boric acid solution. (Bio PhD)

Firstly, numbers of observation in first segment (Nj) and in second segment (N2) are combined. A 'pooled' regression is conducted, which is equation (LL-1). *Secondly,* individual regressions of the two periods have been done as well. . . . Then, F test is applied . . . (BS PhD)

In assigning either propositional or metadiscoursal values to items, the distinction between internal and external reference differentiates two writer roles, reflecting Bunton's (1999: S47) view of *research acts* and *writer acts.* The former concerns events which occurred as part of the research in a laboratory, library, or office, relating to the theoretical modelling or experimentation which form part of the subject matter of the text. In contrast, writer acts refer to how these are eventually written up, the decisions the writer makes in fashioning an argument for a particular readership.

The internal/external distinction is analogous to that made in modal logic between *de re* and *de dicto* modality, concerning the roles of linguistic items in referring to the reality denoted by propositions or the propositions themselves. While modality is an interpersonal feature in our model, signalling the writer's assessments of possibilities and his or her commitment to the truth of a proposition, this meaning needs to be carefully distinguished from uses where writers are referring to external circumstances which can affect the outcome of the proposition (Coates 1983: 113; Hyland 1998a: 110c). Thus hedges and boosters are metadiscourse markers which express the writer's logical inference about the likelihood of something, while alternative modal meanings concern the role of enabling conditions and external constraints on its occurrence in the real world.

Palmer (1990: 185) recognizes this distinction as epistemic and dynamic modality, the latter 'concerned with the ability or volition of the subject of the sentence, rather than the opinions of the writer' (1990: 36). The determining feature is therefore the objectivity of the event, and the clearest cases are those where such objective enabling conditions are made explicit. Thus, we recognize (13) as an example of metadiscourse as it comments on the writer's estimation of possibilities, and (14) as propositional as it represents that an outcome depends on external enabling or disabling circumstances.

(13) The diverse insect fauna reported from the reedbeds in Mai Po suggests that the reedbeds *could potentially* be an important habitat for a wide variety of animal taxa.   (Bio PhD)

A possible explanation for this phenomenon *may be* that due to the standing orders of floor traders . . .   (BS MA)

(14) This statement obviously exploits the Maxim of Quantity at the expense of the Maxim of Quality because the salesperson *could have* simply said: 'This company is also very famous in Taiwan.'   (AL PhD)

Using this scale makes it *possible* to compare results of the present study with those of previous socialization studies.   (BS PhD)

In some cases the co-text allows for both an epistemic and a dynamic reading, referring to either the writer's assessment of possibility or the appropriate circumstances, but coding is rarely problematic.

This distinction between external and internal relations, or more precisely between matters in the world and those in the discourse, is not always observed in the work on metadiscourse. It is, however, clearly crucial to determining the interpersonal (or metadiscoursal) from the ideational (or propositional). Obviously, if metadiscourse is to have any coherence as a means of conceptualizing and understanding the ways writers create meanings and negotiate their claims in academic texts, then the distinction between internal and external reference needs to be central.

A model of academic metadiscourse

We believe, therefore, that there are good reasons for distinguishing metadiscourse from the propositional content of a text and for conceptualizing it more broadly as an interpersonal feature of communication. In contrast to writers such as Crismore, Kopple, and Williams, we suggest that the textual features they see in texts are actually contributing towards either propositional or interpersonal functions. In contrast to writers such as Mauranen and Bunton who explore 'metatext' as the writer's self-awareness of *text,* we suggest that metadiscourse represents the writer's awareness of the unfolding text as *discourse*: how writers situate their language use to include a text, a writer, and a reader. Metadiscourse thus provides us with a broad perspective on the way that academic writers engage their readers; shaping their propositions to create convincing, coherent text by making language choices in social contexts peopled by readers, prior experiences, and other texts.

In practical terms, metadiscourse is identified as the writer's reference to the text, the writer, or the reader and enables the analyst to see how the writer chooses to handle interpretive processes as opposed to statements relating to the world. At a finer degree of delicacy the distinction between external and internal aspects of discourse provides a principled means of recognizing how the interpersonal dimension of language can draw on both organizational and evaluative features (Hyland 2001a), or what Thompson (2001) calls *interactive* and *interactional* resources. Thompson uses the term *interactive* to refer to the writer's management of the information flow to guide readers through the text (compare Hoey 1988), and *interactional* to refer to his or her explicit interventions to comment on and evaluate material. While our model takes a slightly wider focus than Thompson's by including both stance and engagement features of interaction (Hyland 2001a) and by building on earlier models of metadiscourse (Hyland 1998b, 2000a), it owes a great deal to his clear conception of these two dimensions. Table 7.1 offers a model of metadiscourse developed from these views of language use in academic writing.

Interactive resources, as noted above, refer to features which set out an argument to explicitly establish the writer's preferred interpretations. They are concerned with ways of organizing discourse, rather than experience, to anticipate readers' knowledge and reflect the writer's assessment of what needs to be made explicit to constrain and guide what can be recovered from the text. These resources include *transitions,* mainly conjunctions, which comprise the rich set of internal devices used to mark additive, contrastive, and consequential steps in the discourse, as opposed to the external world. *Frame markers* are references to text boundaries or elements of schematic text structure, including items used to sequence, to label text stages, to announce discourse goals, and to indicate topic shifts. *Endophoric markers* refer to other parts of the text and so make additional material salient and available to the reader in recovering the writer's intentions. *Evidentials* perform a similar role by indicating the source of textual information which originates outside the current text. *Code glosses* signal the restatement of ideational information in other ways.

Interactional resources, on the other hand, involve readers in the argument by alerting them to the author's perspective towards both propositional information and readers themselves. Metadiscourse here is essentially evaluative and engaging, influencing the degree of intimacy, the expression of attitude, epistemic judgements, and commitments, and the degree of reader involvement. This aspect thus relates to the *tenor* of the discourse, concerned with controlling the level of personality in a text. *Hedges* mark the writer's reluctance to present propositional information categorically while *Boosters* imply certainty and emphasize the force of propositions. The shifting balance of these epistemic categories conveys the extent of the writer's commitment to propositions and signals rhetorical respect for colleagues' views (Hyland 1998c). *Attitude markers* express the writer's appraisal of propositional information, conveying surprise, obligation, agreement, importance, and so on. *Engagement markers* explicitly address readers, either by selectively focusing their attention or by including them as participants in the text through second person pronouns, imperatives, question forms, and asides (Hyland 2001a). Finally *Self-mentions* reflect the degree of author presence in terms of the incidence of first person pronouns and possessives.

An orientation to the reader is crucial in securing rhetorical objectives in research writing. Readers always have the option of re-interpreting propositional information and rejecting the writer's viewpoint, which means that writers have to anticipate and respond to the potential negation of their claims. Metadiscourse is the way they do this: drawing on the rhetorical resources it provides to galvanize support, express collegiality, resolve difficulties, and avoid disputes. Choices of interactive devices address readers' expectations that an argument will conform to conventional text patterns and predictable directions, enabling them to process the text by encoding relationships and ordering material in ways that they will find

TABLE 7.1 A model of metadiscourse in academic texts

Category	Function	Examples
Interactive resources	**Help to guide reader through the text**	
Transitions	express semantic relation between main clauses	in addition/but/thus/and
Frame markers	refer to discourse acts, sequences, or text stages	finally/to conclude/my purpose here is to
Endophoric markers	refer to information in other parts of the text	noted above/see Fig/in section 2
Evidentials	refer to source of information from other texts	according to X/(Y, 1990)/Z states
Code glosses	help readers grasp functions of ideational material	namely/e.g./such as/in other words
Interactional resources	**Involve the reader in the argument**	
Hedges	withhold writer's full commitment to proposition	might/perhaps/possible/about
Boosters	emphasize force or writer's certainty in proposition	in fact/definitely/it is clear that
Attitude markers	express writer's attitude to proposition	unfortunately/I agree/surprisingly
Engagement markers	explicitly refer to or build relationship with reader	consider/note that/you can see that
Self-mentions	explicit reference to author(s)	I/we/my/our

appropriate and convincing. Interactional choices focus more directly on the participants of the interaction, with the writer adopting a professionally acceptable persona and a tenor consistent with the norms of the disciplinary community. This mainly involves establishing a judicious, discipline-defined balance of tentativeness and assertion, and a suitable relationship to one's data, arguments, and audience.

Metadiscourse in postgraduate writing

To illustrate the model and show how these resources are used to facilitate effective, disciplinary specific, interpersonal relationships in academic writing, we briefly describe some of the results of our study of graduate research writing.[1] Analysis of the corpus indicates the importance of metadiscourse in this genre, with 184,000 cases, or one signal every 21 words.[2] Table 7.2 shows that writers used slightly more interactive than interactional forms, and that hedges and transitions were the most frequent devices followed by engagement markers and evidentials.[3]

The high use of transitions, representing internal connections in the discourse, is clearly an important feature of academic argument. Transitions represent over a fifth of all metadiscourse in the corpus, demonstrating writers' concerns that the reader is able to recover their reasoning unambiguously. The most frequent sub-category, however, is hedges which constitute 41 per cent of all interactional uses, reflecting the critical importance of distinguishing fact from opinion in academic writing and the need for writers to evaluate their assertions in ways that are likely to be persuasive. In fact, *may, could,* and *would,* used epistemically to present claims with both appropriate caution and deference to the views of reader/examiners, were among the highest frequency items in the corpus. In general, then, these students' use of metadiscourse demonstrates a principal concern with expressing arguments explicitly and with due circumspection.

Because metadiscourse is a rhetorical activity whose use and meaning is relevant to particular socio-rhetorical situations, it is not surprising that it varied considerably across the two groups of dissertations we examined. The Master's theses were balanced overall between interactive and interactional metadiscourse, with slightly more interactional uses, while the doctoral texts, in contrast, contained 10 per cent more interactive forms. Hedges dominated interactional categories (40 per cent in the PhDs and 44 per cent in the masters theses) and transition markers the interactive group (36 per cent and 41 per cent respectively), with evidentials and code glosses the next most frequent interactive devices and engagement markers representing a fifth of both masters and doctoral interactional devices. The PhD dissertations, however, contained far more metadiscourse, with 73 per cent of all cases in the study and 35 per cent more when normed for text length.

The variations in metadiscourse frequencies are partly due to the fact that the PhD corpus is twice as long as the masters corpus, making it necessary for writers to employ more interactive devices to structure more discursively elaborated arguments. However, the higher frequencies in the PhDs also represent more concerted and sophisticated attempts to engage with readers

TABLE 7.2 Metadiscourse in postgraduate dissertations (per 10,000 words)

Category	Master	Doctoral	All	Category	Master	Doctoral	All
Transitions	75.8	95.6	89.0	Hedges	86.1	95.6	92.4
Evidentials	40.0	76.2	64.1	Engagement mkrs	39.7	51.9	47.8
Code glosses	27.4	40.6	36.2	Boosters	31.7	35.3	34.1
Frame mkrs	20.7	30.3	27.1	Attitude mrkrs	20.4	18.5	19.2
Endophorics	22.3	24.0	23.4	Self-mentions	14.2	40.2	31.5
Interactive	186.1	266.7	239.8	Interactional	192.2	241.5	225.0

and present their authors as competent and credible academics immersed in the ideologies and practices of their disciplines.

In the interactive categories, for instance, doctoral writers made far more use of evidentials, with over four times the number of intertextual references. Citation is central to the social context of persuasion, as it helps provide justification for arguments and demonstrates the novelty of the writer's position, but it also allows students to display an allegiance to a particular community and establish a credible writer identity, displaying familiarity with the texts and with an ethos that values a disciplinary research tradition. The writers of masters' theses, on the other hand, are unlikely to be so concerned about establishing their academic credentials. Not only are their texts much shorter, but they are also completed fairly quickly and in addition to substantial coursework, while their writers are normally studying part-time and are looking forward to returning to their professional workplaces rather than taking up a career in academia. Consequently their reading of the literature, and their desire to demonstrate their familiarity with it, may be less pressing.

Similarly, doctoral students employed far more interactional metadiscourse markers, with much higher use of engagement markers and self-mentions. Self-mention is a key way through which writers are able to promote a competent scholarly identity and gain accreditation for their research claims. While many students are taught to shun the use of first person, it plays a crucial interactional role in mediating the relationship between writers' arguments and their discourse communities, allowing writers to create an identity as both disciplinary servant and creative originator (Hyland 2001b). The points at which writers choose to metadiscoursally announce their presence in the discourse tend to be those where they are best able to promote themselves and their individual contributions. Engagement features, particularly imperatives and obligation modals which direct the reader to some thought or action, are important in bringing readers into their text as participants in an unfolding dialogue.

There were also substantial variations in the use of metadiscourse across disciplinary communities. Table 7.3 shows that the more 'soft knowledge' humanities and social science disciplines employed more metadiscourse overall (56 per cent of the normed count) and over 60 per cent of the interactional features. The interactive figures were more balanced across disciplines, but generally formed a much higher proportion of the metadiscourse in the science dissertations.

These distributions across broad interactive and interactional fields closely reflect those in university textbooks (Hyland 2000a) and research articles (Hyland 1998a), where interactional forms also tend to be much higher in the soft knowledge disciplines. Although boosters and engagement features were fairly evenly distributed across fields, hedges were well over twice as common in the soft fields and self-mentions almost four times more frequent (before norming for text length). These figures generally reflect the greater role of explicit personal interpretation of research in the humanities and social sciences and the fact that dealing with human subjects and data is altogether more uncertain and allows for more variable outcomes. The writer is unable to draw to the same extent on convincing proofs, empirical demonstration, or trusted quantitative methods as in the hard fields, and must work harder to build up a relationship with readers, positioning them, persuading them, and including them in the argument to turn them from alternative interpretations.

Evaluative judgements and hedges are found in all academic writing, for instance, but are particularly important in the more discursive soft fields where interpretations are typically more explicit and the criteria for establishing proof less reliable. Applied linguistics, business, and social studies all rely on the careful interpretation of qualitative analyses or statistical probabilities to construct and represent knowledge, requiring more elaborated exposition and greater tentativeness in expressing claims. Self-mention also plays a far more visible role in the soft disciplines. Students are often exhorted by style guides and supervisors to present their own 'voice' and display a personal perspective, suitably supported with data and intertextual evidence, towards the issues they discuss, weaving different kinds of support into a coherent and individual argument. In the hard fields, and particularly in the more 'pure' sciences as represented by biology in our corpus, the community tends to value competence in research practices rather than those who conduct them, and so a personal voice is subsumed by community knowledge and routines.

TABLE 7.3 Metadiscourse in postgraduate dissertations by discipline per 10,000 words

Category	Applied Linguistics	Public Admin	Business Studies	Computer Science	Electronic Engineering	Biology
Transitions	95.1	97.8	89.1	74.3	76.9	86.6
Frame markers	25.5	29.5	25.3	35.4	24.7	22.5
Endophorics	22.0	15.5	19.6	25.9	43.1	23.0
Evidentials	82.2	55.6	60.7	31.1	20.1	99.5
Code glosses	41.1	36.6	30.0	32.3	30.7	36.0
Total interactive	265.9	240.5	224.7	199.0	195.5	267.6
Hedges	111.4	109.7	93.3	55.8	61.5	82.1
Boosters	37.9	39.5	29.8	29.4	28.0	30.5
Attitude markers	20.3	26.1	20.7	16.2	10.6	15.5
Engagement markers	66.1	42.0	35.8	59.2	32.7	15.4
Self-mentions	50.0	22.4	31.6	29.3	18.1	5.7
Total interactional	285.7	239.8	211.1	190.0	150.9	149.2
Totals	551.6	474.9	435.8	389.0	346.5	416.8

Computer Science tended to differ from this general picture of impersonality in scientific discourse, displaying relatively high frequencies of self-mentions and engagement markers. While essentially a hard field dealing with impersonal computational calculations, computer science is also very much an applied discipline, practical in its orientation and concerned with its relevance to operations in a range of disciplines, including internet marketing, machine translation, and e-business. Thus, unlike the other two hard fields in our corpus, where emphasis is often directed to the development of discipline-internal theories, techniques, and applications, research in computer science tends more to the everyday world and as a result its metadiscourse has evolved, like those in the soft applied fields, to speak to both academics within the discipline and to practitioners outside it.

The findings for interactive metadiscourse in this study represented less stark contrasts between hard and soft fields and greater variation between disciplines within these categories. Transitions tended to be more carefully marked in the soft fields, perhaps reflecting the more discursive nature of these disciplines, and the hard disciplines employed a relatively higher number of endophorics, especially in engineering, thus emphasizing their greater reliance on multi-modality and arguments which require frequent reference to tables, figures, photographs, examples, and so on.

Perhaps the greatest surprise here is the extremely high use of evidentials in the biology dissertations. Evidentials are metadiscoursal features which provide intertextual support for the writer's position, a frame within which new arguments can be both anchored and projected, and as such they tend to play a more visible role in the soft disciplines where issues are more detached from immediately prior developments and less dependent on a single line of development (Becher 1989). The fact that new knowledge follows more varied routes means there are less assured guarantees of shared understandings and less clear-cut criteria for establishing claims than in the sciences. Because of this, writers often have to pay greater attention to elaborating a context through citation, reconstructing the literature in order to provide a discursive framework for their arguments and demonstrate a plausible basis for their claims.

But although it is a 'hard' science, biology has the greatest density of citations in the corpus, a finding which mirrors their use in a study of research articles across a similar range of disciplines (Hyland 2000a). The evidence from both these corpora suggests that significant recognition is given to the ownership of ideas in Biology, making it unusual among other hard disciplines in giving greater weight to *who* originally stated the prior work. The biology style guides make it clear that it is important for writers to show how their current research relates to, and builds on, the preceding work of other authors (for example Council of Biology Editors 1994; Davis and Schmidt 1995; McMillan 1997) and this suggests both a considerable emphasis on proprietary rights to claims and an interest in how particular research contributes to a bigger scientific picture.

This brief description of metadiscourse use in postgraduate dissertations clearly shows that disciplines are not only distinguished by their objects of study. The fact that academics actively engage in knowledge construction as members of professional groups means that their decisions concerning how propositional information should be presented are crucial. It is these decisions which socially ground their discourses, connecting them to the broad inquiry patterns and knowledge structures of their disciplines and revealing something of the ways academic communities understand the things they investigate and conceptualize appropriate writer–reader interactions. In other words, their use of metadiscourse, how they choose to frame, scaffold, and present their arguments and research findings, is as important as the information they present.

Conclusions

While there is often an unfortunate tendency in the metadiscourse literature to focus on surface forms and the effects created by writers, especially in pedagogic materials and college essays, metadiscourse should not be seen as an independent stylistic device which authors can vary at will. The importance of metadiscourse lies in its underlying rhetorical dynamics which relate it to the contexts in which it occurs. It is intimately linked to the norms and expectations of particular cultural and professional communities through the writer's need to supply as many cues as are needed to secure the reader's understanding and acceptance of the propositional content. Central to our conception of metadiscourse, then, is the view that it must be located in the settings which influence its use and give it meaning.

The framework we have suggested offers a comprehensive and pragmatically grounded means of investigating the interpersonal resources academics deploy in securing their claims. But while we believe this provides both a theoretically more robust model and a more principled means of identifying actual instances, we recognize that no taxonomy can do more than partially represent a fuzzy reality. The imposition of discrete categories on the fluidity of actual language use inevitably conceals its multifunctionality. Pragmatic overlap is a general feature of discourse motivated by the need to accomplish several objectives simultaneously. Writing effectively means anticipating the needs of readers, both to follow an exposition and to participate in a dialogue, and occasionally devices are used to perform both functions at once. A classification schema nevertheless performs a valuable role. Not only does it help reveal the functions that writers perform, but it also provides a means of comparing generic practices and exploring the rhetorical preferences of different discourse communities.

Metadiscourse is thus an aspect of language which provides a link between texts and disciplinary cultures, helping to define the rhetorical context by revealing some of the expectations and understandings of the audience for whom a text was written. Differences in metadiscourse patterns can offer an important means of distinguishing discourse communities and accounting for the ways writers specify the inferences they would like their readers to make. Put simply, the significance of metadiscourse lies in its role in explicating a context for interpretation, and suggesting one way in which acts of communication define and maintain social groups.

Notes

1   A more detailed discussion of the findings of this study can be found in Hyland (2004b).

2   The fact that metadiscourse often has clause or sentence length realization means that our standardized figures are not meant to convey the overall amount of metadiscourse in the corpus, but simply compare different patterns of *occurrence* of metadiscourse in corpora of unequal sizes.

3   Because a corpus of this size generates thousands of instances of high frequency items such as modals and conjunctions, we counted all the returns of these high frequency items and then generated fifty example sentences of each one in each discipline and degree corpus. We then carefully analysed each of these fifty randomised cases in turn to identify, in context, which items were functioning as metadiscourse. With this figure from fifty, we then extrapolated the number of metadiscourse functions of each item as a percentage of the total number of cases of that item overall to give a proportion of metadiscourse uses. We then added all the figures in that functional category (e.g. all transition markers) and normed them to occurrences per 10,000 words to facilitate comparison across corpora of different sizes.

8

Change of attitude?
A diachronic study of stance

Ken Hyland and Feng (Kevin) Jiang

Conceptions of stance

The ways that writers and speakers express their opinions is an important feature of all interaction, and researchers have long been concerned with describing how stance is linguistically marked. A range of terms have been used to conceptualize the idea. Some are umbrella conceptions such as *posture* (Grabe, 1984), *attitude* (Halliday, 2004), *appraisal* (Martin, 2000), *evaluation* (Hunston & Thompson, 2000), and *metadiscourse* (Hyland, 2005a). Others focus more specifically on the linguistic realizations of judgments, feelings, or viewpoints by looking at *intensity* (Labov, 1984), *disjuncts* (Quirk, Greenbaum, Leech, & Svartvik, 1985), *hedges* (Hyland, 1998a), and *modality* (Palmer, 1986).

Two related ideas have been particularly influential in current conceptions of stance: *evidentiality* (Chafe, 1986; Chafe & Nichols, 1986) and *affect* (Besnier, 1990; Ochs & Schieffelin, 1989). *Evidentiality* refers to the status of the knowledge contained in propositions and concerns its reliability, implying its source, how it was acquired, and the credibility we can invest in it (e.g., Chafe, 1986). In contrast, the term *affect* is used "to include feelings, moods, dispositions, and attitudes associated with persons and/or situations" (Ochs & Schieffelin, 1989, p. 7). So, affect markers express the intensity and particularity of personal feelings and attitudes rather than evaluations of knowledge. Most discussions of stance seek to combine these two elements into a single model, Thus, Biber and Finegan (1989), for example, identified how this is expressed in various spoken and written genres, while Martin's model of *appraisal* lists the possible realizations of attitudinal stance meanings as indicating affect (emotional responses), judgment (moral evaluations), and appreciation (aesthetic evaluations) whereas *engagement* aligns broadly with epistemic stance meanings (Martin, 2000; Martin & White, 2005).

Another recent approach to stance stresses its evaluative aspects (e.g., Hunston, 1994; Hunston & Thompson, 2000), seeking to unpack the judgments writers make in expressing a position on what they discuss. Hunston and Thompson (2000, pp. 22-26), for example, suggest four main dimensions of evaluative language that can convey a writer's stance:

1 Value—or variations along a good-bad scale

2 Status—relating to the degree of certainty the speaker/writer invests in information

3 Expectedness—referring to how obvious or expected the information is to the hearer/reader

4 Relevance—judgments of significance or relevance to the listener/reader

Finally, considerable work has been done under the umbrella of *metadiscourse,* or linguistic material referring to the evolving text and to the writer and imagined reader of that text. Hyland's work has been influential here, distinguishing interactional and interactive types, broadly, language that conveys the attitudes of writers to their material and readers and that is used to create a more accessible and persuasive text (Hyland, 2005a).

All of these approaches are based on a view of writing (and speaking) as social engagement and attempt to reveal the ways writers project themselves into their discourse to signal their attitudes toward the content of their talk. In addition, all draw on a distinction between meanings that indicates a speaker/ writer's personal attitudes and assessments, and those that comment on the truth of a proposition. In line with Biber (2006) and Hyland (2005b), we see *stance* as the writer's expression of personal attitudes and assessments of the status of knowledge in a text, and in the next section we discuss the relevance of these meanings to academic writing.

Stance in academic writing

First, we need to acknowledge that overt stance expressions are far less common in academic prose than in other registers (e.g., Gray & Biber, 2012). Biber and Finegan's (1989) cluster analysis of 12 lexicogrammatical stance devices, for example, statistically categorized 75% of the written academic texts as "faceless." Biber (2006) found that spoken registers are more heavily stance-laden, even in academic contexts, and Hood (2010), using an appraisal model, found little affect in research texts. As a result, and until fairly recently, much of the research into stance tended to concentrate on mass audience texts, such as journalism, politics, and media discourses, which seemed most likely to yield a richer harvest of explicitly author-avowed positions.

We also need to grant that stance, like other features of disciplinary discourses, is not a static and unchanging marker of professional research writing. A considerable literature has addressed the emergent nature of the article genre (e.g., Bazerman, 1984). Thus Atkinson (1999), for example, tracked changes in articles published in *Philosophical Transactions* between 1675 and 1975, finding a relentless growth in "informational" features until the latter date, a change that Atkinson describes as a move from a less "author-centred" rhetoric to a highly abstract and "object-centred one." The genre has also seen a decline in the frequency of *be*-passives compared with transitive actives through the 20th century (Seone, 2013) and a movement toward more compressed, phrasal expressions over elaborated, clausal expressions that allow for faster, more efficient processing by expert readers (Biber & Gray, 2016). Academic genres, however, appear to change only slowly, so that Hundt and Mair (1999, p. 221) suggest that they reside at the more conservative end of "a cline of openness to innovation ranging from 'agile' to 'uptight' genres."

We know little about changes in the expression and frequency of stance features however, although it essential to recognize the importance of authorial-investment in research texts. Successful academic writing depends on the individual writer projecting a personal assessment of research entities and claims, although this needs to be done within a shared professional context. In claiming a right to be heard, and to have their work taken seriously, writers must simultaneously take a stance toward what they discuss and display competence as disciplinary insiders. This competence is, at least in part, achieved through writers constructing a dialogue with readers that situates both their research and themselves. This is crucial in establishing community-approved interactions and highlights the centrality of authorial "positioning": adopting a point of view in relation to both the issues discussed in a text and to others who hold points of view on those issues (Hyland, 2012, 2015). Stance in this sense is a consistent series of rhetorical choices that allow authors to conduct interpersonal negotiations and balance claims for the significance, originality, and plausibility of their work against the convictions and expectations of their readers.

Considerable research interest has therefore been invested in discovering precisely how writers persuade readers of their claims and lead the academic

TABLE 8.1 Corpus Size and Composition

Discipline	1965	1985	2015	Overall
Applied linguistics	110,832	144,859	237,452	493,143
Biology	244,706	263,465	237,998	746,169
Engineering	92,062	97,545	235,681	425,288
Sociology	149,788	196,232	262,203	608,223
Total	597,388	604,556	973,334	2,272,823

community to "adjust its network of consensual knowledge in order to accommodate those claims" (Hunston, 1994, p. 192). Studies have shown the lexical and grammatical choices that seem most productive expressions of stance in academic genres such as undergraduate essays (Aull & Lancaster, 2014), theses (Charles, 2006), abstracts (Hyland & Tse, 2005), and research articles (Hyland, 2012a) as well as in L2 student writing (Hyland, 2004b) and in disciplines as diverse as geology (Dressen, 2003) and art history (Tucker, 2003). Stance is also a key feature of readers' assessments of text quality. In an interview study, for example, Lancaster (2014) found that tutors judgments of critical reasoning and analytic rigor in argument papers by economics students were influenced by valued configurations of stance, while Uccelli, Dobbs, and Scott (2013) discovered that hedges functioning as an epistemic stance marker significantly predicted assessments of writing quality in the essays of high school students.

An important consideration here is that stance is always expressed in relation to some set of expectations so that personal judgments are convincing, or even meaningful, only when they contribute to and connect with a communal ideology or value system concerning what is taken to be interesting, relevant, novel, useful, good, and so on. Writers' stance options, in other words, are made not from an infinite range of alternatives but from a restricted subset of options that reveal how they understand their communities through the assumptions their stances encode. Readers can always refute claims, and this gives them an active and constitutive role in how writers construct their arguments. Thus any successfully published research article anticipates a reader's response and itself responds to a larger discourse already in progress. Stance choices are, in other words, disciplinary practices as much as individual positions.

Because writers comment on their propositions and shape their texts to the expectations of different audiences, the expression of stance varies according to discipline. In the next section we outline our corpus and approach and then explore whether authors' use of stance features has changed over time.

Corpus, model, and method

We created three corpora taking research articles from the same five journals in four disciplines spaced at three periods over the past 50 years: 1965, 1985, and 2015. The different time spans were chosen to see if any changes were more pronounced in the later or earlier period, although we were concerned with overall changes over the 50 years. Applied linguistics, sociology, electrical engineering, and biology were selected as representative of both the soft applied fields and the hard sciences, and we took six article at random from each of the five journals that had achieved the top ranking in their field according to the 5-year impact factor in 2015.[1] The journals are listed in the appendix, and together the corpus comprised 360 articles of 2.2 million words, as shown in Table 8.1. This indicates the huge increase in the length of articles over this period.

The corpora were then searched for stance features using the concordance software AntConc (Anthony, 2011). In this article we follow Hyland's (2005b) approach to stance in academic writing, which regards it as

> *writer-oriented features* of interaction and refers to the ways academics annotate their texts to comment on the possible accuracy or credibility of a claim, the extent they want to commit themselves to it, or the attitude they want to convey to an entity, a proposition, or the reader. (Hyland, 2005, p. 178b)

While this comprises only half of Hyland's model of intersubjective positioning, it would be hard to also do justice to engagement in a single article. The notion of stance, however, represents a coherent concept and body of research that is worth discussing in its own right. This framework encompasses three main components: *evidentiality, affect,* and *presence.*

- *Evidentiality*—the writer's stated commitment to the reliability of the propositions he or she presents and their potential impact on the reader, expressed through hedges and boosters
- *Affect*—a broad range of personal and professional attitudes toward what is said expressed through attitude *markers*
- *Presence*—the extent to which the writer chooses to intrude into a text through the use of first person pronouns and possessive determiners

Hyland's model recognizes the role that readers play in shaping a writer's stance by understanding texts as community negotiated processes so that these projected reader positions help establish "a virtual dialogue" (Hyland, 2012a). Stance is socially inscribed in academic writing as these shared practices align claims with existing knowledge and sanction what ideas will be legitimated. It therefore allows us to study disciplinary variations. Overall we examined 140 different items, taken from the appendix of Hyland (2005a), searching for both U.S. and British spellings, and manually examined and counted each concordance to establish that the feature was performing a stance function (e.g., only cases of exclusive *we*). Both authors, working independently, coded a 10% sample to ensure reliability with 95% agreement.

Changing patterns of stance

Our analysis of these features shows a steady increase in authorial stance in academic writing over the past 50 years, with a massive 54% rise in raw numbers. When adjusted for the large rise in the length of articles, however, we see a small, but significant, *decrease* in the expression of stance, as shown in Table 8.2. The normed figures show that the increase in published words is accompanied by a substantial dip in authors' explicit stance choices until 1985 and then a slight rise once again to the present.

This overall relative decline in stance features turns out to be not evenly distributed across all fields but indicates changing argument patterns in different disciplines. Table 8.3 shows that despite large increases in the use of explicit stance features in applied linguistics and sociology over the period, writers in these disciplines now position themselves less obtrusively than 50 years ago. Biologists have adopted a slightly more visible presence, but engineers appear to be moving toward a very different stance profile.

We have to admit that these trends were completely unexpected. The attention given to stance and interpersonal positioning more generally might convey an impression that academic writing is now more personally engaging. Our results, at least in the social sciences, however, suggest the

TABLE 8.2 Distribution of Stance Features Over Time

	1965	1985	2015	% change
Total stance items	19,372	20,465	29,783	53.7
Per 10,000 words	324.3	291.6	304.9	−6.0

TABLE 8.3 Changes in Stance Frequencies Over Time by Discipline (Raw Numbers and per 10,000 words)

Discipline	1965	1985	2015	% change
Applied ling	4,934 (351.2)	5,434 (375.4)	7,009 (295.2)	42.1 (−16.0)
Sociology	5,656 (377.6)	6,084 (301.0)	8,493 (323.9)	50.2 (−14.3)
Electrical eng	2,561 (262.4)	2,905 (283.4)	7,632 (323.6)	198.0 (23.3)
Biology	6,221 (254.2)	6,039 (229.3)	6,649 (275.4)	6.9 (8.3)
Total	19,372 (324.3)	20,462 (291.6)	29,783 (304.9)	

opposite: that the additional word limits allocated to authors to discursively explore issues has not been matched with an equivalent degree of personal intrusion and evaluation. Biologists have slightly increased their use of stance features, although this is almost entirely due to a massive 150% increase in self-mention since 1985. Most intriguingly, and surprisingly, is the upsurge in the frequency of stance features in electrical engineering, which in contrast to all other disciplines, showed an increase in all four features examined.

We discuss these changes in more detail below, but there seems to have been a significant shift in how these disciplines understand academic argument and the ways their members seek to persuade peers. Obviously, presenting a self is central to the writing process (Ivanic, 1998), and we cannot avoid projecting an impression of ourselves and how we stand in relation to our arguments, discipline and readers (Hyland, 2004a). Writers cannot, in other words, not take a stance, whether this is a confident and assertive author, a tentative interpreter or a modest scientist, carefully hiding behind empirical findings, or flawless logical deduction. There is no "faceless" writing, and all stance choices are important rhetorical decisions that affect how the message is received and the ways readers react to a text. But while writers in different disciplines represent themselves, their work, and their readers in very different ways, with those in the humanities and social sciences tending to take more personal positions than those in the sciences and engineering, this seems to be changing.

The soft knowledge fields, particularly in the past 30 years in the case of applied linguistics, have been slowly moving toward more "author-evacuated" prose, increasingly mimicking hard science practices. Speculatively this might be attributed to an increase in more empirically grounded and quantitative studies that restrict opportunities for more extensive overt stance-taking. On

TABLE 8.4 Changes in Stance Features by Discipline (per 10,000 Words)

Feature	Applied ling			Sociology			Electrical eng.			Biology		
	1965	1985	2015	1965	1985	2015	1965	1985	2015	1965	1985	2015
Hedges	201.1	169.4	128.6	187.8	148.2	148.7	117.5	122.5	126.4	130.2	132.9	148.4
Boosters	107.7	79.7	67.0	88.8	62.7	57.3	82.6	87.3	95.0	76.3	51.9	51.7
Attitude markers	42.3	38.5	31.2	35.7	32.0	28.0	9.5	8.5	16.4	27.9	24.3	27.2
Self-mention	94.0	87.8	68.4	65.3	67.1	89.9	52.8	65.1	85.8	19.8	20.2	52.1

the other hand, writers in the hard sciences in our sample, and spectacularly in the case of electrical engineering, seem to be edging toward greater visibility, especially through self-mention to create a more evident presence in the text. In the following two sections we explore these results in greater detail by looking at the main components of stance: evidentiality, affect, and presence.

Shifting commitment and increasing presence—What changes and to what extent?

Table 8.4 shows that stance features are not changing in a single direction nor behaving in uniform ways, either across times or disciplines. It indicates rises in hedges in engineering and biology, falls in the use of boosters by all disciplines but engineering, relatively steady uses of attitude markers, and, excepting applied linguistics, substantial rises in self-mention.

Overall, it appears that, over time, the biologists have become more measured in their stance expression and also more "present" in terms of self mentions, intervening more overtly as writers. The electrical engineers are, in general, taking a stronger stance, increasing their expression of attitude, presence, and evidentiality. In contrast, while the sociologists are becoming increasingly "present" in terms of self mentions, they are expressing attitude and epistemic judgments less frequently. The applied linguists are becoming increasingly faceless overall, a directly opposite trend to that we see in the engineering texts. We will look at the changes in stance features in more detail below.

Evidentiality

Evidentiality refers to epistemic aspects of stance and includes meanings of certainty, doubt, actuality, precision, or limitation. Gray and Biber (2012) point to the prevalence of these stance meanings in all academic genres, regardless of mode, and much has been written on their use in academic writing (e.g., Biber, 2006; Hyland, 1998a). These assessments of certainty allow authors to express statements from a particular perspective. They are realized grammatically, through clauses (e.g., *as we all know, it is doubtful that . . .*), lexically through words like *definitely, might, likely, etc.*, and by phrasal forms (*seems obvious that, may be due to*). **Hedges** are often a prudent option for authors, allowing them to open a discursive space and mark statements as provisional to involve readers as participants in their ratification while conveying respect for colleagues' views. **Boosters**, on the other hand, express conviction, seeking to assert claims categorically and shut down alternative voices (Hyland, 1999).

These then are devices that either invest statements with the confidence of factual reliability or withhold complete commitment to imply that a claim is based on the author's plausible reasoning rather than certain knowledge. Both options represent an author's explicit intrusion into a text to convey a personal stance. Because all statements are evaluated and interpreted through a prism of disciplinary assumptions, writers must calculate what weight to give to an assertion, attesting to the degree of precision or reliability that they feel able to invest in it. This means that we see variations across time and disciplines.

Interestingly, both applied linguistics and sociology recorded dramatic falls in the use of hedges and boosters over this 50-year period (when normed for the increase in corpus size). Hedges fell 36% per 10,000 words in applied linguistics and 21% in sociology, and boosters were down 38% and 35%, respectively. *May* and *would* remain the most frequently used hedges in both disciplines but their combined frequencies per 10,000 words fell by almost 50% over the period, and the other modal hedges also declined. *May, would, should, could,* and *might* represented 40% of all hedges in sociology and 48% in applied linguistics in 1965 and 34% and 34%, respectively, in 2015, while *suggest* and *likely* were the only forms to become more common among the most frequent hedges. This seems to represent not only a decline in hedging but a shift away from forms that, according to Coates (1983), express assumption (*should* and *ought*), possibilities (*may, might* and *could*), and hypotheticality (*would*) toward those that carry more speculative judgments, predicated on a reference to the uncertainty of human evaluation.

Boosters have also steadily declined in both soft knowledge disciplines and also have shown a trend toward more verbal uses. The most common form in 1965 and 1985 in both applied linguistics and sociology was *must,* the predominant modal of inferential certainty, but this had disappeared from the top 20 by 2015. Cognitive verbs such as *think, believe,* and *know* have also declined, while *show, demonstrate,* and *find* have all increased. Although cognitive verbs can function as either hedges or boosters, their decline as boosters in favor of more objective choices suggests a move from personal beliefs toward more empirical and data-supported commitments to claims:

(3) That, I **believe** must be sought in an unhappy confusion in the minds of the teachers of composition. (Applied linguistics, 1965)

(4) Let me say at once that I **think** these main tenets are indisputable, and that they have had a tremendous impact on linguistics. (Applied linguistics, 1965)

(5) We **demonstrate** that this inconsistency has resulted from inadequate control, . . . including the basis of norm comparisons. (Applied linguistics, 2015)

(6) In summary, we **find** that females are markedly superior to males in recalling social network information. (Sociology, 2015)

In the sciences, evidential changes have surprisingly moved in the opposite direction. Hedging increased in both fields we studied, albeit only marginally, rising 14% (per 10,000 words) in biology and 8% in electronic engineering, with *may* remaining the preferred form for both fields throughout. This expresses greater certainty than *might* and indicates a 50-50 assessment of possibilities, but also facilitates a certain ambiguity between enabling circumstances which permit X ("it is possible for x") and the writer's lack of confidence in the possibility that X ("it is possible that x"). The fuzziness allows writers to hedge a hedge. Frequencies of this form, however, declined in both fields, as did those for *would* and *should,* as well as approximation words such as *about, estimate,* and *approximately,* perhaps indicating greater precision in measurements. *Could,* which is similar in meaning to *may* in expressing tentative possibility, gained increasing popularity through the period, pointing to cautious interpretations instead of hypothesizing or exploring assumptions.

While engineering and biology displayed broadly similar trends with hedges, they diverged considerably in the use of boosters, with biology following the pattern of the social science disciplines and recording a 32% fall and engineering increasing its use of boosters by 15%. *Show, must,* and *know* were the preferred forms in both disciplines throughout, although engineers came to use a much wider array of expressions as the period wore on, especially *establish, prove,* and *clearly.* These are forms that are used to ensure readers are aware of the strength of results or the claims being made:

(7) We have **established** above that *(e(i)(t),* ω*(i)(t))* are bounded for all $i = 1, \ldots, j \ldots$ (Engineering, 2015)

(8) Our contribution here will be to **prove** this conjecture:

Theorem 1.3. There exists a regular function X0 : [0, 1] $\rightarrow$ R such that, if the sequence (xt)tN is defined according to (1.3), then xt(1) $-$ xt(0) > 2 for all t. (Engineering, 2015)

(9) **Clearly,** the control action has significant impact on the variance of the temperature. (Engineering, 2015)

In summary, authors in the soft knowledge fields in our corpus have exhibited a movement toward considerably less marking of evidentiality over the past 50 years. It seems that research is being reported more categorically in these disciplines without the authors annotating their statements for either truth or precision. We also find changes in the most common forms themselves, with fewer modals and more verbal choices and a trend toward more empirically oriented boosters. In the sciences, on the other hand, there is greater circumspection in the use of hedges, although with a growing tendency to blur personal judgments and enabling possibilities and to hedge quantities rather than interpretations. Both weaken the stance taken. Engineering was the only discipline to see an increase in boosters, using them to underline the strength of results rather than subjective interpretations of them. Together these changes represent a substantial shift to more cautious and author-evacuated positions in the soft knowledge fields and a growing tendency for authors in the sciences to express their results with greater visibility.

Affect

Attitudinal stance indicates the writer's affective, rather than epistemic perspectives, and includes evaluations and personal feelings as he or she comments on the material under discussion (*interestingly, surprisingly*) or on the communication itself (e.g., *honestly, in truth*). While attitude is expressed throughout a text by the use of subordination, comparatives, progressive particles, punctuation, text location, and so on, it is most explicitly signaled by **attitude markers** such as attitude verbs (e.g., *agree, prefer*), sentence adverbs (*unfortunately, hopefully*), and adjectives (*appropriate, logical*). The emphatic expression of affect is relatively infrequent in academic research writing (Biber, Johansson, Leech, Conrad, & Finegan, 1999; Hyland, 2004a) and tends to be implicitly *invoked* rather than openly *inscribed* (Martin & White, 2005). Attitude markers thus bring affect to the surface through forms with positive or negative meanings, as in these examples:

(10) These results were **surprising** in that they showed that a Ca2+ influx apparently is not required for a low level of increase in the rate of transcription . . . (Biology)

(11) Student A2 presented another **fascinating** case study in that he had serious difficulties expressing himself in written English. (Applied linguistics)

By signaling an assumption of shared attitudes and reactions to material, writers both express a position and seek to bring readers into agreement with it.

In our longitudinal data we find an overall decline in the explicit marking of affect with fairly dramatic falls in the soft sciences and, once again, an increase in electronic engineering. In fact, the patterns we observe in attitude markers closely mirror those for hedges and boosters discussed above. Over the 50-year span, frequencies fell by 26% in applied linguistics and by 22% in sociology (normed to 10,000 words), remained almost unchanged biology, and, admittedly with very small raw numbers, rose by 73% in engineering.

The items *important* and restrictive *even* remained the top two choices across all four fields, although their frequencies fell in all disciplines. Both these forms enable writers not only to express a stance toward something but also to align that stance with the interests of their community. Crafting texts, which offers a positive evaluation of a result or entity assumed to be valued by readers, also has the added value of making it difficult to challenge:

(12) The cutoff frequency is **important** for analysing bandwidth of small-signal amplifiers and power gain of power amplifiers. (Electrical engineering)

(13) One **important** feature which differentiates authentic lecture discourse from written text or scripted lectures is the way they are structured at the micro-level. (Applied linguistics)

By categorically asserting statements that assume shared attitudes, the writer constructs a relationship along with a text, but this is a relationship where the writer is firmly in the driving seat.

Similarly, *even* conveys an attitude by focusing attention on the author's assessment of the relative unexpectedness of something being the case, while implicitly drawing on shared knowledge to do so, as here:

(14) Aflagellate cells maintained in CdCl2 did not regrow flagella, **even** after a 24 h period. (Biology)

(15) **Even** Ralph Dahrendorf considered the Nazi regime as "specific" and "unique." (Sociology)

The biologists in (14) takes a stance that squarely positions him with the views of knowledgeable peers who are likely to be equally surprised at this result. In (15) the sociologist positions himself in proximity with an informed audience by drawing on their shared knowledge of the great social thinker's views of enforced constraint the ways groups are controlled by a hierarchy of authority and power. In both cases a clear stance is expressed toward both the topic and the audience.

Other items among the most preferred expressions of attitudinal stance across all disciplines are *expected* and *interesting,* which generally comment on data or the results being presented:

(16) As <u>expected</u> from the border analysis, the unstable H59 line contained a single intact copy of T-DNA. (Biology)

(17) We <u>expected</u> the group with more exposure to the target words in natural context through in-class activities to perform better on the posttest. (Applied linguistics)

(18) It is <u>interesting</u> to observe at this point that (18) lends further credence to our earlier assertion that the extremum implied by (1) is a minimum. [Electrical engineering]

(19) When the distribution of data was examined by a branch of the legal profession, some <u>interesting</u> differences were found . . . (Sociology)

Once again, what might be considered interesting or anticipated is likely to be shared by one's peers, the informed insiders who inhabit the same community and contribute to its research and discourses.

One high frequency item marking attitudinal stance that is more discipline specific is *essential,* which occurs far more commonly in the hard sciences. While the adverbial form *essentially* is often used as a hedge, *essential* underlines the author's assessments regarding the contribution or component of an entity or process. It is a stance expressing the author's view that something is vital to a situation or activity (20) or a fundamental part of the nature of something (21):

(20) . . . openness to strangers, and acceptance of differences are *essential* in establishing urban communal life. (Sociology)

(21) Guessing, top-down processing, and inferencing are all *essential* aspects of effective communication (Applied linguistics)

Interestingly, the two hard science disciplines we studied use *essential* in broadly different ways. Engineers overwhelmingly use the form as a predicative adjective, either characterizing the nominal expression in subject position (22) or, less commonly, following a direct object (23):

(22) System modelling and experimentation are *essential* parts of the engineering discipline. (Electrical engineering)

(23) For control systems to be able to adapt to a variety of unexpected changes, dynamic learning is *essential*. (Electrical engineering)

In all cases, the author gives an evaluation of the material he or she is presenting, marking a clear stance toward its perceived value.

Although this pattern is also common in biology, authors here tend to make far greater use of structures with an extraposed *to-clause* or followed by *for*, with the latter twice as frequent in the 2015 corpus. While both forms express judgments of extreme importance or necessity and are generally considered interchangeable when preceding a noun phrase, *essential to* seems to convey a stronger meaning of *essence* or inherent quality. Thus in (24) the connection the author posits might be seen as intrinsic to the plant's development and so seeks to establish a stronger relationship than that in (25), which might only be a necessary factor.

(24) The continued delivery of ions via the transpiration stream ensures that the leaf is supplied with the mineral nutrients *essential to* its growth and development. (Biology)

(25) Therefore, the crs 1 gene is *essential for* the biogenesis of the ATP synthase complex. (Biology)

The stronger meaning of *essential to* is more common in our corpus and seems to be increasing, conveying a stronger authorial stance. In all uses of *essential,* however, as in many other markers of affect, there is no explicit attribution of the stance to the author, but this overt positioning can easily be inferred.

To sum up the changes we have observed in the use of affective stance in these four disciplines, we can say that they closely reflect those observed for epistemic markers. Normed frequencies for the explicit marking of attitude have seen large increases in engineering, declines in applied linguistics and sociology, and biology virtually unchanged. Once again, it is difficult to account for these changes, but it appears that over the past 50 years there has been a broad movement toward homogenization of stance in academic writing, with authors in the soft sciences being more prudent in their expressions of attitude and scientists becoming less so. However, while the sciences have increased both the frequency and range of the markers they use to convey affect, the most common items remain those that refer to judgments of importance and expectation rather than emotion. They are also items that seek to closely align writers with a disciplinary value system.

Presence

A writer can present a more or less visible stance by choosing to step into a text through self-mention or to use impersonal forms. How we understand writers and their attitudes to their arguments and readers is heavily influenced by their choices of authorial presence, and this, in turn, typically reflects disciplinary practices. In the sciences it is common for writers to downplay their personal role to highlight the phenomena under study, the replicability of research activities, and the generality of the findings, subordinating their own voice to that of unmediated nature. Such a strategy subtly conveys an empiricist ideology that suggests research outcomes would be the same irrespective of the individual conducting it. In the humanities and social sciences, in contrast, the use of the first person is closely related to the desire to both strongly identify oneself with a particular argument and to gain credit for an individual view.

Table 8.4 shows that authorial self-mention has undergone the greatest changes of all stance categories over the past 50 years. Frequencies have increased 38% in sociology (per 10,000 words), 63% in electrical engineering, and a massive 163% in biology, while declining by 27% in applied linguistics. Again, these results, and especially those for biology and applied linguistics, defy our expectations, with the biggest increases in the hard sciences. Table 8.5 shows that the most dramatic changes have occurred since 1985 and that the patterns of change have been uneven, with *I* replacing *we* as the preferred marker in applied linguistics, for example, and *I* becoming more common in sociology after a fall in the middle of the period.

It is obviously much harder to explain these trends than to describe them. While scientific writing remains the prototypical exemplar for representing meanings in an objective and formal way, subordinating the authority of the individual to the authority of the text, authors in the sciences are increasing their presence in the text with far greater willingness than in the past.

This increase, however, is confined to plural forms, which allow authors to create more distance between themselves and their reporting than first person and so temper a more invasive stance. It is nevertheless a significant trend. One reason for it may be the need for scientists to respond to the imperative of "impact" as a measure in annual performance reviews and career assessments, making their work less impersonal and more accessible to audiences in the commercial world outside the university. More likely, however, a more visible presence is a way of ensuring that their individual interpretations and claims do not go unnoticed by those same university human resource panels where applications for jobs, tenure, and promotion are judged. A more personal stance garners more visibility, more citations, and more professional credit. Something of this claim for recognition can be seen in these examples:

(25) From these data we have proposed a model for the distributive regulation of Byr2 by Rasl, Ste4, and Shkl. (Biology)

TABLE 8.5 Changes in Self-Mention by Discipline (per 10,000 Words)

Feature	Applied ling			Sociology			Electrical eng.			Biology		
	1965	1985	2015	1965	1985	2015	1965	1985	2015	1965	1985	2015
we	42.8	42.9	29.5	33.6	29.9	40.2	47.1	65.1	71.0	10.1	12.3	37.9
I	22.1	15.8	29.5	12.8	9.3	28.7	0.1	0.2	0.5	2.2	1.7	1.8
our	16.2	19.1	8.9	12.8	15.8	12.9	3.6	4.1	6.9	5.8	4.1	10.1
us	5.7	6.6	3.2	3.9	2.9	4.0	1.8	3.9	7.2	1.2	0.8	1.8
Other	7.2	3.4	4.1	1.3	1.3	8.0	0.0	0.0	0.1	0.6	0.4	0.4
Total	94.0	87.8	68.4	65.3	67.1	89.9	52.8	65.1	85.8	19.8	20.2	52.1

(26) We suggest that some of these differences may be attributable to changes in the chromatin structure introduced by the fixation procedure itself. (Biology)

(27) . . . we are of the opinion that the first method (co-simulation by the simultaneous solution of the electrical and thermal problem) is superior, since it is applicable to a wider range of electro-thermal problems. (Electrical engineering)

Personal reference is a clear indication of the perspective from which a statement should be interpreted, enabling writers to emphasize their own contribution and to seek agreement for it.

Sociology has followed a similar path with authors increasing their presence in their texts through self-mention, although this is perhaps more understandable than in the sciences. Here research is generally more explicitly interpretative and less abstract, with less "exact" data collection procedures and less control of variables. Readers expect that authors will not write with positivist detachment but will craft a convincing argument using a credible stance:

(28) The political incorporation of women, we argue, is a worldwide process deeply influenced by world models of progress and justice and strongly associated with becoming a legitimate nation-state. (Sociology)

(29) I want to set out a different approach and suggest that this may be able to resolve some of the dualisms found in the previous contributions. (Sociology)

The patterns in applied linguistics, however, are more curious, with a substantial fall in self-mention over the past 50 years. This shift to a more faceless style of prose also underlines an increasing formality we have noticed in the research writing of this discipline (Hyland, & Jiang, (2016)). It possibly reflects an increase in empirically oriented studies, as opposed to perhaps personalized accounts of teaching practices in earlier times, and where authors seek to moderate their personal stance to present research that can withstand the rigors of falsifiability. An alternative explanation may be the influence on the corpus of growing numbers of second language writers, most notably from China and the Middle East, schooled in the virtues of eliminating explicit agency from academic writing (e.g., Hyland, 2012a). More likely, perhaps, is that it simply represents a heightened awareness, and growing self-consciousness, among rhetorically sensitive academics keen to replace an ego-centric stance with a more collectivist one.

These explanations, however, are rather fanciful, as their fragility is accentuated by the fact that both applied linguistics and sociology showed an increase in the use of singular first person over the period, with it doubling in the former since 1985 and increasing threefold in the latter. Quite clearly, the ability of authors to position themselves as reflective agents with important ideas to contribute (30) or insights to convey (31) remains very important in these fields:

(30) I raise the issue of forms of extreme social disadvantage and oppression which do not seem to stimulate a critical social scientific response. In this way, I am confronting fundamental issues of critical social science practice; I show why the experiences of very cognitively disabled people should be a central concern of critical sociologists, social historians and social theorists. (Sociology)

(31) Whatever I decide to do with what I have learned about what originally puzzled me, I will also be aware of another type of knowledge which I take way from the research process. The research process will have taught me something about both of these aspects, in essence what has served my research needs and what has not. If the latter proves to be substantial then I will adopt a different approach as regards the techniques I choose the next time I do research. (Applied linguistics)

While *I* is increasing in the soft knowledge fields, we can see from Table 8.5 that *we* dominates the frequencies in all disciplines, composing 58% of all forms in 2015. This has nearly doubled in electrical engineering and increased by almost four times in biology, making it overwhelmingly the preferred marker of self-mention in the sciences. Clearly one reason for this is the growing trend, and career pressures, toward collaborative writing and coauthorship. Recent decades have seen the proportion of multiauthored articles gradually increase, so according to data from Thomson Reuters, the average number of authors on articles in the Science Citation Index grew by 50% between 1990 and 2010 (Hyland, 2015). Thus, while any issue of a journal such as *Nature* contains around the same number of articles today as one from 1950, it has about four times as many authors (Greene, 2013).

Today, collaboration and teamwork are among the most obvious features of scientific and technological research with a worldwide trend toward more coauthors affiliated with more universities in more countries. This development reflects both the publishing imperative of the performance culture and the growth of increasingly intricate multi-investigator research. Under career-defining review systems the world over, each author can claim each article and each citation as his or her own, so creating a system that rewards heavily multiauthored articles. Coauthorship is also encouraged by the ever-growing complexity of the modern environment and, by extension, of research problems. Thus, almost half (46%) of all U.K.-authored articles in 2010, for example, were coauthored with a non-U.K. researcher, and this created a citation advantage greater than that of nationally authored or coauthored articles (Department of Business, Innovation and Skills, 2011).

Multiple authorship is also growing in the soft disciplines, as can be seen by the increased use of *we* in our sociology corpus. There is a possibility that this reflects a trend toward studies aimed at, and funded by, a wider audience from the political, commercial, and industrial worlds, which prefer their research presented in a more objective fashion. Essentially, however, coauthorship creates research advantages as a result of sharing resources, ideas, expertise, and data, as well as the synergistic creativity that comes from working with others. Perhaps more important, it also provides opportunities to effectively split the work to speed up progress and publish more articles, so that coauthorship is driven by the imperative of visibility and the fact that it enables academics to publish more, in higher impact journals, and to collect more citations for their work.

Final observations and conclusions

In this article we have tracked how the expression of authorial stance has changed in research writing over the past 50 years. Using Hyland's model and looking at the articles from leading journals in four representative disciplines, we have uncovered a somewhat surprising picture, finding that stance features have not kept pace with massive increases in the length of articles and that stance frequencies are increasing in the sciences and falling in the soft fields. These seem to be quite major, but largely unnoticed, recent historical changes that have resulted in a gradual movement toward rhetorical convergence as the hard and soft fields adjust their stance profiles to changing circumstances.

It appears that, from the point of view of authorial stance at least, we are witnessing slow changes in traditional knowledge construction practices. The cumulative and tightly structured procedures of the sciences have generally allowed for succinct communication and relatively "strong" claims that have been presented as emerging from observations in the lab rather than interpretations at the word processor. The relatively clear criteria for establishing or refuting claims have allowed authors to remove themselves from the picture, but research articles in both biology and electrical engineering, and particularly in the latter, display an increased deployment of stance markers, most noticeably self-mention. We speculate this may not be unrelated to the need to address audiences beyond an immediate group of informed insiders to promote both one's research and oneself with tenure and promotion committees and commercial sponsors.

In the more discursive fields we observe a marked trend in the opposite direction, toward less authorial intrusion and a less visible stance. Articles in the top applied linguistics and sociology journals contain far fewer hedges, boosters, and attitude markers (per 10,000 words) than they did in 1965, and those in applied linguistics also contained less self-mention. These changes minimize authorial presence and convey more cautious stances, directing readers to the persuasive strength of data or methodological practice rather than the convictions of the interpreting writer.

Overall, our study supports research that shows an inexorable growth in formality and authorial withdrawal since the inception of scientific writing some 350 years ago, a change that Atkinson (1999) describes as a move from a less "author-centred" rhetoric to a highly abstract and "object-centred one." Using a cluster analysis of features, Atkinson found that articles steadily became less affectively and "narratively" focused and more "informational" and abstract over time, shifting from a discourse based around the experiencing gentleman-scientist to community-generated research problems. Impersonalization is a powerful tool for repackaging research phenomena to downplay the interpreting researcher and "semiotically reconstruct" (Halliday, 2004) experience to realize new ways of seeing the world. Just as the grammatical style of articles has emerged over the centuries from the political establishment of a scientific community, the changes we see in these stance choices similarly reflect changing audiences and material conditions. As Biber and Gray (2016) report in their study of academic research writing over the past two centuries, it appears that stance in academic writing is increasingly coming to be expressed implicitly rather than employing explicit grammatical devices.

It must be remembered, of course, that the stance an author takes, and the creation of an academic persona to which this stance contributes, is always an act of personal choice where the influences of personality, confidence, experience, and ideological preference are clearly important. Ultimately we are all rational agents rather than robotic instruments of our disciplines. However, writers do not act in a social vacuum and knowledge is not constructed outside particular communities of practice. Such communities exist in virtue of a shared set of assumptions and routines about how to collectively deal with and represent their experiences. The ways language is used on particular occasions is not wholly determined by these assumptions, but a disciplinary voice can only be achieved through a process of participating in such communities and connecting with these socially determined and approved beliefs and value positions. In this way, independent creativity is shaped by accountability to shared practices. This is why disciplines do things differently and why the ways they do things change over time.

In the end, effective academic writing depends on rhetorical decisions about interpersonal intrusion that recognize and align with both disciplinary epistemologies and social practices and with wider political and institutional changes. The most significant of these in recent times would seem to concern the ways knowledge is constructed and disseminated to new audiences outside a traditional peer group, including commercial and industrial sponsors who might make use of the knowledge created and personnel boards that make high-stakes decisions regarding the careers of academic writers.

Appendix

Journal List

Applied Linguistics

TESOL Quarterly (1967-)
Language Learning (1948-)
Foreign Language Annals (1967-)
Modern Language Journal (1916-)
College Composition and Communication (1950-)

Sociology

American Journal of Sociology (1895-)
Social Problems (1953-)
British Journal of Sociology (1950-)
American Journal of Economics and Sociology (1941-)
Sociological Quarterly (1960-)

Biology

Quarterly Review of Biology (1926-)
Biological Reviews (1923-)
Radiation Research (1954-)
BioScience (1964-)
Journal of Experimental Biology (1923-)

Electronic Engineering

Proceedings of the IEEE (1963-)

Automatica (1963-)
IEEE Transactions on Automatic Control (1963-)
IEEE Journal of Solid-State Circuits (1966-)
IEEE Transactions on Information Theory (1963-)

Note

1 Two journals, *TESOL Quarterly* and *Foreign Language Annals,* began
   only in 1967, and so article were chosen from issues in that year. College
   Composition and Communication (CCC) is the only one not indexed in
   SSCI, but it is an influential, long-running journal listed in the language and
   linguistics subcategory of Arts and Humanities in the Thomson Reuters
   citation index.

Commentary on Part II

Brian Paltridge

In order to be successful, academic writers need to represent themselves in ways that are valued by their discipline as well as take account of the values and beliefs of their disciplinary community. This involves 'negotiating a self which is coherent and meaningful to both the individual actor and the group' (Hyland 2011b: 11). Academic writers, further, are only successful in doing this if their work is recognized as being legitimate by the discipline and the group. Writers, thus, as Hyland argues in the chapters which follow, need to choose ways of expressing themselves that will resonate with their audience so that their claims to be one of them will be seen to be credible and valid. This kind of writing, however, is highly situated and requires an in-depth understating of the values and ideologies of the discipline in which the writer is working, as well as how writers make connections with their readers through the use of language.

There are a number of ways academic writers represent themselves in their text, their relationship with their readers and their relationship to the knowledge they are discussing. This is often through the use of *metadiscourse* features (see Chapter 7 of this book) such as of *stance*, involving the use of *self-mentions* (*I*, *we*, *my*), *hedges* (*might, perhaps, almost, suggest, believe*), *boosters* (*definitely, in fact, beyond doubt, clearly*) and *attitude markers* (such as *unfortunately, surprisingly, admittedly* and *important*) which express the writer's attitude towards a proposition. Writers also draw on *engagement* strategies such as *reader pronouns*, *personal asides*, *appeals to shared knowledge*, *directives* and *questions* to establish relations with their readers (see chapters 6, 7 and 8 of this book). Through the use of these strategies, Hyland points out, writers both acknowledge and recognize the presence of their readers at the same time as they position themselves in relation to what they are writing. Academic writers, thus, aim to represent themselves in a way that is valued by their discipline as they adopt the beliefs, expectations and identity of a successful academic writer in the particular genre and disciplinary context.

Successful academic writers, then, take up or anticipate the voices and positions held by their potential readers. They do this as they comment on claims that they make, establish solidarity with their readers and represent their credibility in terms of what they are saying. Hyland and Tse (Chapter 7 of this book) examine stance in postgraduate student dissertation writing, arguing that the use of stance (and engagement) features allows writers to frame what they say, to position and engage readers, and to enter into a relationship with their reader/s. An analysis of this can help reveal the expectations and understandings of the audience for whom the text is written which, in turn, reveals 'something of their social practices, values and ways of thinking' (Hyland 2005a: 58).

A number of researchers have drawn on Hyland's framework for examining this aspect of academic writing. Hyland and Diani (2009) and Hyland and Sancho Guinda (2012), for example, provide examples of studies which focus on such genres as book reviews, academic bios, research articles and undergraduate students' final year reports. Hyland (2012b) focuses on stance in undergraduate student writing at a university in Hong Kong. The use of stance features, he points out, is a complex task for second language writers, as well as for beginning L1 academic writers who might often hedge what they write much less than expert writers do (see Aull 2015; Aull and Lancaster 2014). The more tentative stance of an expert writer, Hyland (2012a: 146) argues, is a 'clear signal of insider membership and disciplinary identity'. The use of hedging to express tentativeness, thus, is seen as an indicator of more successful student writing, both L1 and L2 (Lee and Deakin 2016), where findings and results are explained 'with the appropriate degree of precision and caution' (Koutsantoni 2007: 105), reflecting 'the need for negotiation of knowledge before claims are accepted and consensus is reached' (Koutsantoni 2004: 179). McGrath and Kuteeva (2012) show this in their study of published research articles in the area of pure mathematics where tentativeness by the use of hedging is a more frequent stance feature than attitude markers, boosters or self-mentions.

Khamkhien (2014) also finds tentativeness, expressed through the use of modal items such as *can*, *may* and *could*, to express degrees of possibility and confidence, a common feature of published research articles in the area of applied linguistics, especially where authors compare their research to previous studies in the discussion section of their article. Paltridge (2017) examines the stance that reviewers of submissions to academic journals take in the reports that they write. He found the most frequent markers of stance were attitude markers in reviews that made 'accept' and 'minor revisions' recommendations and self-mentions in 'major revisions' and 'reject' reviews. There were, however, when compared to published academic writing, many fewer instances of hedging in the reports that the reviewers wrote. This, coupled with the high use of self-mentions in the reviews, showed a high level of certainty in the views expressed by the reviewers who, by contrast with the Aull (2015) and Aull and Lancaster (2014) studies, were highly expert academic writers. The writers did this, he suggests, not because they had not yet developed a sense of how to hedge in their writing, but because they wanted to show a strong commitment to the views that they were presenting in their texts.

In a review of the use of engagement features in academic writing, Hyland and Jiang (2016) examine engagement strategies in published research articles over the past fifty years, finding very large changes in rhetorical conventions of interaction over that time. One of their key findings is a marked decline in the use of interactional features in soft knowledge fields. At the same time they found a dramatic increase in the use of directives in electrical engineering writing. This, they argue, shows that writers of research articles seem to report on their research more impersonally and 'with less effort to finesse the reader' (p. 39) than they did in the past (see Chapter 8 for a discussion of changes in the use of stance in academic writing over the same period of time).

In sum, then, Hyland's work on the interactive nature of texts as realized through the use of metadiscourse features shows how academic writers interact with disciplinary cultures and audiences, both mediated and influenced by the disciplinary communities to which the readers and writers of the texts belong and their discipline-approved practices (see Chapter 5). Hyland's work has made an important contribution to our understanding of how academic writers' identities are performed through the use of discourse as they take account of the norms, values and expectations of their disciplinary communities. His metadiscourse framework, in particular, enables us to see how, through the use of language, academic writers both present themselves to their readers and interact with them and, in doing this, index membership of their disciplinary communities.

PART THREE

Interactions in peripheral genres

Introduction

In this section I present some of my research into genre and look at three different kinds of text in the academy: dissertation acknowledgements, popular science and academic homepages. Readers may be surprised at these choices, after all, they are not core genres in any way: they contribute little to the processes where knowledge is created, degrees achieved, courses passed or reputations made. They may, in fact, seem somewhat peripheral to the genres where the real work of the university gets done. Textbooks, essays, dissertations, final year reports, abstracts and, of course, research articles are the staples of academic writing research in applied linguistics.

Those 'core' genres, however, have been massively studied and almost every detail of them has been unpicked. We know what they are like and how they work and are now in a far better position to advise students on their writing as a result. But I have not neglected these genres because I think all interest has been wrung out of them. On the contrary, there is more to learn, particularly about genre creativity and innovation (e.g. Tardy, 2015) and genre uptake, or how individuals create genres as a result of their genre knowledge, contextual understandings and sense of self (Freadman 2002). I have, in fact, done my fair share of research on these core genres, but I have always had a soft spot for those texts on the edge of things, forever the 'extras' and rarely playing starring roles.

My workout on the backroads of written academic interaction includes studies of book reviews, graduate prize applications, journal descriptions and article bios as well as the three 'peripheral genres' discussed in this section. Despite their existence on the boundaries of academic life, however, these genres display the characteristics of their more celebrated cousins. They are (a) social, in their conventional, recognizable and consequential ways of organizing texts, activities and social reality, and they are also (b) cognitive, revealing how we recognize and make sense of the situations in which we find ourselves. In these genres we find the same regularity and clustering of conventions which help guide perceptions and actions so they have the same potential for exploring the practices of language-using communities.

This underlines the social nature of texts and the fact that language is embedded in (and constitutive of) social realities. Genres are the property of the communities that use them, and by analysing them we learn more about the ways discoursal practices both reflect and construct those communities. As always, John Swales (1998: 20) puts this better than I ever could:

> In-group abbreviations, acronyms, argots, and other special terms flourish and multiply; beyond that, these discourse communities evolve their own conventions and traditions for such diverse verbal activities as running meetings, producing reports, and publicizing their activities. These recurrent classes of communicative events are the genres that orchestrate verbal life. These genres link the past and the present, and so balance forces for tradition and innovation. They structure the roles of individuals within wider frameworks, and further assist those individuals with the actualisation of their communicative plans and purposes.

It is through the recurrent uses of familiar genres and conventionalized forms that individuals develop relationships, establish communities and get things done, and this can be seen in the genres described in this section as much as in research articles or dissertations. They also show that the more genres we look at, and the more examples of those genres, the more variety we find, the more dimensions of variation, the more invention and the more situations and meanings. We also see something of the diverse ways individuals use interactions to tell their stories and construct professional relationships.

Chapter 9 explores how some of these issues of credibility, engagement and interaction are played out in the genre of popular science articles. Popularizations are interpretations of academic activity recast for the interests, beliefs and preoccupations of an audience without a professional need for information about science. Despite carrying less prestige and importance in the academic genre hierarchy, they are not merely examples of superficial infotainment. This is a genre written not only by specialist journalists but also by leading scientists such as Richard Lewontin and Stephen Hawking; it is also a commercially successful genre which appeals to a huge audience. In speaking to a general audience, however, authors of these texts cannot assume that readers will always recognize the significance of information in the same way as experts, and so relevance has to be supplied in the text itself rather than presupposed in the context. This means that the original scientific claims are often 'boosted' in popularizations and new information is added to adjust to readers' assumptions and values.

The purpose of the scientific literature is to persuade other specialists of the validity of conclusions and the effectiveness of methods, and to do this they use a standard format and give prominence to results and to tables, figures and diagrams to represent physical evidence. Popular science, in contrast, attempts to convince scientific outsiders of the significance of data and to celebrate the results, with their validity largely taken for granted. In this chapter I look at one way this is achieved by drawing on the idea of *proximity,* or how writers rhetorically display both authority as experts and personal positions towards issues in their texts. This is an earlier, less fully formed, conception of *proximity* than that discussed in Chapter 2. There the term referred to features which shape a text to make sense to readers and was paired with *positioning,* or how writers locate themselves in a dialogue with others. In Chapter 9 proximity is asked to perform both these jobs, by demonstrating *membership* and a personal *commitment,* towards what is being discussed. To highlight these features I compare them with the ways the same functions are performed by scientists in research articles.

Chapter 10 looks at *Dissertation acknowledgements* and how postgraduate students create a professional expertise and personable character while thanking those who have contributed to the accompanying dissertation. The corpus was extracted from the 240 dissertations Polly Tse and I collected as part of the metadiscourse project and was analysed as an afterthought once we had completed the funded work. It is never a good idea to let good data just sit neglected on the hard drive. Once again, this is a somewhat disregarded 'Cinderella genre', sometimes considered an expression of vanity or careering, but it is one of the few opportunities students have to present a social and scholarly self disentangled from academic discourse conventions. It therefore provides some interesting insights into academic interactions, the creation of credibility and the management of the author's relations to the disciplinary community. Once again, we see that such interactions are mediated by disciplinary preferences and strategic professional choices, reflecting the way in which postgraduate writing represents a situated activity.

The final 'peripheral genre' in this section is the academic homepage (Chapter 11), now a ubiquitous genre of scholarly life, with perhaps 80 per cent of academics having at least one. This venture into the ways interactions are accomplished through a combination of textual, visual and hyper-textual choices was a new departure for me and was part of a growing interest in how academics construct a public identity. But while the personal homepage is a valuable tool for supporting a particular version of the author, representing how he or she wishes to be seen by others, it can also serve institutional ends: functioning to promote the employing institution and marginalizing the individual. Because of this, academics often reclaim something of their self-representation by creating their own personal page and this study compares similarities and differences between the two, examining fifty personal and ffity university-hosted pages of the same people in two contrasting disciplines.

Through an analysis of what is said and how it is presented, as text, design and hyperlinks, the study suggests how individuals are positioned by corporate discourses yet manage to carve a sense of self to assert professional credibility.

The section closes by a commentary on the texts by one of the leading writers on specialist genres. Professor Vijay Bhatia, of City University of Hong Kong, has been a colleague, friend and source of inspiration for many years.

9

Constructing proximity: Relating to readers in popular and professional science

1 Interpersonality and proximity

The view that academic writing is interpersonal is no longer news. Over the last twenty years analyses of academic texts have shown us that while they may be more lexically dense, cautious and nominalised than many other kinds of writing, they aren't as completely 'author evacuated' as we had once supposed (e.g. Hyland, 2004a; Swales, 2004). Instead, they are actually comprised of careful evaluations and interactions. Research papers, we have discovered, are sites where academics don't just offer a view of the world, but negotiate a credible account of themselves and their work by claiming solidarity with readers, appraising ideas and acknowledging alternative views. Interpersonality, then, concerns the ways that writers use language to negotiate social relationships by telling their readers what they see as important, how they believe they should select and present material for them, and how they feel about what they write about. It concerns the explicit system of meanings which link the participants in a text as *interactants*, adopting an acceptable persona and a tenor consistent with the norms of the community.

The important interactional norms in research writing involve 'establishing a judicious, discipline-defined balance of tentativeness and assertion, and a suitable relationship to one's data, arguments and audience' (Hyland, 2005a, p. 54), but not all academic writing is the same. Any text is part of a register which spans a range of purposes, audiences and contexts from undergraduate essays to Nobel speeches. In other words, the work of the academy is interpreted using language for different audiences: for experts, for students, for industry, and for practitioners, and in each case, this recontextualization of material offers different ways of understanding academic practices. By extending our interest beyond the mainstream discourses of the academy we access very different scientific discourses and gain insights into how relations between science and society are mediated and the cultural authority of science is promoted. More centrally for this paper, we also learn something of how interpersonality is negotiated with different audiences. We discover the ways writers display who they are and construct a convincing argument drawing on different discoursal conventions to establish proximity with readers.

I use the term *proximity* here to refer to a writer's control of rhetorical features which display both authority as an expert and a personal position towards issues in an unfolding text. It involves responding to the context of the text, particularly the readers who form part of that context, textually constructing both the writer and the reader as people with similar understandings and goals. While it embraces the notion of interpersonality, proximity is a slightly wider idea as it not only includes how writers manage themselves and their interactions with others, but also the ways ideational material, what the text is 'about', is presented for a particular audience. It is concerned with how writers represent not only themselves and their readers, but also their material, in ways which are most likely to meet their readers' expectations.

So proximity captures 2 key aspects of acting interpersonally. The first refers to what might be called the *proximity of membership*: How academic writers demonstrate their authority to *colleagues* through use of disciplinary conventions. What does the writer do to position him or herself as a disciplinary expert and competent colleague? The second concerns the *proximity of commitment,* or how the writer takes a personal position towards *issues* in an unfolding text. That is, what does the writer do to locate him or herself in relation to the material presented? One points to how we position ourselves in relation to our communities, and the other to how we position ourselves in relation to our text. These are, of course, difficult to separate in practice as we can't express a stance towards the things we talk about without using the language of our social groups. But the concept does allow us to say something about how writers take their readers' likely objections, background knowledge, rhetorical expectations and reading purposes into account.

Proximity emphasizes a reader-oriented view of writing and is closely related to Sacks, Schlegloff and Jefferson (1974: p. 272) notion of "recipient design", or how text and talk are shaped to make sense to the current recipient.

In writing, as in conversation, we display an orientation and sensitivity to the particular others who are our co-participants through lexical choice, topic selection, conventions of argument, and so on. In academic writing this means that the process of writing involves creating a text that we assume the reader will recognise and expect and the process of reading involves drawing on assumptions about what the writer is trying to do. Hoey (2001) likens this to dancers following each other's steps, each building sense from a text by anticipating what the other is likely to do. Skilled writers are therefore able to create a mutual frame of reference and anticipate when their purposes will be retrieved by their audiences.

The concept of *proximity* therefore helps us to understand how writers typically position themselves and their work in relation to others, so we can see interpersonality as the rhetorical construction of proximity. In this paper I plan to look at how writers in two very different genres, research papers and popular science articles, create a sense of proximity by textually constructing themselves and readers as having shared interests and understandings. Comparing key features from these contexts, I show how different language choices are employed to negotiate academic claims and construct proximity with two very different audiences.

2 A tale of two genres

Despite competition from electronic publishing alternatives such as e-journals and personal websites, the research article (RA) remains the pre-eminent genre of the academy. The RA is not only the principal site of disciplinary knowledge-making but, as Montgomery (1996) has it, 'the master narrative of our time'. One reason for this preeminence is the value attached to the processes of peer review as a control mechanism for transforming beliefs into knowledge. Another is the prestige attached to a genre which restructures the processes of thought and research it describes to establish a discourse for scientific fact-creation. In this context language becomes a form of *technology*, or a resource for controlling an environment, by presenting interpretations and positioning participants in particular ways to establish knowledge.

Several studies have shown the ways that serial drafts and reviewers' comments guide science writers to rhetorically accommodate their laboratory activity to the concerns of the discipline (Berkenkotter & Huckin, 1995; Casanave & Vandrick, 2003). Through reviewers' recommendations to modify the strength of their claims, provide propositional warrants, and establish a context through citation, writers gradually integrate their new claims into the weave of disciplinary relevance and prior work. This process indicates that new facts are not added piecemeal to the heap of existing knowledge, but are the extension of an ongoing conversation among members, conducted in a shared 'theory-laden' language and particular patterns of argumentation. Professional research writing is therefore seen as a tension between originality and deference to the community, and the language authors use to argue their claims constructs this proximity with readers.

This extract from an article in *The Astrophysical Journal Letters,* a highly-regarded molecular science periodical, helps to illustrate this:

(1) We used *Suzaku* observations to measure the spatial variation of the Fe Kα line with radius in the *Tycho* supernova remnant. The Fe line widths show a significant decrease from a FWHM value of 210 eV at the center to 130 eV at the rim. Over the same radial range the line center energy remains nearly constant. These observations are consistent with a scenario in which the shell of Fe-emitting ejecta in *Tycho* is expanding at speeds of 2800–3350 km s^{-1}. The minimum line width we measure is still a factor of two larger than expected from a single component plasma emission model. If thermal Doppler broadening is the dominant additional source of broadening, we infer an ion temperature of $(1 - 3) \times 10^{10}$ K. (*Astrophysical Journal Letters* 693L61)

This text is written for a professional audience with a high degree of specialised expertise. Information is presented with considerable exactness, foregrounding procedures and using technical jargon, nominalisations, precise measurements, cautious inferences from data, and acronyms. It is a discourse of exclusivity underpinned by a specialised knowledge of methods and of the meanings which results have for insiders.

'Popular science', in contrast, is produced for audiences without a professional need for information about science but who want to keep abreast of developments. In fact, popular science discourses play an enormous role in shaping most people's views of academic research: informing lay understandings of the interests, methods and knowledge that it produces. Their existence underlines that 'science' is not a monolithic entity always understood in the same way, but a social construct created by different groups with different interests. Popular science itself also addresses a range of different audiences (Hyland, 2009). While many popular science books are written by scientists for an elite educated audience, the public gets most of its information about science from specialised magazines like *New Scientist* and *Scientific American.* Most daily newspapers now have specialized science sections and the number of science articles in the press has been increasing (Pellechia, 1997).

These offer interpretations of academic activity recast with an eye for the interests, beliefs and preoccupations of a new readership. But it would be a great oversimplification to dismiss popular science as merely infotainment. This is a discourse related to the academy, its work, and its forms of communication but stripped of its more forbidding rhetorical features. While attempting to wield the authority of science, both scientific facts and the argument forms of professional science are transformed in the process.

The different purposes and audiences of the two genres mean that writers negotiate proximity and represent science in very different ways to those employed by professional scientists in research journals. This can easily be seen by comparing example 1) with the same research reported in the popular journal *Science Daily*:

(2) The most crowded collision of galaxy clusters has been identified by combining information from three different telescopes. This result gives scientists a chance to learn what happens when some of the largest objects in the Universe go at each other in a cosmic free-for-all. The researchers found that four separate galaxy clusters are involved in a triple merger, the first time such a phenomenon has been documented. Galaxy clusters are the largest objects bound by gravity in the Universe. . . . . "In addition to this enormous pileup, MACSJ0717 is also remarkable because of its temperature," said Cheng-Jiun Ma of the University of Hawaii and lead author of the study. "Since each of these collisions releases energy in the form of heat, MACS0717 has one of the highest temperatures ever seen in such a system." (*Science Daily, Apr. 17, 2009*)

In contrast to the research report, the popularization places an emphasis on the actors and their interpretations with short sentences, congruent grammar and use of first person quotes. Priority is given to the potential payoffs of the research and results, rather than the means of obtaining them.

Criticism of this kind of writing as a journalistic dumbing down of science, disseminating simplified and often sensationalised accounts to a passive mass readership, misses the point. Popular science does not just report scientific facts to a less specialist audience but represents phenomena in different ways to achieve different purposes. Most fundamentally, proximity helps science writers transform beliefs into knowledge, producing evidence for claims to persuade specialists of the reliability of their interpretations and the rigour of their methods. Popular science, in contrast, is concerned with establishing the novelty and relevance of a topic to celebrate scientific results, with their validity taken for granted. Popularizers, then, actually transform the products of an academic culture in the process of appropriating them and so influence the nature of elite science itself.

3 Facets of proximity

There are various ways writers achieve proximity and I want to discuss five in the remainder of this paper. Drawing on two corpora, a collection of 120 research articles from four science and engineering fields and a sample of 120 popular science articles, I explore how writers use *Organisation, Argument, Credibility, Stance and Engagement* to negotiate proximity with readers.

3.1 *Organisation*

There are several aspects of presentation which distinguish these genres. One is the role of visuals, for example, which play a key role as *arguments* in the research papers by giving visibility to information and offering a proof for interpretations, and as *explanations* in the popularizations, where they work mainly to attract the reader and elucidate the text (Miller, 1998). The most striking difference, however, is the way that the genres are organised. Instead of finding the main claim towards the end of the paper as in a research article, it is typically foregrounded at the beginning. Nwogu (1991), for instance, found that journalistic accounts typically open with a background move which contextualizes the research issue as a problem for readers and then follows this with the main outcome, often including reference to the scientists themselves. These examples are typical:

> (3) New moms beware: If you want to shed those extra pounds you packed on while pregnant, you better get your sleep. A new study shows that women are more likely to lose baby fat if they get over five hours of shut-eye a night. (*Scientific American,* Nov 2007)

> When we emerge from a supermarket laden down with bags and faced with a sea of vehicles, how do we remember where we've parked our car and translate the memory into the correct action to get back there? New research identifies the specific parts of the brain responsible for solving this everyday problem. The results could have implications for understanding the functional significance of a prominent brain abnormality observed in neuropsychiatric diseases such as schizophrenia. (*Science Daily,* April, 2009)

This deductive rhetorical pattern highlights the novelty and importance of the topic to lay readers rather than the methodological steps taken to get there, and while this provides a hook to bait the uninitiated, it can be confusing for scientists. Myers (1990: p. 141), for example, reports how the editor of the *New England Journal of Medicine* published both an original immunology article and a version rewritten by a Science journalist who gave greater attention to organization, explication and clarity. General physicians subsequently wrote applauding the fact that even difficult topics could be made accessible to non-specialists while immunologists complained that the revised version was harder to read because information wasn't where they expected to find it. Both groups therefore had different views about the best way to write immunology based on their own needs, background knowledge, discourse expectations, and reading purposes.

3.2 *Argument structures*

Arguments are also structured very differently to shape material for the two audiences, reaching out to dissimilar audiences through distinctive kinds of appeals, focuses, and framings.

In terms of *appeals,* both genres are driven by *novelty,* although this is presented differently for the two audiences. Novelty is clearly a key feature of academic advancement and intellectual change in disciplinary communities, a means by which individuals gain credit for themselves, prestige for their field, and growth for their discipline within a shared understanding of what is worth knowing and where to take this. Kaufer and Geisler (1989: p. 286), in fact, refer to academic disciplines as 'factories of novelty, encouraging members to plod towards their yearly quota of inspirational leaps'.

In research papers novelty is negotiated by being the first to synthesize a contribution with the existing weave of community knowledge, situating local research in the broader concerns of the discipline. To be new, work must recognize the knowledge which has already gained consent and against which it makes a claim for change, so innovation is managed by establishing explicit intertextual links to existing disciplinary knowledge. So while claims for novelty are often made (as in 4), they are also backed up by arguments which relate them to the literature:

(4) The assays presented herein illustrate two novel approaches to monitor the intracellular dynamics of nuclear proteins. (Bio RA)

This paper proposes a new methodology that further develops Taguchi's method to incorporate multiple objectives and constraints in product design. (Mech Eng RA)

Novelty thus acknowledges what has gone before and builds on the field's organizational structures, beliefs and current hot topics. Topics, in fact are more than a research focus: they represent resources of joint attention which coordinate activities and mark co-participation in communities. Selecting a topic and arguing for its novelty and relevance is thus critical in securing colleagues' interest and in displaying membership credentials.

Popular texts, on the other hand, transform the novel into the *newsworthy*. Journalistic criteria mean that academic claims are changed into scientific breakthroughs and presented in terms of what is of immediate value or potential benefit to readers. Newsworthiness establishes proximity by suggesting a common enthusiasm for science and a technocratic ideology that our lives are constantly improved through scientific progress. There is a celebration of relentless advancement which science is said to produce and which the current case is just one example:

(5) Very young brains process memories of fear differently than more mature ones, new research indicates. The work significantly advances scientific understanding of when and how fear is stored and unlearned, and introduces new thinking on the implications of fear experience early in life. (*Science Daily*, March, 2008)

In contrast to women, men are fertile throughout life. But new research at the Sahlgrenska Academy has now shown that a fertilising sperm can get help from the egg to rejuvenate. The result is an important step towards future stem cell therapy. "We are the first to show that egg cells have the ability to rejuvenate other cells, and this is an important result for future stem cell research", says Associate Professor Tomas Simonsson, who leads the research group at that has made this discovery. (*Science Today* March 2009)

In both academic and popular contexts, novelty is related to proximity by appealing to what is assumed about readers' knowledge and interests. Newness is not a property of the ideas themselves, but a relation between ideas and communities as professional writers package material for particular readers.

We also find writers managing proximity in the **focus** of the argument, or what writers chose to concentrate on. Quite simply, we can see that science journalism centres on the *objects* of study rather than the disciplinary procedures by which they are analysed. Professional papers construct what Myers (1990) calls a 'narrative of science' which follows the argument of the scientist by arranging time into a parallel series of events and emphasizing the conceptual structure of the discipline in their syntax and vocabulary. The discourse embodies key assumptions of academic practice: impersonality, cumulative knowledge construction, and empiricism, so that readers can relate a current claim to their understandings of the epistemological beliefs, prior findings, and currently approved methods. This is a typical example:

(6) Leukocyte telomere length was positively associated with increasing physical activity level in leisure time ($P < .001$); this association remained significant after adjustment for age, sex, body mass index, smoking, socioeconomic status, and physical activity at work. The LTLs of the most active subjects were 200 nucleotides longer than those of the least active subjects (7.1 and 6.9 kilobases, respectively; $P = .006$). This finding was confirmed in a small group of twin pairs discordant for physical activity level (on average, the LTL of more active twins was 88 nucleotides longer than that of less active twins; $P = .03$). (*Archives of Internal Medicine,* Jan 2008)

A narrative of science, then, traces what is of interest to the discipline, employing rhetorical and linguistic choices which imply anonymity, precision, induction, comprehensiveness, and specialist knowledge of methods.

Popular articles, on the other hand, present a *'narrative of nature'* concentrating on the thing studied. This extract reports the same research in a very different way:

(7) Individuals who are physically active during their leisure time appear to be biologically younger than those with sedentary lifestyles, according to a report in the Archives of Internal Medicine. "Our results show that adults who partake in regular physical activity are biologically younger than sedentary individuals. This conclusion provides a powerful message that could be used by clinicians to promote the potential anti-aging effect of regular exercise." The authors say. (*Science Daily* Jan 2008)

Here the focus is more explicitly concerned with the connection between exercise and aging, which is likely to be of greater interest to lay readers than the precise methodological procedures. Material is assembled to enhance the visibility of information and make the message more convincing through a chronological presentation and highlighting concerns readers may be expected to have about ageing, youthful appearance, and exercise. These different language choices convey meanings which change understandings of both research and of science.

Finally, proximity is achieved in argument by the ways writers *frame* information for their target readers. Framing is achieved by tailoring information to the assumed knowledge base of potential readers, creating proximity for different audiences through language choices which ask readers to recognise something as familiar or accepted. In research papers this is largely accomplished though the use of technical terminology, acronyms, reference to routine craft practices and specialised forms of equipment. This example from physics is typical:

(8) The sample used was 80% H2O with 20% D2O as the lock substance. Two dimensional proton NMR experiments were conducted on a Bruker ARX-500 spectrometer with a 5 mm inverse probe and a 10 mm normal broad band probe tuned at 500.13 MHz using the standard J-resolved and COSY pulse sequences. The transmitter offset was located at 162 Hz up (for SECSY) or down (for COSY) field from the resonance of water for both dimensions. In 2D data acquisition 512 points were used and the number of tl-increments was changed. Data matrix for Fourier transformation was 512X512 with zero filling in the t domain only. (Phy RA)

Appealing to proximity through this abbreviated, highly specialised reference to shared knowledge is possible in the natural sciences because research is typically characterized by linearity and well-defined and agreed upon problems (Becher & Trowler, 2001). New knowledge is generated from what is known and each new finding inexorably contributes to the eventual solution of the issue under study. Specialism is encouraged by this as research often occupies considerable investments in money, training, equipment and expertise, which means individuals are often committed to particular research areas for many years and build their careers through precise contributions to a highly delimited field. Readers are often familiar with prior texts and research, and can see whether procedures and materials have been used appropriately and what results mean, so writers can describe their work economically.

Popularisations, on the other hand, can't assume this degree of shared knowledge and have to make connections to what readers are likely to already know. This involves constantly defining new concepts as they are introduced and making explicit links between entities. They therefore tend to avoid jargon and offer an immediate gloss where this is not possible. Clarifications are often inserted on-the-fly where the writer assumes an unfamiliar usage (9) or where complex processes are related to more familiar everyday events through simile (10):

(9) Prozac, the popular antidepressant, blocks the action of a pump that sucks serotonin, a key mood-regulating chemical, out of the gaps between two neurons. (*Popular Science*, Nov, 2007)

(10) To get into the brain they must be shuttled across the blood–brain barrier by specialized transport proteins. Like passengers trying to board a crowded bus, amino acids compete for rides on these transporters. Not only does tryptophan have paltry representation among the passengers; it also competes with five other amino acids for the same transporter. Aced out by other amino acids, tryptophan thereby has a tough time hitching a ride to the brain. (*Scientific American*, Nov 2007)

The unfamiliar is thus made intelligible by brief definitions and explanations which relate complex processes to everyday events, taking the reader's perspective to present the strange and exotic in the terms of the commonplace and unexceptional.

Non-scientists are also accommodated by the writer's management of cohesion. Cohesion depends on the semantic structure of a text and therefore on the reader's expectations and knowledge, particularly knowledge of lexical relations. However, because scientific texts rarely contain replacement or pronouns for cohesion, non-specialists may struggle to see connections across sentences (Myers, 1991). Journalists, however, make these links explicit by using a variety of cohesive devices to serve as the basis for inferences about the meanings of any unfamiliar terms. In this extract, for example, the writer is careful to ensure that the reader is able to recover the links describing the genetic causes of mental retardation. Through the use of repetition, conjunctive phrases such as 'which means that', determiners (the, those), and synonyms, connections are specified and the passage becomes transparent:

(11) In humans, the disorder stems from a mutation on the X chromosome as a three-base sequence begins to repeat over and over in a section of the fragile X mental retardation 1 gene (FMR1). The portion of the gene where this error multiplies does not code for a protein, which means that several repetitions of the sequence can occur without damaging the fragile X mental retardation protein (FMRP). People who have a gene with a sequence that is repeated 50 or fewer times are considered normal; those with fewer than 200 repetitions are carriers of the disorder. Individuals with more than 200 triplets, however, have disruptions to the promoter region of FMR1 that block the gene from being transcribed into RNA and forming a protein, thereby prompting onset of the syndrome. (*Scientific American* Nov 2007)

Clearly the 'naïve reader' is unable to learn the cultural system encoded in the language of science merely through reading scientific texts, but the representation of scientific knowledge in popularizations at least provides a basis for understanding the products of that culture.

3.3 Credibility

A third way in which popularizations seek to promote proximity with readers and engage with the incomplete knowledge-base of the non-specialist is to emphasize the credibility of the source of the information they report. In professional articles reliability is largely bestowed on findings by the writer's display of craft practice and expert handling of recognized research methods.

Attributions to other scientists mainly function to align the writer with a particular camp or reward researchers who have conducted relevant prior work. Popularizations, on the other hand, can't assume this level of knowledge in readers and so bestow credibility on scientists through their position in an institution, only identifying particular scientists when they are directly relevant to the research being reported:

(12) Animal scientist William <u>Dozier, formerly with the ARS Poultry Research Unit in Mississippi State, has been working with colleagues at the ARS Swine Odor and Manure Management Research Unit in Ames, Iowa, and Iowa State University</u> to find ways to supplement animal diets with glycerin. (*Scientific American*, April 2009)

"By controlling muscle groups instead of individual muscles, we're reducing the variables, but we're not losing efficiency," said Matthew Tresch, <u>assistant professor of biomedical engineering at the McCormick School of Engineering and Applied Science and of physical medicine and rehabilitation at the Feinberg School of Medicine.</u> (*Science Daily*, April, 2009)

Unable to demonstrate the relevance and importance of new work by embedding it in a community generated literature, popularizations import material through the quotes of insiders and give credence to that material by drawing attention to the credibility of its source. Not only is authority given to the research by underlining the status of informants, but scientists are often allowed to tell the story themselves through direct quotes introduced by the verb *say*. These options are hardly ever found in the article genre where imported material is overwhelmingly rewritten as a summary from a single source or as a generalization combining several different studies. In popularizations, then, science is both made intelligible in the characteristics of conversation and brought to life through the voices of those involved.

(13) the two groups compiled drafts of the bovine genome, identifying genes important for fighting disease, digesting food and producing milk. "It gives you a window into what makes a cow a cow," says Harris Lewin, of the University of Illinois at Urbana-Champaign. (*Science News*, April, 2009)

Not only are quotes rare in the articles, but the reporting verb *say* to introduce the work of other researchers is a significant rhetorical choice and almost never found in science and engineering papers (Hyland, 2004a). Not only does the choice of reporting verb allow writers to convey the kind of activity reported and to express an attitude to that information, but it also represents a key way of achieving proximity with peers because it signals shared rhetorical practices among disciplinary members. Figure 9.1 shows the most common forms in a larger corpus of 240 RAs in 8 disciplines, indicating the clear demarcations in the structure of subject-area knowledge systems and the fact that different domains use almost completely different verbs.

So while all writers draw intertextual links to their disciplines, they do so in ways which reflect disciplinary distinctions. The humanities and social sciences tend to use verbs which refer to *Discourse activities* (e.g. *discuss, hypothesize, suggest, argue, etc.*) which involve the expression of arguments and evaluation. Engineers and scientists, in contrast, prefer verbs which point to the *research* itself (like *observe, discover, show, analyse*) which represent actions in the real world.

Scientists try to suggest that results would be the same whoever conducted the research and so rhetorically distance themselves from their interpretations. Impersonality is used to give *objectivity* to conclusions through use of *passive voice, dummy it* subjects, and the *attribution of agency* to inanimate tables or results. By subordinating their own voice to that of nature, scientists emphasise the methods, procedures and equipment used, and not the researcher. In other words, in direct contrast to the personalising strategies of popularizations, credibility is increased in science articles through impersonalization; by writers *downplaying* their personal role.

3.4 Stance

The use of direct quotes, relevant topics, researcher identification and a sense of immediate value all reaffirm the role of personal activity in scientific research reported in popularizations which is usually rhetorically airbrushed in professional academic discourses. In addition, journalistic practices also intrude into popular articles through more emphatic claims about the findings and a fuller expression of personal attitude. In other words, writers also

Discipline	Most frequent forms	Discipline	Most frequent forms
Philosophy	say, suggest, argue, claim	Biology	describe, find, report, show,
Sociology	argue, suggest, describe, discuss	Elec Eng.	show, propose, report, describe
Applied Ling.	suggest, argue, show, explain	Mech Eng.	show, report, describe, discuss
Marketing	suggest, argue, demonstrate, propose	Physics	develop, report, study

FIGURE 9.1 *Most frequent reporting verbs in articles in 8 disciplines (Hyland, 2004a).*

establish proximity with readers by taking a clear stance. Here we draw closer to core notions of interpersonality and the fact that statements don't just communicate ideas, but also the writer's attitude to those ideas and to their readers.

Modality is a much discussed feature of interpersonality (Hyland, 2004a; Martin & White, 2005) and is important in constructing proximity by allowing writers to use language flexibly to adopt positions, express points of view and claim affinity with readers. Hedges and other devices which allow writers to comment on the factual status of propositions are therefore abundant in research genres, indicating the degree of caution or assurance that can be attached to a statement. Writing for a peer audience, academics must carefully handle their claims to avoid overstating their case and risk inviting the rejection of their arguments. By withholding complete commitment to a proposition, hedges imply that a claim is based on the writer's plausible reasoning rather than certain knowledge while opening a space for readers to dispute interpretations. This example from a research paper is typical of this recognition of alternative voices:

(14) An early flowering response to high temperature is maintained in *pif4* mutants, <u>suggesting</u> that architectural and flowering responses <u>may</u> operate via separate signaling pathways. The role of PIF4 in temperature signaling does not, however, <u>appear to</u> operate through interaction with either phytochrome or DELLA proteins, <u>suggesting</u> the existence of a novel regulatory mechanism. We conclude that PIF4 is a <u>potentially</u> important component of plant high temperature signaling and integrates multiple environmental cues during plant development. (Current Biology, 2009; 19 (5))

The frequent use of hedges therefore marks out a modest and careful researcher trying to keep interpretations close to the data and unwilling to make overblown claims.

But scientists see their work as far more tentative and mediated than journalists, who take a very different view towards facts. The process of transforming research into popular accounts involves removing doubts and upgrading the significance of claims to emphasize their uniqueness, rarity or originality. This can be seen in the way that the same research is reported in a popular science journal, with the tentativeness removed in favour of unmodified or boosted assertions which amplify the certainty of the claims and, in so doing, the impact of the story:

(15) Researchers at the universities of Leicester and Oxford have made a <u>discovery</u> about plant growth which <u>could have an enormous impact</u> on

crop production as global warming increases. Dr Franklin said: "This study provides <u>the first major advance</u> in understanding how plants regulate growth responses to elevated temperature at the molecular level. This <u>discovery will prove fundamental in understanding</u> the effects of global climate change on crop productivity." (*Science Today* April 2009)

For the science journalist, hedges simply reduce the importance and newsworthiness of a story by drawing attention to its uncertain truth value, but in glamorizing material for a wider audience, popular science texts do not help readers to see how scientific facts can be questioned.

In addition to using hedges to manipulate proximity to an audience and distance from a text, scientists also reduce their use of explicit attitude. My science RA corpus contains one expression of attitude every 350 words in engineering and one every 200 words in the sciences. This works to increase the objectivity of claims and the persuasiveness of the argument, and where they do occur it is typically to express significance (16) or to comment on results (17):

(16) The interaction of light with the scanning tunneling microscope (STM) junction should provide an <u>important</u> tool for the characterization of surfaces. (Phy RA)

ATP plays a <u>central</u> role not only in the energy status of the cell, including microorganisms, but also as a regulator of enzyme activity. (EE RA)

(17) This result was <u>unexpected</u> as the degree of internal colonization by the mycoparasite was greater in sclerotia of the latter fungus. (Bio RA)

There is one <u>important limitation</u> in working with systems of only a few hundred particles. (Phy RA)

In contrast, the popularizations are littered with attitude markers, indicating the writer's affective responses to material, pointing out what is important and encouraging readers to engage with the topic. Unlike their role in research papers, however, these markers do not signal the writer's affiliation to shared disciplinary attitudes and values. Instead, they help to impart an informal tone and underline the accessibility of the material. In fact, the attitudes are often not the writer's at all, but those which the interested lay reader might be expected to hold:

(18) In May this year astronomers announced the first weather report for an alien world. Although fairly crude, it hints that <u>some pretty wild weather is blowing</u> on a giant Jupiter-like planet. (*New Scientist,* Sept 2007)

> Physicists at the University of Toronto have cracked the mystery behind the <u>strange and uncannily well-ordered</u> hexagonal columns found at such popular tourist sites as Northern Ireland's Giant's Causeway and California's Devil's Postpile. (*Science Daily*, Dec 2008)

> After digging their way out and molting into adults, billions of the <u>big, clumsy, red-eyed</u> insects will sing their <u>ear-splitting</u> love songs. (*Scientific American*, 2004)

Proximity with readers is, then, achieved by writers attributing attitudes to them. This is what a lay audience might think and believe on the basis of a community-endorsed common sense.

3.5 Reader engagement

Finally, proximity is negotiated through the ways writers explicitly address their readers. *Engagement* (Hyland, 2005b) is an alignment dimension of interaction where writers acknowledge and connect to others, recognising the presence of their readers, pulling them along with their argument, focusing their attention, acknowledging their uncertainties, including them as discourse participants, and guiding them to interpretations. I only have space to mention two aspects of engagement here, reader pronouns and questions, both of which are far more common in the popular texts.

Reader pronouns offer the most explicit ways of achieving proximity by bringing readers into a discourse, and while *you* and *your* are actually the clearest way a writer can acknowledge the reader's presence, these forms are rare in research papers. Instead, inclusive *we* is prominent. There are several reasons for using this form, but most centrally it identifies the reader as someone who shares a point of view or ways of seeing with the writer. It sends a clear signal of membership by textually constructing both the writer and the reader as participants with similar understanding and goals, as we see here:

> (19) <u>We would expect</u> that over time, plant genotypes that maximize mycorrhizal benefits would be at a selective advantage. (Bio RA)

> At the moment <u>we tend to accept that</u> the incident light, or at least most of it, is bounced twice before returning and re-emerging outside the eye. (Phy RA)

This emphasis on binding writer and reader together through inclusive *we* is also a feature of popular articles, where it functions less to claim community solidarity than to insinuate a shared, taken-for-granted way of seeing the world. This helps guide readers towards a perceived relevance of the reported research, aligning the reader with the writer, and perhaps against an institutional 'they':

(20) On the list of things we're supposed to do but generally don't, nothing ranks higher than eating well. And no wonder, considering that they keep changing the rules on us. Margarine was once supposed to be better than butter — until it turned out to be worse. Low-fat eating was supposed to be the way to lose weight — until it was low-carb, then back to low-fat again. (*Popular Science* April, 2009)

While this shades into a positioning of the reader, the considerable use of the second person does this more explicitly, directly involving readers in the topic of the article through common experiences or drawing them into a credible, but unknown, world:

(21) You're just about ready to buy a pair of tickets on Ticketmaster, but before you can take the next step, an annoying box with wavy letters and numbers shows up on your screen. You dutifully enter in what you see— and what a robot presumably can't—in the name of security. But what you may not know is that you also have helped archivists decipher distorted characters in old books and newspapers so that they can be posted on the Web. (*Scientific American* Aug 2008)

One look at the effects of a bomb blast suggests that you'd have to be extremely lucky to emerge from one unscathed. If you were not burned by the explosion or blasted by shrapnel, the chances are you'd be hit by the shock wave. Travelling at several hundred metres per second, this causes massive fluctuations in air pressure which can knock you unconscious, rupture air-filled organs such as eardrums, lungs and bowels, and stretch and distort other major organs. (*New Scientist,* April 2009)

A final way that writers build a connection with readers is through the use of *questions.* These are a key strategy of dialogic involvement in many registers but are almost non-existent in science and engineering research papers. Their role in manufacturing immediacy with readers is viewed with suspicion by professional scientists, although they are common journalistic devices, representing an important way of managing proximity:

(22) Solix must struggle for answers before it can sell a thing: Which species of algae will produce the most oil? What's the best way to grow it? And not

least, how do you extract the oil from the algae once it's grown? (*Popular Science*, April 2008)

Presenting the researcher's problems as questions achieves proximity with readers by engaging them in the scientific enterprise, bringing them closer to the concerns of the scientist at the same time as making the science real and intelligible.

4 Conclusions

Science journalism illustrates the ways proximity (and interpersonality) work as writers set out material for different purposes and readers. Popularizations represent a discourse which establishes the uniqueness, relevance and immediacy of topics which might not seem to warrant lay attention by making information concrete, novel and accessible. Findings are therefore invested with a factual status, related to real life concerns, and presented as germane to readers with little detailed interest in the ways that they were arrived at or in the controversies surrounding them. Readers, in fact, experience the academic world and its discourses as a succession of discoveries in the relentless advance of inductive science. In sum, science journalism works as journalism rather than science. It is written in ways which make the research accessible and allow non-specialists to recover the interpretive voice of the scientist.

There are several advantages for EAP practitioners in considering how proximity operates across genres. First it helps us see how different features combine to orchestrate interpersonality. Writers position themselves through rhetorical choices which emphasize a relationship to content and to readers, displaying an audience sensitivity through the way the material is organised argued and attributed as well as choices concerning writer and reader visibility in the text. Simply, both scientists and journalists have to evaluate their audiences as they write, and this means that texts tell us something about how writers see their readers. Understandings of readers' goals, interests, knowledge and processing capabilities are indexed in rhetorical choices so that context is constructed in text.

Second, genre comparisons suggest that proximity is important not only to analysts of academic writing but also to teachers of it. Clearly, analyses caution us to take care in using popular texts as models for scientific writing as differences in constructing proximity mean that they will not help students see how scientific facts can be questioned or modified. Comparisons, however, can have an important consciousness raising function by highlighting features of scientific discourse for learner noticing. Their study, moreover, may help students see something of the importance of audience. In recontextualizing academic research for a lay audience, much popular science portrays research as an immediate encounter of a scientist with nature. Scientists become actors and claims become a discovery event; jargon is evicted or roughly glossed; certainties replace tentativeness, and nouns regain their verbal status. Such a reconfiguration of discourse reminds us that science is a communicative activity so science once again becomes ideas to be discussed rather than information to be received.

Most importantly, the study of proximity in these genres helps us to see how each group appropriates and transforms science by presenting the same material for different purposes and readers. In popular science proximity is achieved by making research accessible and allowing non-specialists to recover the voice of the scientist which is absent in professional papers. In research science writers position themselves as competent colleagues by displaying familiarity with methods and a disciplinary literature, presenting research with caution, and supporting claims with evidence. Ultimately the insights into academic writing which these kinds of contrasts reveal makes the study of interpersonal proximity a highly productive and valuable activity for researchers and teachers of academic discourses.

10

Dissertation acknowledgements: The anatomy of a cinderella genre

In recent years, a burgeoning literature has illuminated our understanding of the written discourses of the academy as well as consolidated the importance of genre as a means of investigating the situational and cultural influences which operate within academic communities. This work has led to an increased appreciation of the interpersonal role of writing and an understanding of the fact that academic writers do not simply produce texts that plausibly represent an external reality but use language to situate themselves in a discipline and to construct and negotiate social relations. Yet while research has explored the varied ways that writers offer a credible representation of themselves and their works in a range of genres, interest in the interpersonal has tended to focus on its role in argument. What is perhaps the most personal genre of all, that of acknowledgements, has been relatively neglected.

Swales' (1996) categorization, for instance, divides the academic genre system into three groups according to the kind of audience addressed: *primary* or research-process genres for peer communication; *secondary* or pedagogic genres; and relatively private or *occluded* genres dealing with the exchange of material and information between academics which support the research process. Acknowledgements seem to fall through the cracks here and represent something of a "Cinderella" genre. Like the heroine in the children's fairy tale, acknowledgements are a taken-for-granted part of the background, a practice of unrecognized and disregarded value deserving of greater attention. Neither strictly academic nor entirely personal, acknowledgements stand outside the research record but have a considerable sociopragmatic relevance which makes them integral to it. They are central to the academic practice of reciprocal gift giving and for this reason are particularly important to students. Acknowledgements can act as a means of demonstrating academic credibility, recognizing debts, and achieving a sense of closure at the end of a long and demanding research process. In this article, I explore these issues by examining the acknowledgements accompanying 240 Ph.D. and M.A. dissertations written by nonnative speakers of English.

Acknowledgements in published scholarly texts

While student acknowledgements have received little research attention, slightly more is known about the genre in published texts. Acknowledgements in books date back to a time when public gratitude for the generosity of the powerful was almost a precondition for publication, yet the relative political and economic autonomy of modern times does not appear to have dented their popularity. Academic tomes have always contained expressions of gratitude, and in journal publishing, early scientific articles often featured acknowledgements in an introductory cover letter (Atkinson, 1999). Their emergence as a textual feature was uneven until the 1940s, only becoming common during the 1960s (Bazerman, 1988), and while they are still to be found in book prefaces or article footnotes, the compulsion to recognize colleagues and funding bodies is now more likely to receive institutional endorsement and editorial prominence in a separate textual space.

Although sometimes considered to be a minor feature of research reporting, and therefore unrelated to the substantive issues of establishing claims and reputations, the persistence of this optional genre confirms its usefulness to disciplinary communities. Acknowledgements are commonplace in the scholarly communication process today and have, in fact, become both longer (Caesar, 1992; Cronin, 1995) and more common, with perhaps over half of all published articles including this section (Cronin, McKenzie, & Stiffler, 1992). Their value is also attested by survey data. Cronin and Overfelt (1994), for example, found that over 50% of their survey of 280 academics generally read acknowledgements when scanning a new article, often to make a preliminary relevance assessment of the article, and 90% were aware of having been acknowledged themselves, a few even keeping a formal record for institutional evaluation.

Yet acknowledgements are much more than a simple catalogue of indebtedness. They offer insights into the persona of the writer, the patterns of engagement that define collaboration and interdependence among scholars, and the practices of expectation and etiquette that are involved. In a study of anthropological ethnographies, Ben-Ari (1987) observes that acknowledgements are

> formulations that take on an intermediate position between the internal contents of the ethnography and the people and relationships outside it: they are both an introduction to an intellectual product and a reconstruction of the external contributions that have gone towards its realization. (p. 65)

Acknowledgements often play a metadiscursive role in being physically set apart from the main social and textual product yet functioning to both facilitate the construction of this product and to comment on it. They point inward to the text and its author and outward to the factors which help construct them both. Acknowledgements thus provide a valuable space for writers to encode a representation both of themselves and of those they wish to publicly recognize as influencing the project.

It would, however, be wrong to see acknowledgements as entirely idiosyncratic and personal. Studies have found that they reflect both disciplinary proclivities and the dialogic processes of academic research. Cronin, McKenzie, and Rubio (1993) discovered disciplinary variations in the frequency of acknowledgements, suggesting a continuum across the soft-hard spectrum, with virtually all articles in the hard sciences carrying one (McCain, 1991). These patterns tend to mirror the recognized disciplinary working practices and the ways knowledge is constructed in different fields. Philosophers, for instance, are more likely to inhabit dispersed communities with little reliance on close interaction with others and so have low acknowledgement rates. Hard scientists, in contrast, recognize financial support and their engagements in highly developed webs of mutual transmission and exchange, trading the materials and preprints upon which their research depends. The structural properties of acknowledgements also exhibit disciplinary variations, with writers in the humanities and social sciences writing more elaborate texts (Giannoni, 2002; Hyland, 2004c).

Acknowledgements, self representation and gift-giving

The fact that acknowledgements are one of the few available gestures for conveying appreciation increases their interest to discourse analysts and to bibliometricians seeking to trace complex genealogies of interaction. Through them, we learn something of the forces shaping the accompanying text. But although academics appear to subscribe to a governing etiquette of acknowledging practices (Cronin & Overfelt, 1994), expressing thanks to others is not an entirely altruistic practice, and it is this potential for flattery and self-promotion which has attracted criticism to acknowledgements.

Book acknowledgements, in particular, have been criticized for the "twin vices of fawning and vanity" with a suggestion that this genre provides authors with

> an excuse for long, rambling essays, in which they flatter the powerful, gurgle over their families, and otherwise boast to the world what happily married, highly-educated, well-connected and generally right-on people they have the good fortune to be. ("Acknowledgements," 1996)

This judgment perhaps seems rather harsh on those who have been recognized for support provided at some social, emotional, or financial cost. The role of acknowledgements in providing a representation of the writer as a social person, however, is common, particularly in book acknowledgements, and reflects the emerging importance of the writer in society. It reveals the writer as someone with a life beyond the page in whom readers may be interested.

Ben-Ari (1987) also comments on the role of acknowledgements in creating a professional as well as personal identity: showing how the formulation of acknowledgements act as strategic choices in "careering," achieved through the author's management of his or her relations in the disciplinary community and the creation of authorial credibility. The inclusion of references to those only marginal to the research is common and, in some cases, has drawn fire from journal editors for its excessiveness. After receiving a manuscript with a five page acknowledgement section listing 63 institutions, 155 physicians, and 51 members of seven different committees, the editors of the *New England Journal of Medicine* were moved to officially limit their space for acknowledgements (Kassirer & Angell, 1991).

This kind of gift-giving is a core feature of academic communities but has mainly been studied in its manifestation as citation (Gilbert, 1977; Hyland, 2000a). Latour and Woolgar (1979), for instance, see citation as central to the cycle in which academics engage to maximize their credibility, accumulating recognition which they can convert into research grants, further data, and fresh publications. Bibliometric research also shows that the number of citations an individual receives correlates strongly with other forms of career recognition (Garfield, 1998). While not tabulated by bibliometricians or promotion boards, the gift bestowed by acknowledgement may be no less influential. Yet this is clearly a more personally marked form of credit through which the author both indicates an individual tribute for some service and directly indicates an affiliation. In short, acknowledgements are "not trivial, meta-textual flourishes, rather they are formal records of often significant intellectual influence" (Cronin & Overfelt, 1994, p. 183) which point to strong networks of association between researchers.

It is this direct and personal aspect of acknowledgements moreover which make them important to research students. The dissertation is a high stakes genre; a formidable task of intimidating length and exacting expectations, demanding heavy investments in time, money, and labor. Acknowledgements can therefore provide a means of reconciling the student's individual achievement with the interpersonal debts incurred in completing the study. Yet despite this importance, published advice to students about acknowledgements is sketchy and, at best, lukewarm. Lester (1993), for example, tells novice researchers that acknowledgements are an unnecessary encumbrance, while Day (1994) cautions that they may suggest shaky authorship. Only Swales and Feak's (2000) academic writing guide appears to recognize the significance of the genre and provides anything like a serious pedagogic treatment of the topic.

For analysts, postgraduate acknowledgements are of interest for the insights they offer into disciplinary environments for research and patterns of academic exchange and patronage. In writers' metatextual reflections, we can discover personal histories of collaboration, patterns of affiliation, demonstrations of academic credibility, and glimpses of a more contingent world. The remainder of this article explores the ways these issues are played out across disciplines and degrees through analysis of who is acknowledged, what they are acknowledged for, and the ways this is realized.

Corpus and procedures

The acknowledgement data were collected as part of a larger study of postgraduate research writing in Hong Kong comprising 240 dissertations and interviews with supervisors and students. The text corpus consists of the acknowledgement sections in 20 M.A. and 20 Ph.D. dissertations from each of six academic fields written by nonnative English speaking students at five Hong Kong universities totaling 35,000 words. The disciplines were chosen

TABLE 10.1 Acknowledgement Corpus (20 dissertations in each discipline)

Discipline	master's			Doctoral		
	Texts	Words	Average	Texts	Words	Average
Applied linguistics	18	2,402	133.4	20	7,718	385.9
Biology	15	1,825	121.7	19	3,864	203.4
Business studies	6	810	135.0	19	2,512	132.2
Computer science	18	1,483	82.4	20	3,470	173.5
Electronic engineering	20	1,427	71.4	19	2,771	145.8
Public administration	19	3,289	173.1	20	3,594	179.7
Totals	96	11,236	117.0	117	23,929	204.5

to represent a broad cross-section of academic practice, namely, electronic engineering (EE), computer science (CS), business studies (Bus), biology (Bio), applied linguistics (AL), and public administration (PA).

The two corpora were initially coded inductively, developing categories from recursive passes through the texts looking for classes of people to whom debts might be owed and terms designating behavior that might attract credit. The corpora were then searched more systematically using *MonoConc Pro*, a concordance program, which yielded further search items and enabled a detailed study of the cotexts to discover who was acknowledged, what they were acknowledged for, and the naming practices employed. All acknowledgements were coded for these features and studied for frequencies of occurrence. *Winmax Pro* was also used to code all cases to ensure that unpredictable individuals (such as "my twins"), which would be missed by machine searches alone, were recovered. A sample of 60 texts was coded independently by a second rater for reliability with 90% agreement.

In addition, two M.A. students and two Ph.D. students from each discipline were interviewed by the project research assistant, herself a recent linguistics graduate. We were interested in participants' understandings of the meanings of acknowledgements and their thoughts on disciplinary practices as a way of gaining insights to the text data. The interviews followed a semistructured format (Cohen, manion, & Morrison, 2000, p. 270) which allowed peripheral topics to be followed up if important.

Postgraduate acknowledgements:
Differences of degree

Both the qualitative and quantitative data confirm the importance of acknowledgements to these students despite its optional, nonexaminable status, with around 90% of the texts, and almost all the Ph.D.s, containing one (see Table 10.1).

The length of the texts ranged enormously. This almost cursory 38 words was the briefest:

> (1) I would especially like to thank Dr. Douglas Vogel,[1] my dissertation supervisor, who gave me his advice and guidance throughout the preparation of the dissertation. Also my colleagues and friends are thanked for sharing their experience in outsourcing. (CS M.Sc.)

But most were much longer, up to an applied linguistics opus of 1,085 words. The average however was 160 words, about 3 times the average of 55 words that Giannoni (2002) found in his sample of research article acknowledgements.

The Ph.D. students were more conscientious in acknowledging assistance. Twenty two percent more doctoral theses contained acknowledgements and these were on average about twice as long as those by master's students. This is perhaps because doctoral students are generally anticipating an academic career and are often already apprenticed to a scholarly community. It is during doctoral research that individuals take on the cultural frame that will define a greater part of their academic lives, and these novitiates are immersed in the practices of their chosen disciplines as much as their research fields. For these students, an acknowledgements section is an important courtesy guided by scholarly norms, a means of publicly recognizing the role of mentors, the sacrifices of loved ones, and sharing the relief of completing the process:

> It is a must to write an acknowledgement as it is an important channel to express our thanks to those who helped in our project. It is an important section for the completeness of the thesis. (CS Ph.D. student)
>
> This section is meaningful for myself to show my gratitude, but I don't think people who read my thesis would care much. (Bio Ph.D. student)
>
> It is a very important section as it gives me an opportunity to express my gratitude. It is a very personal thing. (PA Ph.D. student)

Masters students, on the other hand, typically study on a part-time basis and are looking forward to returning to their professional workplaces. Their theses tend to be much shorter (averaging only a third of the length of Ph.D.s), constructed fairly quickly, and completed in addition to substantial coursework. Not surprisingly, therefore, acknowledgements were afforded less significance by these master's students and some saw acknowledgements as merely a convention:

> I think acknowledgement is not an important section, but rather a formality. I've discussed with my classmates on whether to include one or not, and we agreed at the end that it seems to be a must to write one. (AL M.A. interview)

> There is no need to write an acknowledgement. I don't think it is an important section at all, I think my supervisor would appreciate it more if I ask for his opinions and discuss more with him than thanking him in an acknowledgement. (BS M.A. interview)

> I once considered omitting this part, but I feel pressured to include one as almost everyone did so. Personally, I think this section is not important, but just a convention. (EE M.Sc. interview)

Allocating credit: Some disciplinary patterns

The main purpose of acknowledgements is to allocate credit to institutions and individuals who have contributed to the dissertation in some way. There were 1,400 separate acts of acknowledgement in the corpus, 70% of which were in the Ph.D. texts. Of those, 1,276 different individuals and 138 institutions were acknowledged, with supervisors appearing in all acknowledgements, in fact, almost half of these acknowledging acts were to academics, a quarter to friends, and 14% to family members. These figures were remarkably consistent across the Ph.D. and M.A. corpora, although there were considerable disciplinary variations. Table 10.2 shows the distributions as a percentage by discipline and degree.

Bibliometric studies of acknowledgement patterns have classified some of the functions which are credited by writers, showing that they are not simply a miscellany of thanks (Cronin, McKenzie, & Rubio, 1993; McCain, 1991). The student dissertations contained six main reasons for giving credit: academic support, access to data, moral support, clerical services, financial resources, and technical help, with three quarters being for academic and moral assistance. Academic assistance attracted the most thanks, representing 45% of each corpora, with moral support comprising 30%. Table 10.3 summarizes the results.

TABLE 10.2 Participants Acknowledged in Dissertations (in percentages)

	M.A. Dissertations						Ph.D. Dissertations					
	Academic	Friends	Family	Orgs	Other	Total	Academic	Friends	Family	Orgs	Other	Total
AL	41.7	17.7	21.9	3.1	15.6	100	36.2	24.8	15.1	13.3	10.6	100
BIO	46.0	30.0	8.0	16.0	0.0	100	55.2	21.5	8.7	13.4	1.2	100
BUS	43.8	31.2	0.0	12.5	12.5	100	57.4	21.8	5.9	11.9	3.0	100
CS	39.6	22.7	11.2	11.4	15.1	100	36.5	29.1	23.4	8.8	2.2	100
EE	60.4	16.7	4.1	18.8	0.0	100	44.4	26.8	13.9	13.9	1.0	100
PA	35.7	28.7	21.8	6.9	6.9	100	48.4	21.8	12.1	10.5	7.2	100
Total	43.1	23.4	14.9	9.7	8.9	100	45.3	24.3	13.5	12.1	4.8	100

NOTE: Other = clerical, technical, and financial support. AL = applied linguistics, BIO = biology, BUS = business studies, CS = computer science, EE = electrical engineering, and PA = public administration.

TABLE 10.3 Activities Acknowledged in Dissertations (in percentages)

	M.A. Dissertations						Ph.D. Dissertations					
	Academic	Access	Moral	Other	?	Total	Academic	Access	Moral	Other	?	Total
AL	35.5	24.5	31.9	3.6	4.5	100	40.3	18.0	31.4	6.1	4.2	100
BIO	52.5	23.7	18.7	5.1	0.0	100	48.2	7.6	25.9	11.7	6.6	100
BUS	47.1	11.8	17.6	11.8	11.7	100	54.0	8.7	25.4	5.6	6.3	100
CS	50.0	23.2	21.4	0.0	5.4	100	42.9	1.3	42.2	8.4	5.2	100
EE	60.7	3.3	21.3	4.9	9.8	100	48.0	6.4	28.8	9.6	7.2	100
PA	40.5	11.1	42.4	3.0	3.0	100	43.1	13.8	29.2	8.3	5.6	100
Total	45.5	17.2	28.9	3.7	4.7	100	45.3	10.2	30.6	8.2	5.7	100

NOTE: Other = clerical, technical, and financial support. The symbol ? = unclassifiable, AL = applied linguistics, BIO = biology, BUS = business studies, CS = computer science, EE = electrical engineering, and PA = public administration.

It is interesting that these writers tended to make explicit the kind of help they received, with only 5% of the thanking acts unclassifiable from the context. This open disclosure points to the public nature of these texts and suggests that writers are not just addressing those they are acknowledging, who are presumably aware of their contributions. Instead, they are conscious of a wider professional audience of academics and examiners with the power to influence the reception of the dissertation and perhaps the future of its writer. Here, we see genuine gratitude tinged with impression management as the writer represents himself or herself as both a plausible researcher and a sympathetic individual.

Acknowledging scholarly support: The construction of a professional identity

In all disciplines, academics received the most mentions while academic assistance and access to data were the most mentioned forms of assistance. Overwhelmingly, such acknowledgements are offered to senior academics, not only dissertation supervisors, but others within students' professional communities who had mentored or believed in them, taught them, provided intellectual guidance, assisted with conference papers, or contributed in other ways. They were teachers, members of the student's dissertation committee, and occasionally even examiners. Mentioning these people clearly foregrounds the activities which structure the student's intellectual and academic experiences in undertaking the research, but they also represent strategic choices related to "getting done" with the thesis by crediting influential academics and favorably representing the writer.

Supervisors

Supervisors play a major role in graduate research at both masters and doctoral levels and are always mentioned, and almost always before others, giving them a primacy which reveals the intellectual, and often emotional, obligation writers often feel to their supervisors:

> My supervisors really helped a lot in my project, I think it's not just a formality or politeness. It is like we went through all the difficulties together in these so many years and it is not an easy task. (PA Ph.D. interview)

> I'd definitely include my supervisor. His ideas has contributed a lot to my project. (AL M.A. interview)

Occasionally, these acknowledgements appear rather double-edged (2), but they are generally sincere, ranging from the succinct (3), through the blandly formal (4), to the eulogistic (5):

(2) I would like to thank my advisor, Dr. Mohammed Khalifa for his relentless supervision and guidance. (CS M.Sc.)

(3) I am profoundly indebted to my supervisor, Dr. James Kaising Kung, who was very generous with his time and knowledge and assisted me in each step towards the completion of the thesis. (PA Ph.D.)

(4) I, Chui Mei Ling, would like to express my sincere thanks to my supervisor, Dr. Richard Y. H. Cheung, for his continuous guidance and support. (Bio Ph.D.)

(5) It is with great pleasure that I offer the most heartfelt thanks to my supervisor, Prof. Fu-Shiang Chia, for his unconditional and infinite support throughout my Ph.D. study. Prof. Chia always demonstrate a positive attitude towards my research and career goal. He trusts me, believes my abilities and gives me all the autonomies. I deeply apologize to him for all the stupid things I did and all the troubles I caused in the past three years. Thank for his forgiveness and understanding. Thank you very much, Prof. Chia, for sharing with me your wisdom and experiences in science and daily life. It certainly benefits me in the rest of my life. You are my teacher, my friend, my example and my idol. (Bio Ph.D.)

Academics received the highest proportion of thanks from students working in biology and engineering, and this was particularly marked in the masters' texts. The doctoral acknowledgements revealed a considerable skew to supervisors. Over 70% of all mentions to them were in the Ph.D. texts in the hard knowledge disciplines as writers often thanked them for several kinds of support. The predominance of supervisors in these acknowledgements is perhaps due to their greater involvement in the students' experience of graduate research in the sciences. In hard knowledge fields, for instance, it is not unusual for a graduate student to be taken on as a member of the supervisor's research team and to carry out tasks allocated and watched over by the supervisor which leads to a Ph.D.

The professor's role in selecting the research topic, determining the methodology, providing resources, and overseeing the direction of the research is crucial. Even where the student has more independence, supervisors have the breadth of knowledge of fast moving fields to guide novice researchers to doable problems related to current developments (Belcher, 1994; Prior, 1998). Lab-based research groups tend to be more closely-knit and supervisors have more regular contact with their students and so have greater involvement in guiding and overseeing the doctoral student's work. In softer domains, on the other hand, such contextual imperatives tend to be more relaxed, so the choice of topic is both more open and, once chosen, more amenable to a range of different approaches. Research for students may be a lonely and autonomous endeavor which is frequently conducted at a distance or in circumstances where the supervisor's assistance is mainly restricted to bursts of involvement at the beginning and end of the process (Becher, 1989). In between, students are often left more to their own devices with solitary reading and infrequent meetings with supervisors. These different working practices were reflected in the some of the interview responses:

> I'd include my supervisor to thank him for his ideas on the project, because in our department, it is the professor who proposed some topics and let us choose which one we are interested in, so my project actually come from my supervisor's ideas. (EE M.Sc. interview)

> In my research I work closely with my supervisor. He has funding for what I do and he is responsible for the lab. The research cannot be done without the professor. (Bio Ph.D. interview)

> I think most of the help is at the beginning. Now I get feedback on drafts from my supervisor who would comment on language and ideas. I doubt, however, if we should entirely depend on our supervisors at this level of study, supervisors are busy after all. As there are not many people who would like to read the thesis draft, so I think we should be mature enough and to be considerate. (AL Ph.D. interview)

Other Academics

In addition to supervisors, recognition to other academics is common in these acknowledgements. Once again, these largely tended to be senior scholars who may have played a teaching or advisory role, but they also included help provided by fellow students, colleagues and other peers who provided feedback and critical comments, discussed ideas, or assisted with analyses:

> (6) My colleagues, Ms HO Sze-Man, Ms Yu Wing-kam, and my classmate, Mr. Wai Kam-hung, are also very kind indeed for forming the odour panel in assessing the selected refuse collection points. (Bio M.Sc.)

> I would like to thank K. K. Choi, statistics consultant in the Statistical Consulting Unit of the Management Department at this university, for suggestions on part of the statistical analyses involved. (AL Ph.D.)

Special thanks must also be given to Dr. Edwin L. C. Lai. He showed great interest in my topic and concern about the progress of the dissertation. The books or articles he recommended me to read are very conducive to modifying the model. (Bus Ph.D.)

Significantly, however, those mentioned are more often individuals in whose patronage may lie the seeds of a career strategy. This acknowledgment of senior professionals is more prevalent in the Ph.D. texts and in the hard disciplines, where the mentoring of junior researchers has a stronger tradition, and career paths more directly emerge from the visibility of a research group, the prestige of a supervisory panel, and the patronage of key figures in close-knit professional networks. Acknowledgements can be a coin in the reciprocal dynamic of debts and obligations which Ben-Ari (1987) suggests is central to the continuity and cohesion of the professional community. Acknowledgements potentially announce a relationship binding author and acknowledgee in a mechanism of mutual indebtedness which can benefit both parties over a longer term: the supervisor offering the guidance and benevolence of an established academic and the writer the esteem and loyalty of a grateful mentee. So while master's students may see an acknowledgement to their supervisors and committees as a closure, a ritual signing-off on a relationship; for the Ph.D. graduate, this relationship may be just beginning.

Because of the increasing specialization of both research and funding, an engineer or scientist anticipating an academic career depends heavily on the protection and goodwill of established figures for gaining postdoctoral grants, a lab to work in, or an initial teaching position. Mentioning key figures can therefore gain the writer important credit:

The acknowledgement is an important section for creating good impression. (EE Ph.D. interview)

Though the panel did not really give any help in doing the paper, there are political reasons in thanking them. Therefore, I just make up something to thank them, like thanking them for reading my paper. (EE Ph.D. interview)

Some of the comparative results are from other labs and I will put these people in the acknowledgements. Some of these are from important people in the field and it is a good idea to include them. (Bio Ph.D. interview)

Who the writer studied with, however, also remains an important status marker for those in the soft fields, where it may be immediately useful to appropriate some of the status of influential figures by associating the text, and its writer, with them. These examples of apparent "name-dropping" in applied linguistics of individuals only marginal to the research help to illustrate the point:

(7) I would like to thank Prof. Chris Candlin for his support and time. I would like to thank Prof. Vijay Bhatia for his kind help as my reliable consultant. (AL M.A.)

Thanks to Professor Jack Richards for admitting me into the Department, as it has been a most vibrant and healthy environment for doing academic work. (AL Ph.D.)

This strategic maneuvering becomes more obvious, of course, when extended to those who have a direct influence on how the thesis itself might be received, such as examiners:

(8) I would like to thank Dr. K. P. Chui, my internal examiner, and Dr. D. Y. Lai, my external examiner, who toiled through my thesis. (CS M.Sc.)

My special thank goes to my external examiner, Dr. R. Dorfman for her kindness and patience in going through my manuscript. (PA Ph.D.)

I also wish to express my gratitude to Professor Nelson Fok, Professor Ping Shui and Dr. Ho C. Law for their kind acceptance to sit on the examination committee. (EE Ph.D.)

Symptomatic of this complex interplay of the interpersonal and the strategic are naming practices as writers seek to simultaneously align themselves with well-known or influential academics and to display their respect for them. The form a name takes in acknowledgements, like those named, helps link the author's private sphere of mentors and social affiliations to the recognition of the individual as a public figure. Of all academics, 96% in the corpus were referred to using their full names with an honorific, even if this was a simple *Mr.* or *Ms.*, and only 2% were mentioned without a title of any kind. None at all were mentioned by first name only, and occasionally, this explicit marking of respect leant to the excessively formal:

(9) The author wishes to thank the dissertation supervisor, Dr. Chen Sze Hui, Associate Professor and Acting Head of Department of Biology, Hong Kong University of Science and Technology (HKUST) for the latter's considerable assistance and insightful comments. (Bio Ph.D.)

The author wishes to express his gratitude to Professor S. Y. King, B.Sc. (ENG.), Ph.D., C.ENG., F.I.E.E., Sen. Member I.E.E.E., head of Electrical Engineering Department, University of Hong Kong. (EE M.Sc.)

As Giannoni (2002, p. 21) points out, this helps to appease what Brown and Levinson (1987) call the contributor's "positive face," or desire for recognition and approval, by reminding readers of the acknowledgee's status, while also signaling deference to the academic community by recognizing its norms and hierarchies.

Participants and providers

Beyond immediate academic support, dissertation writers are dependent on the cooperation or direct assistance of those they study or who provide clerical, technical, and financial help. Once again, there are disciplinary and degree differences regarding who is acknowledged, with a higher proportion of thanks for access to data in the master's texts, perhaps because these tend to contain a more restricted range of categories. The social sciences are relatively dependent on instrumental support and access to documents and subjects, while in the science and engineering acknowledgements greater mention is made of technical and financial support and of those who provided materials and unpublished data.

Researchers in the sciences and engineering fields, and some students in business studies, are particularly reliant on expert backup and as a result technicians, lab assistants, computer wizards, funding agencies, and employers are frequently mentioned, usually as individuals but often as groups or institutions. Around 11 % of all acknowledgements referred to these kinds of background support which underlie every research project but are infrequently mentioned within the dissertation itself. Unlike the master's students, the Ph.D. acknowledgements rarely addressed clerical assistance, but often mentioned technical and, more frequently, funding support. Once more, we can see the textual construction of an academic self in these apparently innocent appreciations from a grateful graduate. The detailing of thanks for prizes, prestigious scholarships, company sponsorships, or travel grants marks the writer out as an individual whose academic talents have already been recognized and who may therefore be a deserving candidate for further honors.

While the writer may feel obliged to refer to his or her funding agency, an examiner is unlikely to remain unimpressed by the writer's obvious confidence and undoubted worth:

(10) This project was generously supported by funding from Hong Kong Polytechnic University's Staff Development Committee. Support has been forthcoming, too, from Cathay Pacific Airways in the form of complimentary air travel, which has allowed me to attend a number of overseas conferences

and thereby bring the research to the attention of a wider audience. (AL Ph.D.)

I thank the City University of Hong Kong for awarding me a Conference Grant in attending the International Conference on Environmental Contamination, Toxicology and Health during 23-26th September, 1998, where I presented a poster on "Elimination of phytotoxicity in mixture of chicken and green manure by windrow composting," and the International Composting Symposium, 1999 in Halifax, Canada, during 20-23rd September, 1999, where I had an oral presentation on "Co-composting of chicken litter and yard trimmings: effects of aeration frequency and spatial variation." (Bio Ph.D.)

Similar rhetorical intentions perhaps lie behind acknowledging individuals for organizing conferences, reviewing articles, and collaborating on publishing projects. Particularly prevalent in the sciences, where work practices are more likely to transcend the immediate research site and offer greater opportunities for collaboration in research papers and conference presentations, such gratitude clearly serves to enhance the writer's professional credentials:

(11) A special acknowledgment is extended to Y. K. Leung at Stanford University for providing spreading resistance analysis and to Prof. Simon Wong for reviewing my IEDM paper. (EE Ph.D.)

Portions of this thesis represent joint work with Francis Lor, which appear in the following papers: "Linear-Time Algorithms for Unspecified Routing in Grids," in the Proceedings of International Conference on Algorithms; . . . and "Efficient Algorithms for Finding maximum Number of Disjoint Paths in Grids," in Journal of Algorithm. (CS Ph.D.)

The assistance provided in granting access to the data required for successful research displays far more disciplinary variations. The creation of hard knowledge is heavily dependent on the collaborative exchange of materials, information, and unpublished results which enmeshes the researcher into networks of reciprocal obligations. Here, then, credit for "access" tends to be granted to other academics for furnishing preliminary findings, supplying article preprints, making data collection instruments available, and so on. In the human sciences, on the other hand, the most important source of access is that provided by the participating subjects themselves, often acknowledged, usually in final place:

(12) I would like to acknowledge the invaluable help rendered by my subjects, the elderly diabetic patients follow-up at the Alp Lei Chau and

Tang Chi Ngong out-patient clinic, who spent time to participate in this study without immediate benefit to themselves. (PA Ph.D.)

Finally, I am most indebted to the 517 companies that were willing to return my questionnaire with their responses. (Bus Ph.D.)

While subjects themselves are unlikely to read the text, quite subtle rhetorical intimations of professional commitment and academic competence can be communicated to professional readers, hinting at the authority and involvement of the writer and of trials overcome. This example from a qualitative applied linguistics thesis into the discourse of the underclass of Filipina maids in Hong Kong is perhaps more effusive than most, but underlines the point well:

(13) I hope this work has given justice to the voices from the margins. For reasons that they would understand, they would remain anonymous in this work. However, if someday they get the chance to read this work, I have no doubt that they will readily recognise their voices that have enlivened the many Sunday afternoons shared together in the parks, under the bridges and under the trees; in the sun and rain; enduring the heat and cold of the changing seasons. I also include those whose search for life's better promises have led them to the classrooms of the YMCA where I have had the opportunities to share moments, outside and inside classroom sessions, that have been made unforgettable by their laughter and tears. And the many nameless others whom I have met in countless encounters whose lives have touched and enriched mine in ways that I would find hard to articulate. (AL Ph.D.)

This account of the rapport established between researcher and subjects graphically testifies to the intimate relations and a political commitment that might not be possible to include so explicitly in the main text. This not only testifies to the hardships and commitments of ethnographic data collection but to a political stance and the insights of someone who attained the status of an insider.

Doubtless these expressions of gratitude are sincerely meant and help the writer repay some of the assistance which has been crucial to completing the degree, but the analysis shows that acknowledgements are not simply random checklists of useful people or institutions. They also allow writers to represent some of the procedures and practices which have gone into the dissertation and so present a competent professional identity. They portray an academic immersed in a network of research paraphernalia armed with the ability to manage the substantial resources often necessary for academic study. Thanks to participants, academics, and other experts helps to communicate the authenticity and plausibility of the research and the skill of the writer. This, then, is a site where writers can textualize themselves as autonomous intellectuals worthy of respect, familiar with the norms and practices of their discipline, and deserving of the qualification sought.

Acknowledging friends and family: The construction of a social identity

While I have discussed the ways that these students carefully created a disciplinary situated persona in their acknowledgements, the fact that almost 40% of the thanks in the corpus were to friends and family members suggests that the genre is not simply an opportunity for political strategizing. Acknowledgements also provided these students with the chance to mention what they considered to be decisive influences on the processes of completing their research. These influences extend beyond the public worlds of the academy to the private forces shaping their responses to those influences, often alluding to the tensions and difficulties experienced in graduate study:

> (14) I should thank my dear parents and husband. Though they are thousands of miles away from me, their continuous encouragement, silent concern and endless love converge to my momentum to work hard and achieve the best I can. (Bus Ph.D.)

> Last, but definitely not the least, I am greatly indebted to my family. It was my parents' unconditional love, care, and tolerance which made the hardship of writing the thesis worthwhile. (PA Ph.D.)

> I want to thank my girlfriend, Ms Grace Chang. Without her support, I do not think that I could overcome the difficulties during these years. (AL Ph.D.)

The category of "moral support" in Table 10.3 includes all expressions of thanks for encouragement, friendship, sympathy, patience, and care. Such expressions are unevenly distributed across the master's texts and are overwhelmingly found in the more liberal social sciences of applied linguistics and social work. These dissertations tend to be considerably lengthier than those in the other disciplines and perhaps involve greater isolation from other students in their creation, factors which might make personal support more important. The Ph.D. students are far more likely to recognize the significance of human concern in sustaining their long hours of study and writing. Sometimes this recognition extends to the encouragement provided by supervisors and mentors, but friends and family predominate, providing an opportunity for writers to inject the personal into this public space:

> I think we should not only be thankful for intellectual help, but also for moral support. Therefore, I'd include whoever helped in my project, including those who helped in collecting data, librarians, as well as family members. (AL Ph.D. interview)

> It is also appropriate to thank for spiritual support, so I'd also include my friends in church and family members. (Bio Ph.D. interview)

Friends and family members tended to be thanked succinctly, a brevity often in stark contrast to the lengthy tributes offered to supervisors and academics, and to be mentioned after academic thanks (Hyland, 2004c). There was also a surprising use of full names for both friends and nonacademics, with over 90% in each category identified in this way, occasionally with an honorific. Family members too were often mentioned using the full form of their names, with only two cases of first-name reference in the entire science corpus. In the Ph.D. dissertations, the full name pattern accounted for three quarters of all acknowledgements to family:

> (15) My personal thanks to Miss Katherine Ng for friendship and help. (EE Ph.D.)

> Last, I would like to say thanks to my lovely spouse Melody, Wong Wai Fong for her patience during my study and research period. (PA M.A.)

> The last, but not the least, I thank from the bottom of my heart, my parents, Mr. Lun Wai-Kan and Ms. Yeung Nga-Mong, for their wholehearted supports of my scholarly endeavors and for always being there when I need them. (Bio Ph.D.)

These naming practices mirror Swales and Feak's impressions of U.S. practices, with First Name + Family Name predominating. The decision to refer to family and friends by full names in this way is perhaps related to writers' perceptions of the formality of the genre, a desire to make a *formal* and almost ritually ceremonial acknowledgement for services received. But it also once again draws attention to the public nature of this discourse and an intention to ensure that the recipient is clearly identifiable by a professional readership for the credit owed to him or her. This is in marked contrast to the few intensely personal acknowledgements in the corpus which carry the obvious sincerity of the thanks by addressing an actual reader directly by his or her first name only:

(16) I owe immeasurable thanks to my best friend. Thank you very much, Stanley, for your infinite supports, round-the-clock services, emergency help, intellectual contributions and all other things that I forgot. (Bio Ph.D.)

Thanks to all my friends, especially for Vivian who is my best friend in the graduate school. Her friendship is much cherished. (PA Ph.D.)

However, for some students, the acknowledgements is a place to address only scholarly readers for academic things. It is because of this public and perhaps rather ritualistic character of the genre which made them reluctant to include family and friends at all:

I thanked mainly those who have actually helped in my project. I didn't include my family. I don't feel comfortable to thank my family. I'm not sure why, but I just feel uneasy to do so, maybe it sounds insincere. (PA Ph.D. interview)

I think it is ok to thank family and friends in writing a thesis, but I didn't include them. I think it is a personal style and I expect my family and friends to understand I'm grateful to them even if I don't put them in. (Bio M.Sc. interview)

I think we should only thank our family in writing a book, I mean, writing a thesis is not a great achievement. (Bus M.A. interview)

However, in thanks to family and friends we see the writer's expression of a personal self and a recognition of contributions beyond the academic. Here, writers are able to present themselves as individuals with lives and relationships outside the pages of their manuscripts. They become more human and sympathetic to readers and therefore more deserving of consideration. Equally, these acknowledgements allow writers to demonstrate their recognition of ethics and ideals shared by the reader, affirming their commitment to values such as modesty, generosity, and gratitude which are prized by academic communities as the public face of their disciplines.

If encouragement and guidance are the main themes of acknowledgements to academics, underlying those to family and friends is human care and concern, although thanks was not limited to human sources:

(17) I would like to thank God for giving me the strength and sustaining me through this whole course. (CS M.A.)

First, I thank God's guidance so that I can have the opportunity to study and complete this thesis. During the course of my study, I have lost hope and felt disappointed many many times. It was God who granted me serenity, courage and wisdom to continue. (CS Ph.D.)

Again, there is an obvious earnestness and honesty in these statements, but there is also an awareness of the genre and of the readership which is assessing both the writer and the dissertation, and which is able to influence the future of them both. References to God, Church fellowships, friendships, and social associations occur frequently, and one *Written Communication* reviewer mentioned that it is not unusual for students in the United States to acknowledge their pets. So genuine gratitude for the sacrifices and support of loved ones, human or otherwise, also projects a social persona as the strategic choices regarding the textual construction of gratitude are related to the strategic choices involved in careering.

The invocation of spiritual help in writing a dissertation perhaps testifies to the kinds of pressure research students often feel subject to, but more broadly, as with thanks to friends and family, it also contributes to a more rounded picture of the writer. Here we see acknowledgements as the bridge between the professional and the personal. References to those outside the academic world allow writers to break free of the constraints of the anonymous research rhetoric which has enmeshed them for the duration of the study to convey a sense of themselves as social individuals. So the apparent anomaly of recording thanks to those outside the academy in a text directed exclusively to those within it is explained by the fact that here writers can express what was hidden in the pages of the thesis: a view of a person not limited to an academic persona.

Some observations and directions

Acknowledgements are sophisticated and complex textual constructs which bridge the personal and the public, the social and the professional, and the academic and the lay. Their widespread use in postgraduate dissertations across different fields reflects their importance to students and underlines their considerable significance in scholarly discourse. This is perhaps the most explicitly interactional genre of the academy, one whose communicative purpose virtually obliges writers to represent themselves more openly. It is also a genre which allows readers to peer behind the carefully constructed façade of research texts to see a human writer with a real identity enmeshed in a network of personal and academic relationships. Here the writer can present a self disentangled from the complex conventions of powerful academic discourse types and reveal a real individual coping with the perplexing demands of research and overcoming a myriad of contingent issues which conspire to overwhelm the project.

Yet despite this relaxation in the authorial roles, purposes, and writer-reader relationships of the research genre, the choices available to writers of acknowledgements are not entirely arbitrary. Acknowledgements are not mere lists of thanks to a random group of people for miscellaneous services. There are clear patterns in the texts and discernable constraints acting on their writers. As in the dissertation proper, the problem for students is to demonstrate an appropriate degree of competence and intellectual autonomy while recognizing readers' greater experiences, knowledge of the field, and influence over the fate of the text. As a result, acknowledgements can play an important rhetorical role in promoting a competent, even rhetorically skilled, scholarly identity while signaling important professional connections and relationships as well as the valued disciplinary ideals of modesty, gratitude, and appropriate self-effacement.

The analysis suggests that the textualization of gratitude reveals social and cultural characteristics, an intimation of disciplinary specialization within a broad generic structure. It remains to be seen whether there are distinct Hong Kong features at work here but, much like the dissertation itself, disciplinary field rather than national characteristics is likely to be the strongest determinant of the shape this genre takes (Johns & Swales, 2002). The public display of thanks is shaped by larger forces and interests than simple thanks and is mediated by disciplinary preferences, personal gratitude, and strategic career choices. We can see here one way which postgraduate writing represents a situated activity. Writing a dissertation does not stand alone as the discrete act of a writer but emerges as a stream of activity which weaves together the personal, the interpersonal, and the institutional, and which often continues beyond the completion of the text, through the patronage and loyalty signaled in this genre, to a future career.

Acknowledgements, then, are intimations of the shared ways of understanding experience, representing a window into the personal worlds of student writers and the processes of engaging in the disciplines. In these texts, we glimpse students' disciplinary life-worlds, the ways their experiences of community work patterns and organizational affiliations are reflected in those they choose to recognize and what they recognize them for.

While this article begins to suggest something of these rhetorical characteristics, the picture is incomplete. This has been a textual study and there are opportunities to dig deeper into the meanings acknowledgements have to those who write and read them and to those they include. If there are social consequences of the rhetorical choices students make in their acknowledgements and in whom they choose to include and omit, then we need to examine writers, readers, and texts in greater detail to tease out the limits of personal choice and the kinds of effects they can have. It is also possible that exploring the potential impact of other situational factors, such as the author's age, gender, seniority, and publishing experience, may reveal interesting differences. More also needs to be done to investigate what Swales and Feak (2000) call "elegant variation" and the ways that gratitude is nonrepetitively realized. Finally, and most importantly, we need to see this genre as a significant site of rhetorical engagement; a place where writers' draw on linguistic resources to promote themselves and their contributions, and where analysts can learn how students understand their participation in the research and discourse practices of their disciplines.

Note

1 Except where permission has been given by acknowledgees to use their real names, all names in the corpus examples have been changed to protect anonymity.

11

The presentation of self in scholarly life: Identity and marginalization in academic homepages

1 Introduction

The connection between writing and identity has been a subject of academic interest for some time (e.g. Ivanic, 1998). While issues of agency and conformity, stability and change, remain controversial, there is some agreement that identity is created from the texts we engage in and the linguistic choices we make, thus relocating it from hidden processes of cognition to its social construction in discourse (Benwell & Stokoe, 2006). In academic contexts identity has often been seen as the ways that individuals work to reproduce or challenge dominant practices and discourses, often producing conflicts in academic and other ways of knowing (e.g. Lillis, 2001). The process of constructing an identity, however, most clearly involves selecting materials to present to others, a process of summation which might be seen most directly in personal homepages.

However, while creating a personal page encourages self-dialogue and reflection about how to create a public presentation of self, academic homepages often reveal the purposes of the university as much as the individual. Faculty members construct an identity online in the context of a university managed space and so reveal an individual supported by institutional infrastructures and linked into networks of colleagues, publications, interests, courses and students, all of which are carefully selected to assert both the professional credibility of the subject and the status of the employer. However, precisely what this selection comprises, the extent to which individual choice is eliminated by this process, and how far the construction of identity differs by seniority, gender and discipline remains unclear. I seek, therefore, to discover how academics use various semiotic resources to construct an identity online in the context of the university as a workplace. First I offer a brief sketch of this genre, its relationship to identity and its importance in academic life, then go on to explore the discursive construction of identity in 100 homepages of academics in two contrasting disciplines sub-divided by gender and rank.

2 Online identities

Following the pioneering work by Goffman (1971), there is now considerable interest in how individuals self-consciously manage the impression they give of themselves in different social contexts. Among these contexts, the homepage is recognized as an increasingly significant means of self-representation and as a result is fast becoming an object of academic study. Analyses have emerged, for example, into the homepages created by such diverse groups as adolescents (Stern, 2004), members of the House of Representatives (Adler, Gent, & Overmeyer, 1998), Fortune 500 companies (Liu, Arnett, Capella, & Beatty, 1997) and lonely Chinese seafarers (Tang, 2007).

The homepage owes its rapid rise, at least in part, to the fact that no other medium is better suited to fulfilling the present-day demands of identity work which seeks to balance the differentiation of modern life with the construction of coherence and meaning. The postmodern view of identity, understood as a patchwork (Kraus, 2000) or pastiche (Gergen, 1991) of independent and partially contradictory sub-identities, finds its expression in a medium which is always "under construction" and which can be regularly updated to reflect the latest self-conceptions. It is also a genre which facilitates the joining of diverse and diachronic self-aspects and sub-identities with links in its hypertext-structure (Chandler, 1998). Representing one's patchwork identity on a personal home page can therefore help foster the feeling of self-integration and self-effectiveness, but more than this, and at the same time, it is especially well suited for an elaborate strategic self-presentation (Chandler, 1998; Wynn & Katz, 1997).

The homepage therefore allows analysts to capture a more authentic and meaningful projection of self than through what seems to have become the default methodology for researchers in recent years: the narrative interview. The interview is an autobiographical reconstruction which allows subjects to reconceptualise their actions as representing a coherently motivated picture of continuity and sameness. It reflects Giddens' (1991) view that self and reflexivity are interwoven, so that identity does not involve the possession of particular character traits, but the ability to construct a reflexive narrative of the self. These interviews thus emphasize the subject's continual interpretation and reinterpretation of experience through a cultural lens, but because they focus on what people say about themselves in formal interviews with complete strangers, it is essentially a contrived, formal, anonymous and low-stakes genre. In contrast, the homepage reveals the same kinds of deliberately constructed identity claims but in a context where the elicitation is not the motive for the telling. It allows us to see such claims in a context which has the potential for a genuine projection of self.

Drawing on various modalities, with an attractive, information-rich presentation, homepage authors can strategically manage the impression they make on people who have never met them. For Parks and Archey-Ladas (2003, p. 4), then, the homepage provides 'an incredibly flexible and unencumbered stage for the construction of identity'. Chandler (1998), in fact, believes the medium offers possibilities both for the presentation and shaping of self which are shared neither by paper texts or by face-to-face interaction. As he observes:

> On the Web, the *personal* function of 'discovering' (or at least clarifying) one's thoughts, feelings and identity is fused with the *public* function of publishing these to a larger audience than traditional media have ever offered. (Chandler, 1998, p. 10).

The psychological need for cognitive integration is therefore married to the sociciological need for self-representation and group membership. Personal home pages convey an impression of how one would like to be seen by others, thus offering opportunities for contacts and networking (Chandler & Roberts-Young, 1998) and also for the kind of public relations work which can help create a certain collective identity (Hyland, 2011b).

The personal home page therefore supplements our means of impression management through interpersonal and public relations communication, supplemented by new ways of signifying meaning such as hyperlinks to show social connections and images to project visual presentations of the self. When online, moreover, we inevitably draw on previous social knowledge of how to present ourselves, so that many of these representations involve similar conventions to those found in offline spaces. In personal homepages all these linguistic, content and design features are typically under authors' control, allowing them to construct representations which will be recognized and valued by significant others and so revealing something of how they understand their communities. When located in a university context, however, the situation is less clear.

3 The academic homepage

The homepage is now firmly established as an ubiquitous feature of scholarly life. A recent study of the social networking practices of over half a million academics, for example, found that 71% of the researchers had at least one homepage (Tang, Zhang, & Yao, 2007). It is a genre of interest to a range of groups seeking information about individual scholars, not the least researchers, students, conference organizers, editors, publishers, university administrators, and recruitment head-hunters. It has therefore become almost obligatory for academics to maintain some kind of online presence. It is equally true, however, that university managers are also alert to the value of these homepages to themselves: in offering online resources for students, displaying credentials of the scholars who comprise their faculty, and functioning to advertise and promote departments, courses, and the individual instructor. As Hess (2002, p. 172) observes, 'As a faculty member designs her homepage, it becomes an advertisement for the faculty member, yet it is also an advertisement for the department and university'.

So while part of the allure of maintaining a homepage is to control one's representation of self online, the fact that it is located on a university server and accessed through a university website means that it can act as a means of institutional governance. From a Foucauldian perspective, of course, this is not surprising. In Foucault's (1972) account, identities are regarded as the product of dominant discourses tied to social arrangements, so that identity is inscribed in powerful discourses which shape and direct the individual. When we look at academic homepages in this way, we might see them as products of university design teams which machine them to conform to current models of university commercialism. So, Thoms and Thelwall (2005), for example, argue that academics relinquish their power to the university for reconstructing the self as the institution seeks to legitimate its own credibility and attract ever more resources.

Creating this identity involves what Chandler and Roberts-Young (1998) call *bricolage*, where the author doesn't so much *write*, but *assembles* the culturally valued attributes of his or her trade: the set of symbolic resources from which identity is constructed for public approval. In academic homepages, it is clear that the inclusion, omission, adaptation and arrangement of elements is not entirely under the author's control, so that the author is often not the owner of the page and his or her identity claim is incorporated into the symbolic eminence of the institution. What is a platform for global visibility and a declaration of academic credibility, therefore, is appropriated into the representation of the institution itself which claims ownership over the expertise of its employees. The individual, immersed in the institutional paraphernalia of research groups, publications and students which comprises the public persona of an academic, is simultaneously constructed in terms of the university's electronic shop window.

The extent to which academics are denied any autonomous subjectivity in the construction of their online identities is, however, uncertain. There have been only a handful of studies which have addressed the construction of academic identities on the web (e.g. Hess, 2002; Thoms & Thelwall, 2005) and these have largely focused on simple analyses of content (Dumont & Frindte, 2005) or power relations (Thoms, 2005). These suggest that academics tend to offer a 'focal-self' (Miller, 1995) which emphasizes an academic identity and the expense of more personal aspects of the self. Dumont and Frindte (2005), for example, found that information about research activities dominated the contents of psychologists' homepages, while women's pages, in particular, appear to conform to institutional representations and focus more on their credentials as academics (Arnold & Miller, 2001; Hess, 2002). This study seeks to show the ways faculty members construct their online identity in a university managed space and to reveal something of how these homepages are mediated and shaped by social factors.

4 Data and method

I studied 100 academic homepages with 50 each from philosophy and physics, representing a cross-section of practice from the humanities and hard sciences respectively and email interviewed five webpage subjects from each field. To ensure some institutional standardization, I selected pages from universities in the top 25 of the *QS-Times Higher world University Rankings for 2009* (http://www.topuniversities.com/university-rankings/world-university-rankings/2009/results). The QSTHS listings rank over 600 universities based on six indicators: the proportion of international students and staff, citations per faculty, staff-student ratio and employer views, but weighted heavily (40%) towards an opinion survey of 10,000 academics world- wide. It was thought that this list would provide a broad international base of comparable institutions from which to collect data for the study. This selection was then stratified according to rank and gender, with 25 full professors and 25 Assistant Professors and 25 men and 25 women from each discipline. The authors were affiliated with institutions in over 10 countries.

The decision to identify the authors by their academic status and gender was, like the categorisation of the corpus by discipline, based on assumptions about the possible significance of these factors in mediating a self representation and perhaps challenging institutional representations. My hypothesis was that the ways writers present themselves might be influenced by their experiences in their profession as women or men and by their seniority.

I then analysed the homepages looking at their texts, visual design and hyperlinks. I used *NVivo 8,* a qualitative software programme, to code the text by what was mentioned. The position, size and content of photographs were also recorded and the hypertext links were followed and categorised. The content that writers included was considered in terms of how it was structured as rhetorical units, each unit seen as a distinctive communicative act projecting a particular aspect of the self. After several passes through the corpus, I settled on this coding scheme which embraced all the data:

- Achievements and awards
- Community service
- Education background
- Employment

- Personal details
- Publications and conferences
- Research interests

- Teaching
- Contact details

These categories provide some detail for those searching for the person behind the conference presentation, academic article, or course description, but they reveal that representations were largely restricted to an *academic* identity. I followed this up with email interviews with randomly selected academics from the corpus to explore choices further and to better understand the extent of agency available to individuals. In what follows I discuss what is said about the self and how it is presented – as text, design and through hyperlinks. In this way I hope to draw these threads of authorship and identity together.

TABLE 11.1 Move frequencies by social characteristics (% of all academics)

Moves	Contact	Education	Employment	Personal	Research	Publications	Teaching	Award	Service
Overall	84	62	96	14	91	68	34	14	12
Professors	82	52	96	17	92	80	39	17	20
Assist Prof.	86	70	96	7	90	54	27	7	4
Male	82	61	96	16	91	60	35	9	11
Female	87	60	97	6	91	76	31	16	13
Physicists	72	57	92	11	90	57	26	21	10
Philosophers	96	69	100	19	93	79	43	6	14

5 Text choices in the academic homepage

The interviews offer some insights into agency in this genre and confirm that options are often restricted by institutional design. While only one of my, admittedly small, sample of respondents mentioned having to 'write to a template', academics often felt constrained in creating their profiles by space or the attentions of department webmasters. These comments are typical:

> We are given guidelines about what we should include and word limits for how much we should write. I'm not sure how strict these are, but they just relate to publications and what topics we can supervise.
>
> (Philosopher interview)

> I guess it is important to present a common front as it makes the department look slick and professional. That kind of thing seems to be important these days.
>
> (Physicist interview)

These constraints perhaps help account for the general uniformity in content choices shown in Table 11.1, which gives the percentages of individuals who included each identity move in their homepage.

Employment was the highest frequency move with all but a handful of physicists stating their current and past positions. Together with a statement of research interests, this comprised a substantial portion of the text of the corpus. Employment histories and research concerns are clearly the staples of the curriculum vitae and are available to everyone from the Nobel Professor to the lowly lab rat and so find their way into most constructions of the academic self. This information was generally accompanied by an email address, telephone number and office address, details which invite interaction and provide opportunities for readers to cross-check the given information (Parks & Archley-Landas, 2003). The figures are unsurprising in that employment credentials and research interests promote both the individual and the institution, linking the academic into a network of credible associations and advertising the university to potential graduate students and employees.

Less frequent than expected in this corpus, perhaps, are moves which relate to the author's research and teaching. In only two thirds of the homepages studied did author's mention their publication output and just a third referred to their teaching and supervision on their main page. Clearly there is more going on here than academics simply following instructions to legitimate the authority of their employers. In fact, decisions about what to include are not entirely determined by university marketing objectives but are negotiated with individuals, as these respondents confirmed:

I gave them some info for the webpage but didn't fill in all their boxes. I thought some information was more important about who I was and how I saw myself as an academic. It was more important to include it than some of the other things they wanted.

(Physicist interview)

I've published a lot in the last couple of years as we had a major project come to an end and I wanted to highlight that on my page. Nobody seemed to mind that I left out the courses I was teaching.

(Physicist interview)

An important mediating factor in the ways academics portray themselves online seems to be the status of the author, with assistant professors falling back on their qualifications and education in the absence of the publication and teaching records of their senior colleagues. For many writers, however, they used the opportunity to locate themselves in the academic milieu through association with a particular institution and individuals (1) and by stating their research interests (2):

(1) I got my Ph.D. from the Philosophy Department at UC Berkeley in 2003; my advisors were Richard Wollheim, John Searle, and John MacFarlane.

I did my Ph.D. at Caltech from 1999 to 2003 at Caltech, under the supervision of Prof. Kip S. Thorne. I graduated from Peking University (China) in 1999; my thesis advisor was Prof. Tan-sheng Cheng (I also did some research under Prof. Zheng-xiang Gao).

(2) I currently work on problems related to (or motivated by) the effort to detect gravitational waves.

Areas of Specialization: Ethics; Philosophy of Action; Philosophy of Mind; Philosophy of the Social Sciences.

Specialist interests represent a key way of establishing a claim to expertise and professional credibility for all academics, but senior scholars have a greater repertoire of to draw on in constructing their online identity and so were able to include far more on publications, teaching, awards and service.

Gender is a less significant influence on this data. As with other academic genres such as book reviews (Tse & Hyland, 2009) and bio statements (Hyland & Tse, 2012), the choices made by males and females is broadly similar. But while the actual frequencies are not large, there are small differences in that females were more likely to mention their publications (21% more moves) and awards (43%), and males to give more personal information (45%). Both groups mentioned their research interests, but females, especially those in philosophy, were largely alone in specifying particular titles, often with a summary for non-specialists, as in this example:

(3) She is the author of four books. The Sources of Normativity (Cambridge 1996), an expanded version of her 1992 Tanner Lectures, examines the history of ideas about the foundations of obligation in modern moral philosophy and presents an account of her own. Creating the Kingdom of Ends (Cambridge 1996) is a collection of her essays on Kant's Ethics and Kantian Ethics. The Constitution of Agency (forthcoming Oxford fall 2008), is a collection of her recent papers on practical reason and moral psychology. And Self-Constitution: Agency, Identity, and Integrity (2009) is a book about the foundation of morality in the nature of agency.

The most obvious potential indicator of institutional control over the ways these academics represented themselves online is the overwhelming predominance of professionally related information. Some individuals, however, found space to develop a more multifaceted subjectivity by including information on hobbies, interests, family and so on which reaches beyond the trappings of academic life. Two academics commented on this:

I wanted to put up something a bit more personal – what I like to read and the music I listen too. I didn't want to be seen as just a one-dimensional swot.

(Philosopher)

The brief from the department said we should include our research and publications and so on, but didn't say anything about what we couldn't put up. Of course most of what is there is going to be about our professional activities, after all, it is a university site, but there is more to life than that. I added some links to my favourite places.

(Physicist)

So at least some subjects sought to convey a more personal self among the academic identity constructs of the genre. Males, and especially senior males, tended to present this aspect of themselves rather more than females (links underlined):

(4) In the summer of 2007 my wife and I cycled the length of Great Britain (Land's End to John O'Groats). Here are descriptions and pictures of the trip.   (Physics professor).

*New house in San Francisco * Painting * Sailing * Wooden boat modelling. *Carpentry and furniture design * Mountain climbing * Cooking * Japan

(Philosophy professor).

Whether this more personal aspect of the self is deliberately suppressed by university marketing ideologies is a moot point as we also find this reluctance to convey personal information in the self-created personal homepages of these authors. This suggests that relative avoidance of non-academic subjectivities may, at least in part, be a deliberate choice rather than the effect of institutional branding.

Where activities outside academia were included, however, it was once again men who were more likely to reveal information about their personal lives. It is unclear why this is, although Tannen (1990) might suggest that it has something to do with males predilection for collecting, listing and imparting facts:

(5) I really enjoy music. My favourite genres tend to be progressive metal (e.g. Mastodon, Isis) and progressive rock (e.g. Yes, King Crimson), but many of the artists I like do not fit into either of these categories (e.g. John Coltrane, Allan Holdsworth, and Steve Reich. I also like to watch ice hockey, especially the Pens and the Flyers. I have a great appreciation for the beers of the Chouffland.

Outside the Department, I enjoy the complete suite of dweeb passtimes: listening to The Archers, bird watching, politely refusing alcoholic beverages. Unfortunately, having been born too late for The Age of Steam, I never developed into train-spotting. My favourite things include: hamsters , Rocky Mountain National Park, Here and There in The Observatory magazine.

The mention of "dweeb pastimes" in the second example is a tongue in cheek acknowledgement of this tendency, perhaps; a self-deprecating reference to a rather over-studious and perhaps socially inept individual, but the inclusion of these interests offers a picture of a more complex and rounded character than that found on the university homepage.

The biggest impact on homepage content appears to be discipline, with Philosophers making greater use of moves in most categories except awards. Physicists were far less likely to post details of their teaching and publication outputs, and often simply listed a few recent references or a hyperlink without comment. The philosophers, on the other hand, generally provided a commentary on the book and a sketch of its argument or impact:

(6) Professor Pereboom works in free will and moral responsibility, philosophy of mind, history of modern philosophy, and philosophy of religion, and has published articles in each of these areas. In his book on free will and moral responsibility, *Living Without Free Will,* he argues that we human beings do not have the sort of free will required for praiseworthiness and blameworthiness, but conceiving ourselves as

lacking this sort of free will this does not undermine what is important for morality and meaning in life.

Even where frequencies are similar in the two fields, moves are often expressed differently. Writers in both disciplines had a lot to say about their research interests, for example, but philosophers usually expressed this in the first person:

(7) My research focuses on the medical and life sciences as they existed from the early sixteenth-century to the beginnings of the nineteenth-century. With an emphasis on historical accuracy, I try to provide philosophical accounts of the ideas and arguments of past thinkers as they understood their predecessors and attempted their own experimental and theoretical innovations.

I am mainly interested in how scientific method could possibly lead us to true generalizations about Nature; generalizations that extend infinitely beyond our current, finite perspective.

Physicists, on the other hand, tended to employ the third person and refer to their lab membership as a way of presenting their interests and establishing their research credentials. Membership of a prestigious work group can imply status by association at the same time as it informs readers of one's professional commitments. Interestingly, these descriptions are often written for a non-specialist audience, speaking to a wider scientific community than the webpage author's particular field:

(8) The broad focus of Prof. Baranger's group is the interplay of electron–electron interactions and quantum interference at the nanoscale. Fundamental interest in nanophysics – the physics of small, nanometer scale, bits of solid – stems from the ability to control and probe systems on length scales larger than atoms but small enough that the averaging inherent in bulk properties has not yet occurred. Using this ability, entirely unanticipated phenomena can be uncovered on the one hand, and the microscopic basis of bulk phenomena can be probed on the other.

Her research program on galaxy formation and evolution complements the Penn astrophysics research program, whose main focus is on issues related to dark matter and dark energy. Her research will benefit from Penn's membership in the Dark-Energy Survey (DES) and the Large-Scale Synoptic Telescope (LSST).

A description of the writer's interest therefore both advances the expertise and competence of the homepage owner as well as promotes the specialism and the research group to prospective graduate students and other academics.

In short, while university ideologies undoubtedly exert control over what writers include in their pages, this is not the full story. In particular, we can see in these distinctive representations how disciplinary activity can encourage the performance of different kinds of professional identities. Here there are different orientations to community and epistemology. The philosophers tend to characterize themselves in terms of the more individualistic ethos of a discipline which sees interpretations and arguments as the creative insights of the author, while the physicists position themselves as players in a domain where results are the collective endeavours of a team. Here knowledge is shared and part of a cooperative context rather than personally owned.

6 Formatting and images

University governance is most striking in the visual design of academic homepages. The aesthetics of websites contribute not only to their attractiveness, but also to their effectiveness in how long users stay and what they remember later (Bonnardel, Piolat, & Le Bigot, 2010). But more than this, they contribute to the way the subject is positioned. The importance of design cannot be overestimated as online texts are never purely linguistic but involve different semiotic resources which, in combination, create new meanings. For Kress (2010) and other theorists, the screen is organized by the logic of *image*, not text, which orders the various 'systems of meanings' that are found there, so that the visual does not just superficially embellish text, but plays a central semiotic role (Jewitt, 2005). Unlike the D-I-Y feel of many personal pages then, the glossy, polished product which greets the visitor to an academic site represents academics in a very different way as text is integrated with images, page layout, colour, and hyperlinks.

Page formatting and colour, however, are almost always determined by the employing institution. University and department names and logos tend to dominate academic homepages with banner text along the top, departmental links down the side and webmaster information along the bottom. The repetition of these features across the homepages of all the members of a department have been seen as symbols of ownership which help position the individual as an employee (Thoms & Thelwall, 2005). While this largely serves the corporate interests of the employing institution, it also squeezes the space available for the individual to represent themselves without requiring further work by the viewer to link to other screens. The two homepages reproduced as Figs. 11.1 and 11.2 are typical of this.

While located in different disciplines, departments and universities, these homepages share a uniform appearance. The pages have a compositional unity deriving from a horizontal and vertical-grid structure, and this unity is

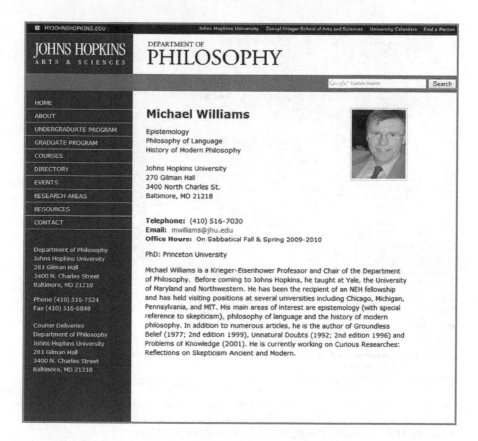

FIGURE 11.1 *Philosopher's homepage at John Hopkins University.*

reinforced by the choice of two main colours, sidebars and banners. The pages are dominated by banners proclaiming the respective departments and universities of the two academics, with the area available to the individual restricted to a third of the screen. In the space remaining, the subjects construct themselves through a brief paragraph, contact details, and a photograph.

Homepage screens ask the reader to perform different semiotic work than page-based texts, offering different entry points and different reading paths from the order of words in a sentence, but this does not mean that the ways we 'read' image-based texts is entirely unconstrained. For multi-modal semioticians, the placement of elements is significant, so that putting something above others can signify superiority or importance, for example. In these texts, the department banner is clearly given prominence at the top of the page and following convention, the reader assumes the text on the left, providing access to information on departmental and institutional matters, is intended to be scanned first. While reading paths are relatively open in images

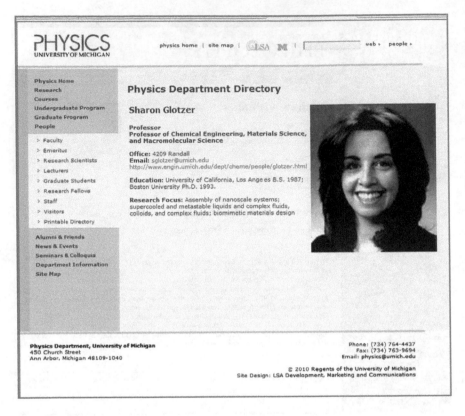

FIGURE 11.2 *Physicist's homepage at University of Michigan.*

compared to texts, the left-side text is also given prominence through being set in a contrastive saturated darker colour and shape.

We do, however, distinguish headers and sidebars from main text. Following Kress and van Leeuwen's (1996) analogy with written texts, it is what occurs on the right which is regarded as the viewer as key information. The right side is where the reader must pay particular attention to find the 'message' while the left is 'already given', or what the reader may be assumed to 'know already' as part of the context. The institutional links are presented as a familiar and agreed upon point of departure for what is not yet known: the representation of the homepage subject. Compared with the relatively dense lists on the left of the screens, the uncluttered white space available to the academics themselves draws the eye to the subjects' information and, in particular, their photographs. In so doing it facilitates and supports an alternative pathway: the modern world's preference for image in online domains. The photos, in fact, are quite salient in this space and therefore demand attention, inviting viewers to form a positive identification of the academic.

All but a dozen homepages in the corpus contained a photo of the author. These were positioned alongside the written text and almost always towards the top of the page, on the right, and were always of the subject alone, irrespective of discipline, gender or rank. While photographs allow viewers another way of understanding the subject and an opportunity to scrutinize the 'carrier's possessive attributes' (Kress & van Leuwen, 1996), or parts of the whole, most gave little away. Many were obviously taken by in-house photographers and over two thirds were a tightly cropped portrait with no articulated background, betraying little context and so reducing our ability to see the individual as anything more than a generic academic filling a vacant place on the homepage. This absence of setting not only reduces personal information about the subject, but disconnects him or her from time and place. Where a background could be discerned it was likely to be an office, although 25% of all authors were represented in a leisure context. Interestingly, this suggests some independence to chose their own photographs and depict themselves in a more relaxed way:

> We had our photographs taken by the guys at Media Resources but we all looked a bit stiff and uniform so I gave them one of me on a camping trip which they put up instead.   (Physicist)

The representation of subjects in muted hues and narrow colour range, one actually in monochrome, helps to realize what Kress and van Leeuwan (1996) regard as 'high modality': cues which impart credibility to the message. On the basis of such modality markers as low colour variation and saturation and the absence of context, the university web designer encourages us to invest the homepage subject with integrity and accept the accompanying text as fact. While unable to control design decisions, however, subjects were able to choose how they faced the camera, and while physicists tended to look serious, most subjects chose to show themselves smiling. The smile invites a connection of social affinity and it is the subject's gaze which dominates, here as in most of these portraits. The academic looks directly at the viewer, bringing about a closer (imaginary) relationship between the subject and those who are interested in discovering more about him or her. A contact is established which augments the bare ideational information presented in the accompanying text and perhaps even subverts it by fleshing out the purely professional persona depicted by the institution.

There are, in other words, a range of meanings in these pages, for while the universities seek to enhance their own standing by representing subjects as academics of substance through uniform formality, some individuals exploit what options are available to offer a more personal self-presentation. The multiplication of meaning and alternative reading paths in these homepages thus offers a complex semantic arena where there may be divergent (and even conflicting) meanings (Liu & O'Halloran, 2009). In fact, there is no reason to assume a coherent semantic integration of semiotic choices in multimodal spaces and this tension seems to exist here, where individual and institutional purposes interact.

7 Hyperlinks and connections

A final aspect of these homepages worth commenting on is the potential interactivity and meanings they establish with the reader through the use of hypertext links. Hypertext is the presentation of information as a linked network of nodes which readers are free to navigate in a non-linear fashion. In webpages it allows for interactivity and multiple reading paths. Just as we make meanings across many paragraphs that we do not make within a single paragraph, so we can make meanings in hypertext through traversals to other sites. Each hypertext link thus provides what Lemke (2002) calls an 'intertextual meaning relation' and in academic homepages these are essentially presentational, tying one topic-specific set of semantic relationships to another. Hyperlink networks seem to offer the reader many potential paths to follow, but choices are actually pre-designed and limited by the page author. For universities this offers opportunities to channel visitors to pages of their choosing, revealing a subject enmeshed in a web of institutional programmes, events, news and so on. For academics, it offers a chance to engage with the reader by indicating social allegiances in a process which Danath and Boyd (2004, p. 2) refer to as "public displays of connection".

Potentially, hyperlinks are key aspects of an online identity construction as they can be seen as flagging topics, stances and people which the author regards as significant (or significantly absent). Turkle (1996, p. 258) observes that 'one's identity emerges from whom one knows, one's associations and connections', while Miller (1995) puts this more directly: 'Show me what your links are, and I'll tell you what kind of person you are'. Through links to the pages of research groups, labs, friends or departments elsewhere, authors can construct a 'virtual community' (Rheingold, 1995). Hyperlinks therefore offer authors an opportunity to individualize their homepage through networks of associations, explicitly claiming membership of particular groups, approval for particular ideas, and relationships with particular others. Unsurprisingly, perhaps, links were overwhelmingly institutional.

The 100 homepages in the corpus contained just under 1400 links, ranging from a handful per page to over 40. The examples in Figs. 11.1 and 11.2 are fairly typical, both in terms of the frequency of links (Williams containing 20 and Glotzer 27) and in their targets. As can be seen from Table 11.2, academic

TABLE 11.2 Target pages of links on University personal homepages

	University	Dept.	Discipline	Students	Publication	Research	Personal	Totals
Male	88	296	24	133	134	54	32	761
Female	86	237	22	123	53	87	25	633
Physics	86	238	11	112	40	116	31	634
Philosophy	88	295	35	144	147	25	26	760
Professor	86	271	22	136	147	93	32	787
Asst Prof.	88	262	24	120	40	48	25	607
Total links	174	533	46	256	187	141	57	1394
Percent links	12.5	38.2	4.1	18.4	12.6	10.1	4.1	100
Total authors	100	99	29	86	28	69	55	

homepages represent a complex source-user relationship with a range of implied offers of further information and demands for action (encouraging the user to click to follow a link). Departmental information dominates the frequencies and together with links to university pages, these connections to the staff, facilities, events, alumni and research comprise half of all the links in the corpus. Information about and for students is the next most common link and occurs on all but six homepages in the sample. Institutional links therefore overwhelm those of the subject.

Links associated directly with the identity of the authors themselves, relating to their publications and conferences, their research, and their personal interests, not only largely served to discursively construct academics in the university's image, but comprised just 27% of all hyperlinks. Disciplinary links, which take readers to affiliated labs and centres elsewhere, journal websites, professional associations, and so on, serve both personal and institutional interests and comprise 4% of links. We can, however, observe a distinction in the university and individual connections, for while the institutional links might be seen as a directory of further information, author links are the equivalent of verbal invitations, encouraging the viewer to discover more about the individual. This type of author-user relationship in hypertext gives rise to a sense of interactive dialogue, allowing the author the opportunity to escape monologism by creating a pseudo-dialogue with the user.

We can also see mediating factors in individual choices which suggest that there is some control by individuals in the selection of these links. Table 11.2 shows that the categories of links do not always correspond with the content categories discussed above. Males, professors, and philosophers all created substantially more links and these were mainly to their departments, students and publications. Professors, unsurprisingly, linked almost four times more to their publications, for example, and males provided one and a half times more links to their publications; females, in contrast, directed readers to their research. Those mentioning research, however, were about three times more numerous for both gender groups and the high figures for publication links can be accounted for by the reporting practices of some dozen senior male philosophers averaging about 12 or 13 links each.

While similar numbers of individuals in each discipline posted links in each of the categories, philosophers used more hyperlinks in almost all of them except research. Virtually all physicists had links to their research interests or, more often, also to their research group, with five times more links. These examples are typical (links underlined):

(9) Research activities.
KTeV Experiment at Fermilab, KTeV: *University of Chicago.*
Braidwood Experiment: Reactor Neutrino Measurement of Theta_13.
APS Neutrino Study: Reactor Working Group.

The 1st phase of the XENON program is well underway, with a 10 kg TPC (XENON10) currently operating underground at the Italian Gran Sasso Laboratory (LNGS). The detector, built at Nevis Labs, was moved in March 2006 and it has been running continuously and stably for the past 5 months in dark matter search mode. First results from this experiment are expected by Spring 2007. The Picture Gallery has a selection of images on the XENON10 experiment.

Interestingly, some physicists saw this as an opportunity not only to a present a representation of themselves through the labs and groups they were associated with, but to promote their fields of interests to outsiders and non-specialists. In the corpus of physics homepages, links are often to sites for the public or online encyclopaedia:

(10) Our group is a member of the Compact Muon Solenoid (CMS) experiment, which has constructed a massive detector for the Large Hadron Collider (LHC). The LHC is a 14 TeV proton–proton colliding beam accelerator, which has been constructed at the European accelerator center, CERN, in Geneva, Switzerland. To find out more, visit the CMS web page and CERN's web page.

Research Interests.
Quantum Information Science <http://en.wikipedia.org/wiki/Quantum_information_science>.
Entanglement <http://en.wikipedia.org/wiki/Quantum_entanglement>.
Quantum Optics <http://en.wikipedia.org/wiki/Quantum_optics>.

Links therefore seem to be less amenable to the university's editorial control than content. But while 55 academics linked their page to a 'personal' target such as a curriculum vitae, few linked to a non-academic site. Fifty three individuals, however, chose to link to personal site. Calling the visitor to a personal homepage enables the author to reclaim some control over format, layout, links and content and, if he or she chooses, to represent a more rounded identity as someone with a life outside the academy. These opportunities, however, were seldom taken and while the *appearance* of the pages differed from the university pages, the content – and the links – often remained much the same.

As I noted when discussing the moves of these homepages, academic material predominated on the personal pages and this was also true of the targets of their links. Just 7% of the links on the personally-constructed pages referred to more private aspects of the self such as family pictures, hobbies, interests, or other icons (see example 5 above). More usually, links followed similar patterns to those on the university pages (again, underlining indicates links):

(11) WEL COME and thank you for visiting my website.

I am a University Lecturer in the <u>Faculty of Philosophy</u> at the <u>University of Cambridge</u>, and a Fellow in Philosophy at <u>Jesus College, Cambridge</u>.

My field is contemporary political philosophy. I am particularly interested in:

1 <u>Contemporary liberalism</u>, including autonomy, equality, multiculturalism and global justice.

2 <u>Feminism</u>, including the body, appearance norms and personal relationships.

3 Theories of <u>social construction</u>, including those of Michel Foucault and Pierre Bourdieu.

On these pages you will find more information on my research, publications and teaching.

Clearly, people who create their own homepage want to present a certain version of themselves to the world, or at least that part of the world interested enough to search for them, and links to others can help this creation. Who we publicly wish to be associated with speaks of how we want to be seen, the communities we belong to, the interests we have and the people we value. It seems, however, that even when freed from institutional constraints, these academics rarely stray beyond links to information related to an academic persona.

8 Conclusions

Critics complain that personal home pages are often trivial, amateurish and superfluous products of narcissism and exhibitionism, where "assertive confession" and endless lists predominate (Rothstein, 1996) while advocates stress the emancipatory and self-reflexive potential of autonomous portrayals of individuals in a public space. Academic homepages seem to fall outside of these generalisations and perhaps only peripherally qualify as homepages in the sense which it is normally used. Certainly these pages promote the individual academic and their accomplishments, but equally they discursively construct the individual as a member of an institution which profits from these accomplishments. Clearly, universities have different policies and different tolerances of deviance while different academics will have different responses to institutional control. It is therefore difficult to make generalisations, but homepages are of interest to EAP practitioners, I believe, and their study can contribute to theory building, language understanding, and pedagogy.

First, the analysis of homepages offers insights into the working of an important, and previously under-investigated genre, revealing the workings of language and how it interacts with other semiotic resources for particular social purposes. In what is included and the ways it is assembled the homepage reveals the negotiation of impression management within specific relations of power. The *content* of personal home pages, whether the text, design, visuals or links, draws on a palette of conventional paradigmatic elements which not only make information about subjects accessible to a potential world-wide audience, but which promote a version of them and their university to that audience. The fact that we find mainly professional biographies and references to research interests and publications; that the design reveals the uniform repetition of a university brand; that the visuals are restricted to cropped passport style portraits; and that links largely take us to places which reinforce the competent, accomplished academic, all reveal a genre which enhances the status of the author and subjugates him or her to its homogeneity.

Second, studying homepages adds to our understanding of identity, its construction in academic contexts, and the ways that factors such as discipline, rank and gender potentially impact on this construction in the face of institutional pressures towards conformity. Community Discourses, and their social ideologies, assist the performance of identities by providing broad templates for how people see and talk about the world and about themselves. Our identities draw heavily on these schema as they both shape and enable particular versions of who we are and how we want to be regarded by those whose opinions we value. In an important sense, identity means constructing credibility, it involves negotiating a self which is coherent and meaningful to both the individual actor and the group. In the ways individuals seek to achieve specific promotional goals through the presentation of themselves in homepages, then, writers are also managing a considered and rhetorically machined identity likely to appear credible to others.

In these constructions of online identities, we find something of the perceptions and values of academic communities as writers assemble a persona which draws on insider perceptions of what their disciplines find meaningful and important. We also, however, see something of the tensions, compromises and negotiations in this enactment of an online self. The authors in this sample were largely complicit in what can appear to be attempts by universities to marginalize faculty members as simply institutional functionaries. Turning to personal homepages situated outside direct departmental governance, we find a more multifaceted self, but this still places a professional expertise at the centre. The individual's attachment to a particular position and his or her commitment to perform a role as an academic clearly serves as the most powerful platform for all their other facets.

Finally, the study of homepages might have pedagogical value in describing a genre useful to students and professionals. This is a self-promotional as much as an informational genre and by creating such pages, academics can, within limits, strategically manage something of their self-presentations. A better understanding of the workings of this genre might assist individuals to reclaim a personal subjectivity from the juggernaut of institutional branding and further their careers by presenting themselves in the most effective way. Perhaps more importantly, however, the process of constructing a homepage can itself be a valuable experience. It encourages individuals to think about which aspects of their identity should be presented and how language works in combination with other modalities, so promoting both genre awareness and a better understanding of the relationship between texts and achieving social purposes.

This has been a relatively small-scale study of 100 homepages and three main variables and there is, of course, more work to be done in this area. It would, for example, be interesting to look at other disciplines, to interview more authors on their writing decisions and readers on the impact of those decisions. It is also worth exploring non-university pages more extensively than the 50 or so examined briefly here to see how authors subvert or conform to institutional requirements, or to examine more thoroughly the impact of national academic cultures on the *bricolage* of academic homepages. I hope, however, to have established an agenda for this research and to have shown that the personal homepage has to be recognized as a rich set of options for portraying an academic self.

Commentary on Part III

Vijay K. Bhatia

Genre analysis, although with a history of just about three decades, has become established as a privileged framework for the study of English for academic and professional communication. Swales, with his pioneering work on academic English, has stood the test of time, and is rightly regarded as the father of Genre Analysis. His work on academic research genres, and its pedagogic applications in research writing, including academic publishing has been viewed as one of the most significant developments in applied linguistics. Some two decades ago, he indicated three kinds of academic genres; research genres that are meant to address fellow academics, pedagogic genres that generally address newly initiated members of the academic community, especially students, and occluded genres, which deal with the private academic communication with other colleagues, such as editors, reviewers and other academic gatekeepers.

However, much of genre analytical literature, in its initial conceptualization, as represented in the works of Swales (1993, 2004), was primarily inspired by applied concerns on the part of a wider linguistic community interested in English for Academic Purpose (EAP) practice. As a consequence, some of the major mainstream analytical investigations focused on key academic research genres, such as journal article introductions and research abstracts, which were considered most crucial for applied pedagogic purposes. The trend continued for several years, and we noticed an unprecedented growth in the analysis of several major academic and professional genres, often at the expense of analytical interest in many other genres that were considered peripheral by the larger academic and professional communities, of course with a few notable exceptions, such as Swales (1996), Connor and Upton (2004), Brown (2004) and Ding (2007). However, some of the most significant contributions have come from Ken Hyland, which seem to bridge this significant gap in the analysis of some of the under-investigated and peripheral genres used in academic contexts. Three of his chapters in this section seem to be quite representative of his contribution to academic genres, in addition to highlighting his contribution to his unique approach to genre analysis, which seems quite distinctive in some respects.

In the first of the three chapters, 'Constructing proximity: Relating to readers in popular and professional science', Hyland offers an interesting study of a rhetorical activity involving two of the well-known and closely related genres with a specific focus on the interactions between writers and readers, which is not often highlighted in conventional genre analytical literature. Based on the quantitative analysis of, and qualitative evidence from, academic discourse, he focuses on the construction of proximity as rhetorical interaction between writers and readers, in authentic scientific discourse and its appropriations in the form of reframed and reformulated popular versions. One of the distinguishing features of this research, as against what is often presented in much of corpus analytical research, is the way Hyland goes well beyond the insights gained from corpus analysis to identify and discuss issues such as the management of expertise in academic genres, offering much-needed insightful explanation, which is often missing in many corpus analytical studies. An interesting aspect of this chapter is the way he distinguishes the two audiences, lay and specialist, in terms of their approaches and understanding of the two interdiscursively related genres, thus making a claim about their understanding of the more general issue of mediation between science and society, which goes well beyond the routine rhetorical analysis of a corpus.

The second study in this section 'Dissertation Acknowledgements: The Anatomy of a Cinderella Genre' investigates how academics express gratitude as what Hyland views as 'academic gift giving', in particular, based on a corpus of acknowledgements written by dissertation writers from English as a second language background. The most interesting aspect of this study is the extent to which academic expressions of gratitude vary with sociocultural as well as disciplinary characteristics. More generally, Hyland seems to explode the widely observed myth that expressions of academic gratitude through the conventional genre of acknowledgements is predominantly personal. His corpus quite significantly reveals sociocultural and disciplinary preferences, along with individual variations.

The final chapter in this section focuses on the presentation of self in scholarly life, in particular on the issues of identity and marginalization in academic homepages. The academic homepage, as Hyland rightly points out, has become what he calls a 'ubiquitous genre of scholarly life', which is generally exploited to reflect not only the individual academic identity in an interdiscursive public space but also the much sought after visibility of individual scholars in the academic world. In order to achieve this, the scholars make a continual effort to manage and establish self-impression through their homepages, invariably managed within institutional frameworks. It is a complex mediated genre that combines personal stories with public discourses. Although the use of new media for the construction and promotion of self is relatively recent, the way it has acquired a privileged status in establishing academic visibility and reputation is particularly notable in recent times. The mainstream research on the analysis of academic homepages within genre analytical framework has been rather slow to investigate this genre primarily because of a relative lack of conventionalization and hence a perceived lack of generic integrity reflected in academic practice in the prevailing environment. Secondly, scholarly homepages are generally constrained by the individual and varied requirements of the institutions that host them. In fact, academic homepage represents a complex hybrid genre that combines key aspects of self-promotion within primarily informative genre, and especially the way both of them are strategically situated in the institutional framework. In this context, Hyland's attempt to explore how various linguistic and semiotic resources in multimodal configuration are exploited for impression management based on a reasonably sized corpus of academic homepages from different disciplinary fields is very significant. In an attempt to analyse such a dynamic construct, Hyland quite rightly makes use of an integrated multidimensional framework, combining aspects of mainstream rhetorical genre analysis, corpus analytical tools and a selection of ethnographic procedures.

In his studies of academic discourse, although Hyland is well known for his profound insights based on his quantitative analyses generally used in conventional corpus-based procedures, he does not treat this as an end in itself, but goes well beyond it to combine his corpus-based insights with qualitative explanations and clarifications to identify and explain key sociocultural issues that reflect his concerns with the nature and function of the rhetorical activities represented in academic discourses. The three chapters on the analyses of some of the under-investigated academic genres in this respect are a true reflection of this concern. The most interesting aspect of this section thus is his use of multidimensional and multiperspective tools to analyse genres by combining quantitative analytical rigour to help develop some sense of the underlying interactional activities. His work, in this section, and also elsewhere, in this way represents an interesting development in genre analytical research, that is his attempt to combine corpus analysis with genre analysis to offer a richer description of genre-based rhetorical activities.

Most mainstream genre analytical studies often rely on manual analyses of a small corpus to generalize rhetorical–grammatical and organizational aspects of genres, invariably assuming that such insights will be shared by other instances of the genre in question. Hyland, in combining rigorous corpus-based analyses of large sets of corpora, makes sure that such generalizations have sufficient statistical evidence as the basis for such generalizations. However, unlike most studies in corpus-based analyses, Hyland has a reputation of going significantly beyond the conventional mainstream studies, which claim generalizations based on statistically significant surface features of textualization of registers and discourses. For him, more than statistical evidence revealed in corpus-based analysis is the identification and explanation of specific characteristic issues that make such a genre unique. In this kind of treatment, he manages to offer a rich description of textualization of individual genres.

As alluded to earlier, an interesting aspect of Hyland's analytical framework is his attempt to integrate ethnographic aspects of analytical interviews with genre analysis based on a comprehensive and statistically rich analytical insights. So if we see the three chapters in terms of analytical rigour, they are strategically situated in a sequence that illustrates Hyland's progression from somewhat pure corpus analysis to a multidimensional analysis integrating corpus analytical work with ethnographic analytical procedures all within the genre analytical framework.

Finally, the interdiscursivity in the integration of research tools mentioned above is also interestingly seen in the way he manages interdiscursive space in his introductory sections, especially the way he displays his expert management of other voices in available literature, an expertise over which he seems to have his complete control. This is a specific skill that even well-established scholars find it difficult to display in their introductory sections, which they seem to think are not worthy of their effort as they are not always taken seriously by most readers. In the section that follows, Hyland demonstrates his mastery of the art and craft of the management of such interdiscursive space.

Features of academic writing

PART FOUR

Features of
academic writing

Introduction

This section includes papers which analyse some key features of academic texts, looking in particular at citation, self-mention, vocabulary and lexical bundles. Some of these papers are among my most cited and the topics have been of continuing interest to me, re-appearing in different ways in different papers. These studies, once again, focus on the ways interactions are accomplished in negotiating agreement; how rhetorical decisions contribute to the construction of knowledge, of disciplines and of self. They also depend heavily on the use of corpora, and would have been impossible to write without them. Corpora have become invaluable in revealing how participant relationships sit at the heart of academic persuasion and there is now a small industry of corpus research demonstrating how some syntactic or lexical feature or other helps reveal the writer's awareness of the text's context and the readers which form part of that context.

The features I've focused on in these papers are no more important than many others, although they represent high-frequency, and salient, aspects of academic writing and so provide immediate access to its rhetorical construction. They reveal how writers, in pursuing their personal and professional goals, embed their writing in a particular discipline through approved discourses. This has told us a great deal about the workings of the disciplines and how persuasion is accomplished through framing ideas in reader-familiar ways. It has also generated a great deal of improved teaching materials which help learners to become both more disciplinary-sensitive and self-aware writers, better able to construct appropriate authorial selves.

Corpus methods are not without their critics, however. Because it is an approach which privileges community practices over individual preferences, neglecting actual texts to focus on specific features, corpus analysis has come under fire for 'fetishizing' data at the expense of what real people are doing in the world. Critics argue that texts are a concrete communicative engagement with an audience and so we should examine interaction as a series of real, situated encounters. Certainly, dematerializing texts to approach them as a package of specific linguistic features fails to capture the richness of composing, but texts function communicatively when they are read, not at the time they are composed. Genres evoke an institutional frame for writers in creating a text, calling up a social milieu which influences the writer and activates specific responses to recurring tasks. By examining key features through corpora we see better how writers understand their audience and how they attempt to engage in academic argument with them. Preferred choices of language use in a representative collection of texts thus shows how writers anticipate the responses of readers to what is written and so participate in particular disciplines.

Chapter 12 illustrates something of this distributional perspective by providing quantitative information about the relative frequency of citation practices in different disciplines, pointing to systematic tendencies in the selection of meanings. Basing conclusions on a corpus of only eighty articles might be sniffed at today, but it was considered a fairly large database back in the 1990s when this paper was written, and happily large enough to persuade the reviewers of *Applied Linguistics* that something interesting was going on. The paper identifies clear disciplinary differences in both the extent that writers refer to the work of others and in how they report this. It shows that writers in the humanities and social sciences use citations far more often than scientists and engineers and that they are more likely to present these differently. These disciplinary differences are repeated for other features that have been studied, showing how academics actively create knowledge as members of professional groups and the ways rhetorical decisions both reflect and construct the epistemological and social conventions of disciplines.

These differences are also apparent in Chapter 13. Self-mention, the presence or omission of *I* or *we* to refer to the author, has long been a contentious subject in language teaching, with conflicting advice in textbooks, style guides and teachers' recommendations. Taking their cue from the physical sciences, those who dispense conventional wisdom on the topic state that academic writing is an 'author-evacuated' product where the writer leaves his or her personality at the door to better enable facts and data to speak directly to readers. The passive voice is supposed to be the default option; a defining feature of academic prose. Much has been written on the topic, both before and since this paper was published, but once again matters are not so cut and dried. Like citation, the choices which express writer presence are also closely associated with disciplinary practices. They are also tied to authorial identity and authority, offering writers a powerful rhetorical strategy for emphasizing a unique contribution and so influence the impression they make on readers.

This paper was written a couple of years after Chapter 12 and makes use of a larger corpus (now grown to 240 articles) and also extends my fondness for corpus data to interviews with expert informants. Frequency and collocational data provide descriptions of existing practice but are not ends in themselves. They are excellent starting points for raising awareness of uses, for telling us what writers do, but to stop here runs the danger of reifying conventions rather than explaining them. What we can't do with corpora we must do in other ways, and interviewing is perhaps the most productive. This was an approach I have used since my PhD (on hedging) and feel it helps to balance text analyses with an understanding of the production and reception of those texts. Discourse-based interviews are particularly productive as they help subjects focus on the language choices they make and, if I am lucky, lead to how they see the literacy practices of the communities that influence their writing. Fortunately, I've generally been able to find colleagues who are reflective and informative about their work. What academics do for a living is talk, and they are often more than willing to talk about their writing.

Chapter 14 is another paper co-authored with Polly Tse and the only one I've ever managed to get published in *TESOL Quarterly* – despite many attempts. This paper continues to tease out systematic differences in disciplinary writing by questioning the assumption behind Avril Coxhead's *Academic Word List* (AWL) and subjecting the list to empirical critique. While we were impressed by the scale and ambition of the list, we were sceptical of its potential value. Our instincts (and research) told us that as helpful as such a list would be, a blessing to EAP students and teachers alike, there was unlikely to be a core of high-frequency words useful to all students across the university.

This time we were more interested in what the text data had to say rather than users, and so created an ever larger corpus (3.3 million words) across different registers in selected genres. Sure enough, our findings showed the list gave a notable coverage of 10.6 per cent of the corpus, but also that individual lexical items often occur and behave in different ways across disciplines in terms of range, frequency, collocation and meaning. In the last decade the idea that there is a single core vocabulary needed for academic study has been replaced by a growing number of more restricted, disciplinary-based lists, and even the *Academic Vocabulary List* (Gardner and Davies2014) based on the 120 million words of the Corpus of Contemporary American English, provides a great deal of information about the meaning and usage words in each of the nine disciplinary sub-corpora.

Chapter 15 follows the same direction as those preceding it: exploring the rich variation of community language use. However, instead of looking at the functions of particular features (chapters 12 and 13) or at individual words (Chapter 14), it focuses on commonly occurring strings of words. These routinely used chunks of texts not only help shape text meanings, but their ubiquity in particular contexts means contribute to our sense that a text belongs to a specific register. This study of a 3.5-million word corpus of research articles, doctoral dissertations and Master's theses in four disciplines also shows that they also underwrite, once again, our appreciation of discipline.

The section is wrapped up with a commentary on the texts by Professor Diane Belcher of Georgia State University, USA, one of the leading figures in the field of advanced academic writing and language for specific purposes.

12

Academic attribution: Citation and the construction of disciplinary knowledge

Introduction

A cademic knowledge is now generally recognized to be a social accomplishment, the outcome of a cultural activity shaped by ideology and constituted by agreement between a writer and a potentially sceptical discourse community. A substantial literature has shown that the research paper is a rhetorically sophisticated artifact that displays a careful balance of factual information and social interaction (for example Bazerman 1988; Gilbert and Mulkay 1984; Latour and Woolgar 1979). Academic writers do not only need to make the results of their research public, but also persuasive, and their success in gaining acceptance for their work is at least partly dependent on the strategic manipulation of various rhetorical and interactive features. Hedges (Myers 1989; Hyland 1996a and b and 1998a), evaluations (Hunston 1993; Thetela 1997), imperatives (Swales *et al.* 1998), theme (Gosden 1993), and metadiscourse (Crismore and Farnsworth 1990; Hyland, 1998b) are only a few of the elements examined for their contribution to the perilous negotiation of a successful writer-reader relationship.

One of the most important realizations of the research writer's concern for audience is that of reporting, or reference to prior research. This is defined here as the attribution of propositional content to another source. Citation is central to the social context of persuasion as it can both provide justification for arguments and demonstrate the novelty of one's position (Gilbert 1976; Berkenkotter and Huckin 1995). By acknowledging a debt of precedent, a writer is also able to display an allegiance to a particular community or orientation, create a rhetorical gap for his or her research, and establish a credible writer ethos. But while the literature recognizes the rhetorical importance of citation, we know little about its relative importance, rhetorical functions or realizations in different disciplines. In this paper I will explore the conventions of citation behaviour in a range of academic fields and, through interviews and corpus analysis, seek to reveal some of the purposes embedded in those conventions. First, I need to provide a context for my research by establishing intertextual linkages of my own.

Citation, intertextuality, and the construction of knowledge

The construction of academic facts is a social process, with the cachet of acceptance only bestowed on a claim after negotiation with editors, expert reviewers and journal readers, the final ratification granted, of course, with the citation of the claim by others and, eventually, the disappearance of all acknowledgement as it is incorporated into the literature of the discipline. This process of ratification clearly suggests that writers must consider the reactions of their expected audience to their work, for it is ultimately one's peers who provide the social justification that transforms beliefs into knowledge (Rorty 1979: 170). The fact that there is no independent, objective means of distinguishing observation from presumption means there is generally more than one plausible interpretation for a given piece of data. Readers always have the option of rejecting a writer's message and therefore always play an active role in its construction. One consequence of this delicate process of negotiation is that writers are obliged to situate their research in a larger narrative, and this is most obviously demonstrated through appropriate citation.

Myers (1990) and Berkenkotter and Huckin (1995) have traced the passage of research articles through the review procedure and see the process as one of essentially locating the writer's claims within a wider disciplinary framework. This is achieved partly by modifying claims and providing propositional warrants, but mainly by establishing a narrative context for the work through citation. One of Myers's case study subjects, for example, increased the number of references from 57 to 195 in a resubmission to *Science* (Myers 1990: 91). Both Myers and Berkenkotter and Huckin see academic writing as a tension between originality and humility to the community, rhetorically accommodating laboratory activity to the discipline. So while Berkenkotter and

Huckin's scientist subject sought to gain acceptance for original, and therefore significant, work, the reviewers insisted 'that to be science her report had to include an intertextual framework for her local knowledge' (Berkenkotter and Huckin: 59).

Academics generally tend to see research and rhetorical activity as separate. However, appropriate textual practices are vital to the acceptance of claims. Explicit reference to prior literature is a substantial indication of a text's dependence on contextual knowledge and thus a vital piece in the collaborative construction of new knowledge between writers and readers. The embedding of arguments in networks of references not only suggests a cumulative and linear progression, but reminds us that statements are invariably a response to previous statements and are themselves available for further statements by others. Fairclough (1992), extending Bakhtin, refers to these intertextual relations as 'manifest intertextuality', which he distinguishes from 'constitutive intertextuality':

> In manifest intertextuality, other texts are explicitly present in the text under analysis; they are 'manifestly' marked or cued by features on the surface of the text such as quotation marks . . . The constitutive intertextuality of a text, however, is the configuration of discourse conventions that go into its production. (Fairclough 1992: 104)

I will be primarily concerned with manifest intertextuality in this paper although the two are clearly related in academic discourse. Overt reference to specific other texts, and the response of writing to prior writing is an important constitutive feature of research articles, contributing to how we identify and evaluate research writing in different disciplines.

The importance of citation as a constitutive element of the modern academic paper can be seen in its increasingly prominent role in the ways writers seek to construct facts through their communicative practices. Historical research on scientific texts has demonstrated the gradual emergence of the Discussion section to replace Methods as the dominant basis of persuasion (Atkinson 1996; Bazerman 1988; Berkenkotter and Huckin 1995). While partly due to the standardization of experimental procedures, this is largely the result of the increasing contextualization of scientific work in disciplinary problems. Bazerman found, for example, that the number of items in reference lists had risen steadily this century from about 1.5 per article in the *Physical Review* in 1910 to more than 25 in 1980.

Further, references have become more focused, pertinent and integrated into the argument, responding to the fact that 'common theory has become an extremely strong force in structuring articles and binding articles to each other' (Bazerman 1988: 157). Citation helps to define a specific context of knowledge or problem to which the current work is a contribution, and therefore more references are now discussed in greater detail throughout the article. New work has to be embedded in a community-generated literature to demonstrate its relevance and importance and to accommodate readers' scanning patterns as they rapidly search for relevance and newness (Berkenkotter and Huckin 1995: 30).

The ways in which such reports are 'manifestly marked' in academic writing is generally seen as relatively unproblematic due to the highly developed conventions prescribed in the official manuals of such authorities as the American Psychological Association, Modern Humanities Research Association, American Chemical Society, and Council of Biology Editors. But while providing a relatively stable rhetorical context for efficient communication (Bazerman 1988), these guides convey the impression that writing is mainly a matter of applying established rules. Citation however also represents choices that carry rhetorical and social meanings, and a growing literature has revealed the availability of a wide range of signalling structures and reporting forms (for example Thomas and Hawes 1994; Thompson 1996). One strand of research has therefore sought to demonstrate the rhetorical effects of syntactic features such as thematic position, tense and voice on the reported information (for example Malcolm 1987; Shaw 1992; Swales 1990). Swales (1990: 154), for example, suggests that the present simple–present perfect–past scale, which covers over 90 per cent of finite reporting verbs, represents increasing distance of various kinds from the reported finding.

Two important attribution features of interest to researchers have been the distinction between integral and non-integral structures and the role of different reporting verbs. Integral citations are those where the name of the cited author occurs in the citing sentence while non-integral forms make reference to the author in parenthesis or by superscript numbers (Swales 1990: 148). The use of one form rather than the other appears to reflect a decision to give greater emphasis to either the reported author or the reported message. The use of a reporting verb to introduce the work of other researchers is also a significant rhetorical choice (Hunston 1993; Tadros 1993; Thomas and Hawes 1994; Thompson and Ye 1991). The importance of these verbs lies in the fact that they allow the writer to clearly convey the kind of activity reported and to precisely distinguish an attitude to that information, signalling whether the claims are to be taken as accepted or not.

The inclusion of explicit references to the work of other authors is thus seen as a central feature of academic research writing, helping writers to establish a persuasive epistemological and social framework for the acceptance of their arguments. But there are clear demarcations in the structure of subject-area knowledge systems which are reflected in the fact that different discourse communities negotiate knowledge in different ways (for example Becher 1989; Bruffee 1986; Swales 1990). Both the incidence and use of citation might therefore be expected to differ according to rhetorical contexts, influenced by different ways of seeing the world and of tackling research and its presentation. I will now set out to examine this view, briefly introducing my method, before presenting the general findings and going on to show what they reveal about disciplinary behaviour.

Procedure and corpus

The study is based on a corpus of 80 research articles, consisting of one paper from each of ten leading journals in eight disciplines (appendix): molecular biology, magnetic physics, marketing, applied linguistics, philosophy, sociology, mechanical engineering, and electronic engineering.[1] These fields were chosen to both represent a broad cross-section of academic practice and to facilitate my access to a group of specialist informants.

The journals were nominated by subject professionals as among the most important in their fields, and the articles chosen at random from issues published in 1997, taking care to select only those based on original data or new theoretical insights to allow a comparison of linguistic features.[2] The articles were scanned to produce an electronic corpus of just over 500,000 words after excluding abstracts and text associated with tables and graphics. In addition, I interviewed one experienced and well-published researcher from each discipline, typically an associate or full professor, about his or her own citation behaviours and their thoughts on disciplinary practices. The interviews were semi-structured and involved a series of open-ended interview prompts, which focused on their own and others' writing but allowed subjects to raise other relevant topics.

The corpus was computer searched for canonical citational forms such as a date in brackets, a number in squared brackets, and Latinate references to other citations (for example op cit, ibid.). This sweep left a number of citations unaccounted for however, particularly in the philosophy papers where renewed or extended discussion of a previously mentioned author often occurred without the repetition of a reference. A concordance was therefore made on all the names in the bibliographies of these articles, of third person pronouns and of generalized terms for agents, such as 'these researchers'. Possessive noun phrases such as 'Pearson's r' and 'the Raleigh-Ritz procedure' that did not integrate prior content were excluded, while those which legitimately referred to textual objects and attributed material to other sources (for example 'Gricean strictures', 'Davidson's argument') were counted. References to schools or beliefs, such as 'Platonists argue . . .', were only included if they referred to a specific author, as in 'Platonists like Bolzano argue . . .'. Finally, it should be noted that I am only concerned with references to the work of other writers in this paper and have excluded self-citation. The latter is far less central to academic argument than other-citation and, I suspect, differs considerably from it in terms of both motivation and disciplinary distribution.

Next I distinguished how authors were referred to syntactically and examined how citations were incorporated into the article, as quotation, summary or generalization from several sources. Finally I scanned the citations again to quantify the use of all main verbs associated with the authors identified through the above procedures, categorizing cases according to a modified version of Thompson and Ye's (1991) taxonomy of reporting verbs in article introductions. Note that in the following discussion I also adopt Thompson and Ye's useful convention of referring to the person citing as the 'writer' and the cited person as the 'author'.

General findings

The quantitative results show clear disciplinary differences, both in the extent to which writers rely on the work of others in presenting arguments and in how

TABLE 12.1 Rank order of citations by discipline

Rank	Discipline	Av. per paper	Per 1000 words	Total citations
1	Sociology	104.0	12.5	1,040
2	Marketing	94.9	10.1	949
3	Philosophy	85.2	10.8	852
4	Biology	82.7	15.5	827
5	Applied Linguistics	75.3	10.8	753
6	Electronic Engineering	42.8	8.4	428
7	Mechanical Engineering	27.5	7.3	275
8	Physics	24.8	7.4	248
	Totals	67.1	10.7	5,372

they choose to represent such work. Table 12.1 indicates the importance of citation in academic writing, with an average of almost 70 per paper, and also the degree of disciplinary variation. The figures broadly support the informal characterization that softer disciplines tend to employ more citations, with engineering and physics well below the average, although the frequencies for molecular biology appear to differ considerably from this general picture.

Table 12.2 shows that there was far less variation in the ways disciplinary communities refer to sources, with all but philosophy displaying a distinct preference for non-integral structures.

Integral forms tend to give greater prominence to the cited author and only the articles in philosophy, which typically consist of long narratives that engage the arguments of other writers, consistently included the cited author in the reporting sentence:

(1) Davidson ascribes to Dewey the view that . . .   (P5)

    . . . some can be analysed in the manner suggested by Lewiss, . . .
                                                                               (P4)

    Sherin (1990) argues that police agencies establish triage systems whereby . . .   (S9)

TABLE 12.2 Surface forms of citations (%)

Discipline	Non-integral	Integral	Subject	Non-subject	Noun-Phrase
Biology	90.2	9.8	46.7	43.3	10.0
Electronic Engineering	84.3	15.7	34.2	57.6	8.2
Physics	83.1	16.9	28.6	57.1	14.3
Mechanical Engineering	71.3	28.7	24.9	56.3	18.8
Marketing	70.3	29.7	66.9	23.1	10.0
Applied Linguistics	65.6	34.4	58.9	27.1	14.0
Sociology	64.6	35.4	62.9	21.5	15.6
Philosophy	35.4	64.6	31.8	36.8	31.4
Overall Averages	67.8	32.2	48.3	32.7	19.0

In the physical sciences, of course, journal styles often require numerical-endnote forms, which reduces the prominence of cited authors considerably:

(2) However, as has been analysed in a recent paper [17] dealing with the spin-spin . . .   (PY2)

The latter has been the subject of debate [13, 14].   (PY3)

Refs [12–19] work out the theory of spatial kinematic geometry in fine detail.   (ME1)

Within integral sentences, greater emphasis can be given to authors by choice of syntactic position and, once again, there seems to be a broad division between the hard and soft disciplines with the former tending to favour reporting passive or adjunct agent structures (for example *according to* . . .). Philosophers once more differed in their greater use of noun phrases and possessive forms, which are often accompanied by more than a hint of evaluation:

(3) If I guess correctly that the Goldblach conjecture is true, . . .   (P2)

We can usefully start with Stalnaker's pioneering sketch of a two-stage theory.   (P1)

. . . according to Davidson's anomalous monism, our mental vocabulary . . .   (P5)

. . . on a par with Aristotle's famous dictum that . . .   (P8)

These forms are far less common in the sciences, and also differed in function, largely acting as shorthand references to procedures rather than a means of introducing an authors' work:

(4) The Drucker stability postulate in the large regains . . .   (ME6)

Using the Raleigh-Ritz procedure, i.e. making it stationary with respect to . . .   (PY4)

. . . are evaluated with $8 \times 8 = 64$ two dimensional Gaussian quadrature formula.   (ME4)

Matthei's equations [17, 19] were first used as a starting point in the scale model . . .   (E1)

In sum, we tend to find a marked overall disposition towards non-integral and non-subject citation forms in the science and engineering papers.

Another aspect of reporting that has interested researchers is how source material is used in the writer's argument (Dubois 1988; Thompson 1996). Clearly the ways writers choose to incorporate others' work into their own, ranging from extended discussion to mandatory acknowledgement, can have an important impact on the expression of social relationships in the collaborative construction of a plausible argument. Choices here largely concern the extent to which the report duplicates the original language event, the options being use of short direct quotes (three or more words), extensive use of original wording set as indented blocks, summary from a single source, or generalization, where material is ascribed to two or more authors.

A cursory analysis of half the articles in the corpus suggests little disciplinary variation in the treatment of imported messages. Table 12.3 shows that citations were overwhelmingly expressed as summary, with generalization comprising most of the rest. Direct quotation was minimal and did not occur in any science papers. These results are clearly related to the discoursal conventions of journals and the fields they represent, ultimately pointing to the persuasive purposes of academic citation in different traditions. The way information is presented is crucial in gaining acceptance for a claim and so writers will tend to express the original material in their own terms. This means employing the cited text in a way that most effectively supports their own argument. Summary and generalization are obviously the most effective ways of achieving this as they allow the writer greater flexibility to emphasize and interpret what they are citing. In most cases the original author's words

TABLE 12.3 Presentation of cited work (%)

Discipline	Quote	Block quote	Summary	Generalization
Biology	0	0	72	38
Electronic Engineering	0	0	66	34
Physics	0	0	68	32
Mechanical Engineering	0	0	67	33
Marketing	3	2	68	27
Applied Linguistics	8	2	67	23
Sociology	8	5	69	18
Philosophy	2	1	89	8

are only likely to be carried into the new environment when writers consider them to be the most vivid and effective way of presenting their case.

Finally, I examined the choice of reporting verbs in the corpus. Over 400 different verbs were used in citations, although nearly half these forms occurred only once. The seven forms in the Averages row in Table 12.4 represent over a quarter of all citation forms used. The ratio of reporting/ non-reporting structures was fairly uniform across disciplines, although philosophers tended to employ more report verbs and physicists used fewer. As can be seen, there were substantial differences between disciplines, in both the density of reporting structures and the choice of verb forms. The table shows an enormous variation between disciplines and suggests that writers in different fields almost draw on completely different sets of items to refer to their literature. Among the higher frequency verbs, almost all

TABLE 12.4 Reporting forms in citations

| Discipline | Reporting structures | | Most frequent forms |
	per paper	% of citations	
Philosophy	57.1	67.0%	say, suggest, argue, claim, point out, propose, think
Sociology	43.6	42.0%	argue, suggest, describe, note, analyse, discuss
Applied Ling.	33.4	44.4%	suggest, argue, show, explain, find, point out
Marketing	32.7	34.5%	suggest, argue, demonstrate, propose, show
Biology	26.2	31.7%	describe, find, report, show, suggest, observe
Electronic Eng.	17.4	40.6%	propose, use, describe, show, publish
Mechanical Eng.	11.7	42.5%	describe, show, report, discuss
Physics	6.6	27.0%	develop, report, study
Averages	28.6	42.6%	suggest, argue, find, show, describe, propose, report

instances of *say* and 80 per cent of *think* occurred in philosophy, 70 per cent of *use* in electronics, 55 per cent of *report* in biology, and 53 per cent of *examine* in applied linguistics. Verbs such as *argue* (100 per cent of cases), *suggest* (82 per cent), and *study* (70 per cent) were favoured by the social science/humanities writers while *report* (82 per cent), *describe* (70 per cent), and *show* (55 per cent) occurred mainly in the science/engineering articles.

Following Thompson and Ye (1991) and Thomas and Hawes (1994), I classified the reporting verbs according to the type of activity referred to. This gives three distinguishable processes: (1) Research (real-world) Acts, which occur in statements of findings (*observe, discover, notice, show*) or procedures (for example *analyse, calculate, assay, explore*); (2) Cognition Acts, concerned with mental processes (*believe, conceptualize, suspect, view*); (3) Discourse Acts, which involve verbal expression (*ascribe, discuss, hypothesize, state*).

In addition to selecting from these denotative categories, writers also exploit the evaluative potential of reporting verbs. The taxonomy employed here diverges from Thompson and Ye's rather complex system by eliminating the need to make fine distinctions between ten sub-categories of evaluation. Despite this simplification however, it does retain their insight that while all recording of sources is mediated by the reporter, writers can vary their commitment to the message by adopting an explicitly personal stance or by attributing a position to the original author. Thus, the writer may represent the reported information as true (*acknowledge, point out, establish*), as false (*fail, overlook, exaggerate, ignore*) or non-factively, giving no clear signal. This option allows the writer to ascribe a view to the source author, reporting him or her as positive (*advocate, argue, hold, see*), neutral (*address, cite, comment, look at*), tentative (*allude to, believe, hypothesize, suggest*), or critical (*attack, condemn, object, refute*). Figure 12.1 summarizes these denotative and evaluative options.

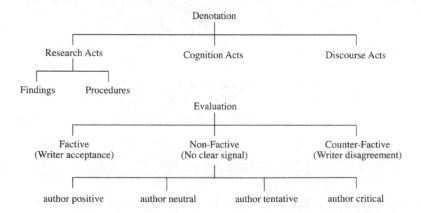

FIGURE 12.1 *Categories of reporting verbs.*

The quantitative analysis shows a fairly clear division in the denotative categories corresponding to the traditional division between hard and soft disciplines (Table 12.5). Philosophy, sociology, marketing, and applied linguistics largely favoured Discourse activity reporting verbs and the engineering and science papers display a preference for Research-type verbs. The evaluative verbs show broadly similar distributions, with verbs indicating the writer's belief in the factual status of a report (Factives) exceeded by those withholding judgement (Non-factives) in all disciplines. Only humanities/social science papers contained counter-factive examples, which represent information as unreliable. The presentation of non-factive material again revealed disciplinary distinctions with writers in the soft disciplines more likely to report authors as expressing either a positive or negative stance. Marketing writers employed a particularly high proportion of author tentative verbs, with *suggest* accounting for over half of all instances.

Discussion: Citation and disciplinary argument

The extensive use of citation in this corpus underlines the fact that, in academic writing, the message presented is always embedded in earlier messages. But while all writers drew intertextual links to their disciplines, they did so to different degrees and in different ways, and these differences reflect clear disciplinary distinctions. The differences are meaningful because citation plays such an important role in mediating the relationship between a writer's argument and his or her discourse community. Differences in rhetorical conventions can therefore suggest characteristic variations in structures of knowledge and intellectual inquiry, the regularities pointing to 'stereotypical social actions' (Miller 1984) and offering insights into the knowledge-constructing practices of disciplinary communities. Broadly speaking, these differences correspond to the traditional distinctions between hard and soft disciplines (sciences/engineering and social sciences/ humanities) (Becher 1989; Kolb 1981). Such distinctions cannot capture the full complexity of disciplinary differences, and may only be acceptable at a general level of analysis, but they do provide a useful basis for identifying dimensions of variability between these fields.

Reference to prior research clearly plays a more visible role in the humanities. Together the articles in philosophy, sociology, marketing and applied linguistics comprised two thirds of all the citations in the corpus, twice as many as the science disciplines. Writers in these fields were also more likely to use integral structures and to place the author in subject position, to employ direct quotes and discourse reporting verbs, and to attribute a stance

TABLE 12.5 Classification of reporting verbs in corpus (%)

Denotation	Biology	Physics	Elec Eng	Mech Eng	Mkting	App Ling	Sociology	Phil	Totals
Research	43.1	56.0	55.2	47.0	31.2	30.5	29.1	23.5	33.5
Cognition	7.2	6.1	2.9	1.7	7.3	10.5	6.9	14.7	8.9
Discourse	49.7	37.9	41.9	51.3	61.5	59.0	64.0	61.8	57.6

Evaluation	Biology	Physics	Elec Eng	Mech Eng	Mkting	App Ling	Sociology	Phil	Totals
Factive	26.7	15.1	16.1	27.3	19.6	20.0	16.3	15.4	19.0
Counter-Factive	0.0	0.0	0.0	0.0	1.8	1.9	3.0	2.8	1.6
Non-Factive	73.3	84.9	83.9	72.7	78.6	78.1	80.7	81.8	79.4
author positive	16.7	8.9	8.2	8.2	29.2	32.2	30.1	31.0	25.7
author neutral	60.9	76.8	69.9	76.5	35.0	48.3	47.7	39.2	49.2
author tentative	22.4	14.3	21.2	15.3	33.5	17.6	16.8	21.0	21.1
author critical	0.0	0.0	0.7	0.0	2.3	1.9	5.4	8.8	4.0

to cited authors. Writers in the hard sciences on the other hand were, with the exception of biology, less extensive in their citation practices and tended to downplay the role of the author. In the remainder of this article I consider possible reasons for these differences, drawing on interviews with expert-writer informants to do so. I suggest that they are closely bound to the social activities, cognitive styles and epistemological beliefs of particular disciplinary communities.

Contextualization and the construction of knowledge

Cognitively it has been noted that a feature of hard knowledge is its relatively steady cumulative growth, where problems are typically seen as determined by the imperatives of current interests and new findings are generated by a linear development from an existing state of knowledge (Kolb 1981; Kuhn 1970). While perhaps an ideological artifact of practitioners, such ontological representations nevertheless have important epistemological and rhetorical consequences (for example Berkenkotter and Huckin 1995). My scientist informants, for example, saw themselves as inhabiting a relatively discrete and clearly identifiable area of study and their research as proceeding along well-defined paths. The conception that their work occurs within an established framework of theoretical knowledge is reflected in scientists' routine discourse practices, as it means writers can presuppose a certain amount of background, procedural expertise, theoretical understanding and technical lexis (Bazerman 1988; Hyland, 1999). In particular, such shared assumptions allow research to be co-ordinated by reporting experiments using a highly formalized and standardized code in place of an extensive system of references to previous work.

This kind of direction and predictability is relatively rare in the humanities and social sciences however. Here new knowledge follows altogether more reiterative and recursive routes as writers retrace others' steps and revisit previously explored features of a broad landscape. In addition, issues are more diverse and detached from immediately prior developments (Becher 1989). Writers draw on a literature that often exhibits greater historical and topical dispersion, being less governed by current imperatives and less dependent on a single line of development. As my informants noted, in these circumstances research cannot be reported with the same confidence of shared assumptions:

I often cite from social psychology or organizational behaviour and other fields that my readers may not be familiar with, so I need to build a basis for what I'm saying. I don't just have to show my reasoning is dependable but also that my scholarship is too. (M interview)

I'm conscious that my work may be read by both academics and teachers and so I often lay out the background carefully. To fit maybe different audiences. (L interview)

More importantly, the literature is open to greater interpretation, findings are more frequently borrowed from neighbouring areas, and there are not the same clear-cut criteria for establishing or refuting claims. Together these differences mean that readers cannot be assumed to possess the same interpretive knowledge. Writers therefore often have to pay greater attention to elaborating a context through citation, reconstructing the literature in order to provide a discursive framework for their arguments and demonstrate a plausible basis for their claims. The more frequent citations in the soft texts therefore suggest greater care in firmly situating research within disciplinary frameworks and supporting claims with intertextual warrants.

These two broad conceptions of knowledge also result in different views about what constitutes a pertinent contribution and, indirectly, who can be cited. Scientific claims, if accepted, are generally regarded as discoveries that augment an orderly and coherent sequence of explanations in a given problem area, each fitting another block in the incremental completion of a research puzzle. This implies the assimilation of prior claims by new. As a result, a reader is unlikely to find Einstein, Oppenheimer, or Planck cited in a physics paper:

This knowledge is assumed. It is not that we reject them but it is just well-known facts. My personal view of science is of a huge volcano and lava is flowing down and I'm at the end of one stream of lava. Nobody cites volcanoes in their papers. (PY interview)

Citation in the hard disciplines is therefore a means of integrating new claims into current knowledge while drawing on it as supporting testimony, situating the new work in the scaffolding of already accredited facts. References, particularly in physics, therefore tend to be tightly bound to the particular research topic under discussion (see also Bazerman 1988: 164), which closely defines a specific context of knowledge and contributes to a sense of linear progression. Intertextuality provides persuasive support by demonstrating the current work as 'valid science': the precedent providing a forceful warrant for current innovation. In the humanities and social sciences on the other hand, the fabric of established understandings has a wider weave. Problem areas and topics are generally more diffuse and range over wider academic and historical territory, and there is less assurance that questions can be answered by following a single path. Thus the substantial differences in citation rates between broad academic fields to some extent reflect the extent to which a context can be confidently assumed to be shared by readers. In soft domains, on the other hand, old ground is re-crossed and reinterpreted rather than suppressed.

The process of coming to terms with the complexity of human behaviour is perceived as less obviously progressive and therefore less likely to discard older ideas as obsolete or irrelevant. As my sociology informant observed, 'good theory doesn't date . . . Durkheim is a cottage industry. People promote or pan his ideas but he's still there because we can't say for sure whether he's right or wrong'. As a result, disciplinary giants are frequently encountered in the soft papers, particularly in 'pure' knowledge fields, where the pathfinders' stocks of relevance are clearly greater:

(5) . . . because it runs counter to the bureaucratic ideal of efficiency (for example Weber 1946).   (S9)

Marx located barricades at the core of conspiratorial movements . . .
                                                                         (S7)

The first is derived from Durkheim's (1938) notion that there is a general
. . .   (S9)

However, both Piagetian and Vygotskian thinking involve constructivist
. . .   (L10)

Wittgenstein insists that what is true or false is what people say, . . .
                                                                        (PL5)

John Stuart Mill taught us to recognize that informal social pressure can restrict . . .   (P10)

Aristotle had a point when he defined humans as language users.
                                                                         (P7)

Agency and epistemology in reporting

The quantitative findings presented above may also be explained in terms of disciplinary dispositions to either acknowledge or suppress the role of human agency in constructing knowledge. As noted above, what constitutes valid claims and admissible reasoning differs between disciplines, and these values and epistemologies are instantiated in aspects of a community's genre conventions. An important aspect of the positivist-empirical epistemology that characterizes a great deal of scientific endeavour is that the authority of the individual is subordinate to the authority of scientific procedure. While these procedures may often be named after their originators (see, for example, 4), this does not directly acknowledge the role of individuals in creating knowledge, but functions as an insider code to situate current work within a framework of shared methodological understandings.

Although many scientists may have perceived the achievement of absolute truth to be an illusory goal, their discursive practices are nevertheless guided by empiricist beliefs. Articles in the hard sciences still suggest that knowledge is accomplished by the correct application of prescribed procedures, and that nature reveals itself directly through scientific method. In this perspective human judgement as a mediating link in the interpretation of data is downplayed, descriptions of phenomena are depicted as representing a reality independent of the observer, and empirical methods are reified in the conventions of scientific narrative. Scientists act as if they see themselves as discovering truth, not making it.

The conventions of impersonality in science articles play an important role in reinforcing this ideology by portraying the legitimacy of hard science knowledge as built on socially invariant criteria. While seeking to establish their own reputations through publication and the recognition bestowed by citation, writers routinely (and often unreflectingly) also subscribe to the assumption that the person who publishes a claim is largely immaterial to its accuracy. The author is merely 'a messenger relaying the truth from nature' (Gilbert 1976: 285). This not only helps to account for the relatively low incidence of citation in the physics and engineering corpus, but also for the predominance of non-integral structures and perhaps also the overwhelming use of the footnote format:

(6) Silicon based methods (Buttenbach (1994)) as well as LIGA (Ehrfeld (1990)) are the major process families in this context.   (E2)

Furthermore, it has been shown [103] that the fundamental dynamic range of . . .   (E7)

As already observed by others [17], T1 was found to be . . .   (PY3)

. . . power gradient linearity in Ref (1) may be partly due to the choice of target fields . . .   (PY5)

. . . Melan theory exhibits the same conceptual structure as the classical one [1–3].   (ME6)

. . . has been summarized by various authors (2, 8, 16) and is still being re-analysed.   (ME3)

The suppression of authors as agents has also blurred Swales' original integral versus non-integral distinction and led to the emergence of a variant form of reporting structure. Examples such as these are fairly common in the corpus:

(7) The angle x must be smaller than x/2 and is described by [3]   (PY3)

According to ref. [11] the coupling parameters in the free electron . . .
    (PY1)

Ref. [9] developed finite formulations and corresponding code.   (PY4)

. . . using the highly efficient techniques described in [22], [23], and [25].
    (E3)

. . . properties of a line trajectory in spatial motion are researched by Refs [21-23], . . .   (ME5)

Reference [20] presents a unified theory of kinematic synthesis to solve the problem.   (ME2)

References [4, 5] reveal points with special kinematic meanings in the main body.   (ME1)

Reference [6] shows a spatial Euler-Savaray analogue based on velocity and . . .   (ME1)

While not all my scientist informants were comfortable with this hybrid pattern, its frequency testifies to a certain acceptability, particularly in physics and mechanical engineering. Its use also underlines the considerable variation in how reporting verbs are used in different disciplines.

Thus, together with relatively low citation rates and high use of non-integral forms, these kinds of patterns help to convey epistemological assumptions that give little space to those whose contributions are cited. Removing the agent helps remove the implication of human intervention, with all the influences of personal interest, social allegiance, faulty reasoning and other distorting factors beyond the empirical realm which that might suggest. These citation practices therefore help maintain the legitimacy of scientific knowledge as built on non-contingent pillars such as strict procedures, replication, falsification, and rigorous peer review in the process of publication.

Molecular biology, however, differs significantly from these fields. Although it is a 'hard' science, and presumably shares the ethos and commitments sketched above, it has the greatest density of citations in the corpus, with three times as many attributions as physics, and also the highest proportion of author subjects among the sciences. While the reasons for these differences are unclear, they appear to reflect the distinctive ways that biology pursues and argues problems and understands the scientific endeavour (cf. Chargaff 1974).

In many ways molecular biology is neither fully established nor prototypically 'scientific'. It is a relatively new discipline, with perhaps less cohesive research networks, and its methods are more descriptive, relying to a greater extent on 'beautiful models' than either physics or chemistry (Kellenberger 1989). The personalities of biology, the creators of its speculations and discoveries, have also tended to assume greater importance than in other hard sciences, both inside and outside the discipline (Judson 1995; Watson 1968). Darwin, Bragg, Pauling, Luria, and Crick are perhaps among the most well known academics of any field. Halloran (1984) has argued that this is the result of an entrepreneurial spirit in the discipline, a notion of scientific knowledge as private property that originated with Watson and Crick's seminal 1953 paper that simultaneously offered a model of DNA and a model of the scientist:

> Both argumentatively and stylistically Watson and Crick put forward a strong proprietary claim to the double helix. What they offer is not *the* structure of DNA or *a* model of DNA, but Watson and Crick's structure or model. (Halloran 1984: 75)

The proclivity for citation in molecular biology, and for the exceptional scientific emphasis on integral reporting structures, might therefore be seen as an indication of a disciplinary ethos that emphasizes proprietary rights to claims. Admittedly this claim is purely speculative and more research is required to account for this difference. However, my data suggests that constructing knowledge in biology seems to involve rhetorical practices that give greater weight to who originally stated the prior work, rather than the traditional conventions of impersonalization still observed in the other hard disciplines studied here.

Writers in marketing also appear to give significant recognition to the ownership of ideas, with high levels of reporting, and subject position author names. A marketing professor admitted that there was a certain self-advocating tendency among practitioners, 'Some people really promote their work and the area of research that they are in, both behind the scenes and in their papers.' While it is difficult to draw strong connections, this may be related to the involvement of large numbers of marketing academics in corporate consultancies, a source of increasing influence on research. 'A lot of the research we do comes from real-world or corporate problems and even if it doesn't originate there, the ultimate goal is that it should end up there' (M interview). Such dual interests increase the possibility of overlap between research and commercial values, with attribution practices becoming influenced by the norms of ownership and competition more typically associated with the marketplace.

The effects of professional and workplace contexts on academic literacy practices are largely unknown, but are clearly pertinent. Writing is collective, co-operative persuasion and occurs within communities bound together by shared assumptions about the nature of the world, how to hold ideas, and how to present them to peers. This social basis of knowledge means its authority originates in the groups who comprise the audience for texts, who both shape this knowledge and render it intelligible. The participation of academics from applied fields such as engineering, applied linguistics, and marketing in their respective public arenas of communication is therefore likely to have consequences for their discursive behaviours. Academic 'forums of competition' (Toulmin 1958), within which new concepts are appraised, become blurred with those of a more applied orientation as members are influenced by the problems, procedures, and criteria of evaluation which emerge from, and are relevant to, workplace concerns and practices. The effects of such interactions on citation conventions remains to be studied.

Another discipline with high author visibility is philosophy, where citation plays a very different role to the one it plays in the hard sciences. Here knowledge is constructed through a dialogue with peers in which perennial problems are recycled through personal engagement:

> Citing allows you to debate with others, the questions have been around a long time, but you hope you are bringing something new to it. You are keeping the conversation going, adding something they haven't considered. . . . You know most of them anyway, you read them and they read you. (P interview)

Bloor (1996: 34) refers to philosophical rhetoric as essentially 'mind-to-mind combat with co-professionals', and the extensive use of citation helps to achieve this high degree of personal involvement among protagonists. Many citations are thus repeated references in a protracted debate or draw on the reader's shared knowledge of an author's views without referring to a specific text. To emphasize the immediacy of the argument and its relevance to current concerns, they are usually presented in the present tense. These few random extracts give some flavour of this:

(8) I disagree sharply with Rawles on the matter of . . . . (P10)

My main critique of Maudlin's solution is that . . . (P6)

Nor can I see how Donnelan's syntactically simpler paradigm and my (for example 3) differ . . . (P1)

Davidson and Wittgenstein are alive to this possibility. (P5)

I have focused on Smith's formulation of the revised sceptical hypothesis
. . .   (P3)

This conventional dispute structure often goes beyond the writer's response
to the paraphrased arguments of an adversary to an imagined dialogue
where claims are provided on their behalf. In the absence of an actual
counterargument, philosophers may strengthen their position by inventing
one and attributing it with a hypothetical citation:

(9) It might be suggested (perhaps by someone like René Descartes) that
     the problem . . .   (P2)

     Wittgenstein would argue that this term expresses . . .   (P5)

     Now Rawls could say that his concern, too, is distributive justice . . .
                                                                             (P10)

     If Churchland intends to say that . . .   (P7)

Clearly these citation practices are not supporting the writer's claim to be
extending the thread of knowledge from what has been previously established,
but helping to position the writer in relation to views that he or she supports or
opposes. Scollon (1994) has argued that citing the work of others is not simply
an issue of accurate attribution, but also a significant means of constructing
an authorial self. Writing in the humanities stresses the individual creative
thinker, but always within the context of a canon of disciplinary knowledge.
Foregrounding the names of those whose work we engage with enables us
to establish a professional persona. This was mentioned explicitly as a reason
for citing by the sociologist:

I've aligned my self with a particular camp and tend to cite people from
there. Partly because I've been influenced by those ideas and partly
because I want them to read my work. It's a kind of code, showing where
I am on the spectrum. Where I stand. (S interview)

Ethos and evaluation: The use of reporting verbs

The distribution of reporting verbs in the corpus also reveals broad disciplinary
differences. There appear to be community-based preferences, both for
specific items and the implications carried by particular semantic categories.
While more work needs to be done in this area, it is possible that these choices
serve to reinforce the epistemological and social understandings of writers by

conveying an orientation to a particular ethos and to particular practices of social engagement with peers.

The finding that the humanities and social science articles contained far more, more varied, and more argumentative reporting verbs is partly a function of their greater need to elaborate a shared context. As discussed above, research in any field has significance only in relation to an existing literature, and citation helps demonstrate accommodation to this community knowledge. In the soft fields, convincing readers that an argument is both novel and sound may often depend on the use of reporting structures not only to build a shared theoretical basis for one's arguments, but to establish a common perspective on the reliability of the claims one reports. Writers have to construct an epistemic as well as a disciplinary context. The different epistemological structure and social organization of the hard sciences on the other hand often allows writers to assume more common ground with readers, requiring less need to demonstrate the relevance and reliability of prior studies using reporting verbs.

In addition to a heavier rhetorical investment in contextualization, the greater use of reporting verbs in the soft fields also reflects the more discursive character of these disciplines. Briefly, reporting verbs are more appropriately employed in an argument schema that more readily regards explicit interpretation, speculation and complexity as legitimate aspects of knowledge. The soft disciplines typically examine relationships and variables that are more numerous, less easily delineated and more subject to contextual and human vagaries than those studied in the hard sciences. Causal connections, conclusive demonstration, and depictions of feature are less easily established in the humanities (Kolb 1981). One reason for the use of a wider repertoire of citation verbs therefore is simply that they facilitate qualitative arguments that rest on finely delineated interpretations and conceptualizations, rather than systematic scrutiny and precise measurement. What appear at first sight to be stylistic proclivities for uniform choices in the hard sciences then, may actually reflect the different procedures, subject matter, epistemological understandings and research perspectives that characterize those fields.

The scientific ideology that perceives laboratory activity as impersonal, cumulative and inductive also helps to explain the relatively high frequency of Research verbs found in the science/engineering corpus. These comprised about half of the denotation choices in biology, physics, and the engineering disciplines, and between a quarter and a third of those in the soft disciplines. This emphasis on real-world activities helps to convey the experimental explanatory schema typical of the sciences, where knowledge is more likely to be represented as proceeding from laboratory activities than the interpretive operations or verbal arguments of researchers.

(10) Edson *et al.* (1993) <u>showed</u> processes were induced only after the cells
were treated . . .   (B1)

. . . linear and non-linear distraction <u>observed</u> in LC delay lines [2]. (E7)

a 'layer' coupled-shot finline structure was <u>studied</u> by Mazur [7] and
Tech *et al.* [8] . . .   (PY4)

. . . <u>using</u> special process and design [42], or by <u>adding</u> [101], or <u>removing</u>
[83] a mask.   (E7)

References [7, 8] <u>developed</u> instantaneous invariance via point
coordinates for the . . .   (ME1)

This emphasis was particularly evident in the physics and engineering
papers, which together contained only nine cognition verbs, thereby
camouflaging the role of author interpretation in the research process.
Applied linguistics and, in particular, philosophy, however, used such verbs
extensively, underscoring the part that reasoning and argument play in the
construction of knowledge:

(11) Acton (1984) <u>sees</u> preparing students psychologically as a . . .   (L4)

Parry <u>concluded</u> that by reading less, this student was encountering
fewer new words . . .   (L2)

Some writers, for example Adams (2) <u>think</u> that . . .   (P4)

Donnelan <u>believes</u> that for most purposes we should take the
demonstratum to be . . .   (P1)

Report verbs however do not simply function to indicate the status of
the information reported, but the writer's own position in relation to that
information. The selection of an appropriate reporting verb allows writers to
intrude into the discourse to signal an assessment of the evidential status
of the reported proposition and demonstrate their commitment, neutrality
or distance from it. There was, however, little difference in how writers
used verbs from the major categories in this corpus, although only soft
disciplines employed counter-factive verbs, representing reported material
critically:

(12) His revisionist interpretation of Twiggy <u>overlooks</u> historical research . . .
                                                                                (M8)

In addition, he <u>fails</u> to fully acknowledge the significance of . . .   (S4)

Lillian Faderman has also probably <u>exaggerated</u> the pervasiveness of . . .
                                                                                (P5)

Churchland did not simply <u>misuse</u> the word 'theory'.   (P7)

Generally writers in all fields tended to indicate their positions to cited material more indirectly, by ascribing an attitude to authors. These imputed positions showed clear disciplinary variations, with scientists and engineers overwhelmingly representing authors as conveying a neutral attitude to their findings. This conveys a detached and impartial reporting style to these papers, reflecting the need to build a convincing argument by simply displaying an awareness of prior or parallel research without appearing to corrupt it with personal judgement:

(13) . . . the relevant theory was <u>developed</u> by Bruno [11].   (PY1)

These ornamentations were <u>described</u> by Schenck *et al.* (1984) as . . .
                                                                              (B2)

Yeh *et al.* [7] <u>reported</u> that a typical force . . .   (E5)

. . . other solutions have also been <u>published</u> [5–7]   (E6)

Paiva and Venturinit (9), <u>presented</u> an alternative formulation . . . (ME4)

·At Liverpool, simulated seismic tests have been <u>performed</u> on plain pipes [9] . . .   (ME3)

Writers in the soft disciplines on the other hand were far more likely to depict authors as adopting a particular stance towards their work, either presenting their view as true, false or tentatively correct:

(14) Baumgartner and Bagozzi (1995) strongly <u>recommend</u> the use of . . .
                                                                              (M9)

Law and Whitley (1989) <u>argued</u>, for instance, that . . .   (L7)

However, both Davidson and Wittgenstein explicitly <u>disown</u> the view
. . .   (P5)

. . . the idea of a characterless substructure is <u>rejected</u> by Aristotle in *Metaphysics*.   (P6)

Tarone (1978) <u>suggests</u> that there is . . .   (L4)

Kubiak <u>hints</u> that the Polish tradition . . .   (S3)

They were also more likely to evaluate this attributed position by adding adverbial comment, although this was not common and was largely restricted to philosophy:

(15) He argues there, correctly to my mind, that . . .   (P2)

Churchland correctly rejected this move . . .   (P7)

As Dipankar Gupta correctly asserts . . .   (S3)

Clearly these explicitly evaluative strategies are better suited to the more disputational style of argument favoured by the humanities, as they allow writers to open a discursive space within which to either exploit their opposition to the reported message or to build on it. Establishing intertextuality by attributing a view to another also allows this to be done in a dialogic way, by engaging the scholarship of the discipline through a discourse with those who have created it. Once again, then, we find that textual conventions point to distinctions in how knowledge is typically negotiated and confirmed within distinct academic communities, facilitating the different ways writers are able to link their local contributions into a wider disciplinary framework of expectations.

Conclusion

Reference to previous work is virtually mandatory in academic articles as a means of meeting priority obligations and as a strategy for supporting current claims. But how writers choose to present information is as important as the information they choose to present. The disciplinary differences discussed here suggest that the imperatives that motivate citations are contextually variable and are related to community norms of effective argument. The fact that academics actively engage in knowledge construction as members of professional groups means that their discoursal decisions are socially grounded, influenced by the broad inquiry patterns and knowledge structures of their disciplines. How an academic community defines its field of inquiry and understands the material it investigates, the ways it conceptualizes and tackles research issues, and the metaphors it employs to characterize knowledge, are all matters of social agreement or contestation. They also contribute to how writers choose to frame their studies for colleagues, relying on a sprinkling of citations to invoke a set of common understandings through to an elaborate scaffold of supporting references.

Such practices cannot, of course, be seen as entirely determined, language users are not simply passive recipients of textual effects. But the impact of citation choices clearly lies in their cognitive and cultural value to a community, and each repetition helps instantiate and reproduces these conventions. The broad distinctions explored in this article therefore provide further support for the view that our routine and unreflective writing practices are deeply embedded in the epistemological and social convictions of our disciplines.

Finally, the study has both raised and neglected a number of issues that provide fruitful ground for further research. Clearly there are attribution practices in other disciplines and genres that we know little about, and it is also probable that closer analyses will reveal variations in sub-genres and sub-disciplines. The comparative study of citation patterns in theoretical and experimental papers, for example, or of established and emergent fields within a discipline, may reveal interesting differences. Similarly, the study of self-citation and how it interacts with both other-citation and writer purposes is likely to be profitable. A further important area that I have not addressed concerns the conventions used to interpret citations, particularly the degree of community or individual variation in determining their scope and specificity.[3] Exactly what part of a proposition is a particular citation understood to refer to? What interpretive understandings do community members employ in their readings of cited material? Are there contextual variations in the licence granted to writers in representing prior work? These questions not only suggest intriguing areas of future study, but may help us to sharpen our understanding of how discourse practices contribute to the construction of academic knowledge.

Acknowledgements

I would like to thank my informants for their time, patience and interest; Chloe Chiu for her help in compiling the corpus; and the anonymous reviewers of *Applied Linguistics* for their insightful comments.

Appendix. Journal corpus

Applied linguistics (L)

1. *Applied Linguistics*
2. *TESOL Quarterly*
3. *Second Language Research*
4. *System*
5. *English for Specific Purposes*
6. *World Englishes*
7. *Journal of Second Language Writing*
8. *Journal of Pragmatics*
9. *Written Communication*
10. *International Journal of Applied Linguistics*

Electronic engineering (E)

1. *Int J. of Microwave and Millimeter-Wave CAE*
2. *Microsystem Technologies*
3. *IEEE Transactions on Microwave Theory and Techniques*
4. *Journal of Microelectromechanical Systems*
5. *Solid-state Electronics*
6. *Microelectronics Journal*
7. *Analog Intg. Circuits and Signal Processing*
8. *J. of Manufacturing Science and Engineering*
9. *International Journal of Production Research*
10. *International J. of Industrial Engineering*

Mechanical engineering (ME)

1. *Mechanism and Machine Theory*
2. *Energy Sources*
3. *J. of Process Mechanical Engineering*
4. *Mechanics and Material Engineering*
5. *Journal of Engineering Manufacture*
6. *Int. J. of Mechanical Sciences*
7. *J. of Mechanical Engineering Science*
8. *Energy Engineering*
9. *International J. of Energy Research*
10. *J. of Energy Resources Technology*

Marketing (M)

1. *Journal of Marketing Management*
2. *International Journal of Research in Marketing*
3. *Journal of Marketing Research*
4. *Journal of Marketing*
5. *Journal of the Academy of Marketing Science*
6. *Journal of Marketing Communication*
7. *Journal of International Consumer Marketing*
8. *Journal of Consumer Research*
9. *Journal of Retailing*
10. *Marketing Science*

Philosophy (P)

1. *Mind*
2. *The Journal of Philosophy*
3. *Analysis*
4. *The Philosophical Quarterly*
5. *Philosophy*
6. *Erkenntnis*
7. *Inquiry*
8. *Political Theory*
9. *Ethics*
10. *Philosophy and Public Affairs*

Cell and molecular biology (B)

1. *Journal of Cell Biology*
2. *Mycological Research*
3. *The Plant Cell*
4. *Plant Molecular Biology*
5. *Plant, Cell and Environment*
6. *Molecular and Cellular Biology*
7. *Mycologia*
8. *The New Phytologists*
9. *Canadian Journal of Botany*
10. *Plant Physiology*

Sociology (S)

1. *American Journal of Sociology*
2. *The Sociological Review*
3. *Current Sociology*
4. *International J. of Comparative Sociology*
5. *Sociology*
6. *International Sociology*
7. *British Journal of Sociology*
8. *British Journal of Criminology*
9. *Criminology*
10. *International J. of the Sociology of Law*

Physics (PY)

1. *J. of Magnetism and Magnetic Materials*
2. *Bulletin of Magnetic Resonance*
3. *Applied Magnetic Resonance*
4. *Electromagnetics*
5. *J. of Magnetic Resonance (B)*
6. *J. of Electromagnetic Waves and Applications*
7. *Journal of Material Science*
8. *Journal of Applied Physics*
9. *Physical Review B*
10. *American Journal of Physics*

Notes

1 Examples and interviews are coded as P = Philosophy, S = Sociology; M = Marketing; L = Applied Linguistics; ME = Mechanical Engineering; E = Electrical Engineering; PY = Physics; B = Biology.

2 I should emphasize here that the analysis is cross-sectional rather than longitudinal, and culturally specific rather than comparative. It offers a snapshot of current practices of writers publishing in English and, while I readily acknowledge the historical and cultural limitations of this snapshot, the data allows me to say little about publishing practices at other times or in other cultures.

3 I am grateful to one of the anonymous reviewers for this point.

13

Humble servants of the discipline? Self-mention in research articles

1 Introduction

Until fairly recently scientific and academic writing was seen as largely objective reporting of an independent and external reality. As long as the researcher used language skillfully and avoided personal bias, then the reader could simply decode the transmitted message to recover the same reality. In contrast to this Shannon and Weaver's (1963) view of communication, a great deal of research has recently shown how professional writers seek to achieve successful interaction with their readers while maintaining the integrity of their data. Hedges (Hyland, 1996a, 1998a), reporting conventions (Hyland, 2000a; Thomas & Hawes, 1994), and evaluation (Hunston, 1993; Thompson & Ye, 1991) are among the features that have been examined for the ways such writer–reader interactions are realised in journal articles. Perhaps surprisingly, the vexed topic of self-mention has received considerably less attention. This issue remains a perennial problem for students, teachers, and experienced writers alike, and the extent to which one can reasonably explicitly intrude into one's discourse, or assert one's personal involvement, remains highly controversial.

In this paper I investigate the extent, forms, and functions of self-mention in a corpus of 240 research articles in eight disciplines to help unravel some of the myths and misperceptions about this topic. In particular, I seek to reveal something of how self-mention is used and perceived as a way of understanding more about writing in the disciplines and the kinds of options available to our students. I begin with a brief outline of the issue, presenting an overview of how impersonality is seen among style manual writers, applied linguists, and composition scholars, then go on to discuss the main features of its realization, examining patterns of exclusive first person pronoun use and self-citation.

2 Impersonality and its discontents

The convention of impersonal reporting remains a hallowed concept for many, a cornerstone of the positivist assumption that academic research is purely empirical and objective, and therefore best presented as if human agency was not part of the process. Albert Einstein (1934, p. 113), for example, wrote, "when a man is talking about scientific subjects, the little word 'I' should play no part in his expositions". A more rhetorically grounded, but closely related, view is that which stresses the persuasive authority of impersonality. Here it is seen as a strategy that maximizes the credibility of the writer and works to elicit credence from the reader. Lachowicz (1981, p. 111) for example, argues that impersonality emphasizes "objectivity, open-mindedness, and the established factual nature of a given activity", it functions to underline the "common share of knowledge with the community", and stresses the collective responsibility of academic endeavour.

Eradication of the self is therefore seen as demonstrating a grasp of scholarly persuasion as it allows the research to speak directly to the reader in an unmediated way. This is a view proposed in many style manuals and textbooks. Rowntree (1991), for example, advises caution in use of the first person, while Spencer and Arbon (1996, p. 26) recommend complete abstention. Similar comments are not hard to find in the pedagogic literature:

> Write your paper with a third person voice that avoids 'I believe' or 'It is my opinion'. (Lester, 1993, p. 144)

> In general, academic writing aims at being 'objective' in its expression of ideas, and thus tries to avoid specific reference to personal opinions. Your academic writing should imitate this style by eliminating first person pronouns . . . as far as possible. (Arnaudet & Barrett, 1984, p. 73)

> Since science is concerned primarily with the objective and the impersonal, the passive point of view, though devoid of human interest and color, is

ordinarily proper for accounts of scientific processes. (Jones & Keene, 1981, p. 125)

Underlying this view seems to be an assumption that academic persuasion is essentially an issue of accommodation, and that humility, towards one's peers, one's reviewers, or the discipline in general, represents the best means of gaining acceptance of one's claims. The research article is thus regarded as a modest, self-effacing genre in which the writer acts as a humble servant of the discipline. There is, of course, some truth in this. Publishing in academic journals demands that the author displays some degree of disciplinarity, that he or she demonstrates a familiarity with the rhetorical conventions and social understandings of the community, and observes suitable patterns of social and rhetorical interactions (e.g. Hyland, 2000a; Myers, 1989). Arguments have to be made in ways that readers find most acceptable and convincing, and research claims framed to project appropriate certainty and maximum plausibility.

However, not all discourse communities employ the same conventions and readers in different fields have different expectations and norms of argument. Thus, because articles are sites of disciplinary engagement, where writers interact with specialist audiences rather than with general readers, admonishments to avoid self-mention are sometimes said to be misguided (e.g. Wilkinson, 1992).

In other words, research writing involves writers in a process of both textualizing their work as a contribution to the field, and in constructing themselves as plausible members of the discipline, competent to make such a contribution (e.g. Ervin, 1993; Pare, 1993). But clearly writers have to do more than display legitimacy; they have to say something new (Berkenkotter & Huckin, 1995; Kaufer & Geisler, 1989). This means that demonstrating solidarity with the community and showing respect for its common goals is only part of the story, and that writers must carefully balance this with vigorous argument for the originality of their claims and by the display of an authoritative professional persona. While claims have to be warranted by appropriate support and reference to existing knowledge by fitting novelty into a community consensus, success in gaining acceptance for innovation also involves demonstrating an *individual contribution* to that community and establishing a claim for recognition for *academic priority.*

So while impersonality may often be institutionally sanctified, it is constantly transgressed. This is generally because the choices which realise explicit writer presence also contribute to a high degree of ego-involvement (Chafe, 1985), and are closely associated with authorial identity and authority. All writing carries information about the writer, and the conventions of personal projection, particularly the use of first person pronouns, are powerful means for self-representation (Ivanic, 1998; Ivanic & Simpson, 1992). Authority, as I noted above, is partly accomplished by speaking as an insider, using the codes and the identity of a community member (e.g. Bartholomae, 1986, p. 156). But it also relates to the writer's convictions, engagement with the reader, and personal presentation of 'self'. Cherry (1988) uses the traditional rhetorical concepts of *ethos* and *persona* to represent persuasiveness as a balance between these two dimensions of authority: the credibility gained from representing oneself as a competent member of the discipline, and from rhetorically displaying the personal qualities of a reliable, trustworthy person.

Presenting a discoursal self is central to the writing process, as Ivanic (1998, p. 32) has made clear:

Writing is a particularly salient form of social action for the negotiation of identities, because written text is deliberate, potentially permanent and used as evidence for many social purposes (such as judging academic achievement).

Writers cannot avoid projecting a particular impression of themselves and how they stand in relation to their arguments, their discipline, and their readers, and this can have an important impact on the outcome of their discoursal purposes.

Kuo (1999) points out that the strategic use of personal pronouns allows writers to emphasize their own contribution to the field and to seek agreement for it. Personal reference sends a clear indication to the reader of the perspective from which their statements should be interpreted. For this reason, self-mention is often seen positively in the literature. The *Manual on Scientific Writing* (1993), for instance, encourages writers to employ the first person, as does the authoritative *Council of Biology Editors Style Manual* (1978, p. 5), which advises writers to shun 'the passive of modesty' and suggests that 'the first person (I, we) is natural for relating what you did'. Several influential style guides also echo this view:

I herewith ask all young scientists to renounce the false modesty of previous generations of scientists. Do not be afraid to name the agent of the action in a sentence, even when it is 'I' or 'we'. (Day, 1994, p. 166)

. . . the scientific attitude is not achieved by either the use or the avoidance of a particular pronoun. Rather, is achieved through the qualities mentioned earlier: honesty, care in handling facts, dignity, and restraint in manner. (Mills & Water, 1986, pp. 32–33)

Because of this conflicting advice, the 'voice' writers choose to employ, the position they adopt to their claims, their readers, and their communities, is a perennial source of difficulty for both native speaker and second language students (e.g. Cadman, 1997; Connor, 1996). Chang and Swales (1999, p. 164), for instance, observe that "feelings and reactions can be both strong and unpredictable" on the use or avoidance of first person pronouns. They found that a group of 37 graduate students felt decidedly uncomfortable with the first person, and comments like "nobody likes to use it in a formal paper" and "only usable for senior scholars" were typical in their data. Cadman (1997, p. 8) argues that the central problem for international students lies in the gap between the epistemological orientations of different cultures; students are unclear about "who they are expected to be" and are unable to establish their own position in their writing. The absence of clear direction in their pedagogic texts, and conflicting expectations among their supervisors and teachers, simply exacerbates these difficulties.

In sum, the linguistic choices writers make not only affect the conceptual or ideational meaning that they convey, but can also influence the impression they make on their readers. The decision to adopt an impersonal rhetorical style or to represent oneself explicitly would seem to have significant consequences for how one's message is received. Indeed, the intrusion of authorial authority to limit claims, enhance plausibility, and promote personal credibility can play an important role in securing acceptance for academic arguments. I will now try to throw some light on the role of personal intrusion in academic writing and to provide an empirical basis for the advice we might wish to give our students.

3 Method and corpus

The study employs both qualitative and quantitative approaches, comprising frequency counts and text analysis of a corpus of published articles and a series of interviews with academics from the relevant discourse communities The text corpus of 240 research articles consists of three papers from each of 10 leading journals in eight disciplines selected to represent a broad cross-section of academic practice. These are mechanical engineering (ME), electrical engineering (EE), marketing (Mk), philosophy (Phil), sociology (Soc), applied linguistics (AL), physics (Phy) and microbiology (Bio). The journals were nominated by expert informants as among the leading publications in their fields, and the articles chosen at random from 1997 and 1998 issues.

The texts were scanned to produce an electronic corpus of 1,400,000 words and searched for expressions of self-mention using Word Pilot, a text analysis programme. The search items were the first person pronouns *I, me, my, we, us,* and *our,* cases of self-citation and references to work conducted elsewhere by the same authors, and examples of self-mention terms such as *this writer* or *the research team*. All cases were examined in context to ensure they were exclusive first person uses and to determine their syntactic position and pragmatic function. All forms of *we, us* and *our* which referred to participants other than the writers were eliminated. The interviews were conducted with experienced researcher/writers from the target disciplines using a semi-structured format of open-ended interview prompts which focused on their own and others' writing, but allowed subjects to raise other relevant issues (cf. Cohen & Manion, 1994, p. 293). Subjects could therefore respond to texts as readers with insider community understandings of rhetorical effectiveness, while also discussing their own discoursal preferences and practices.

4 Frequencies and forms of self-mention

The two most striking features of the corpus are the saliency of self-mention in the articles and the variety of its disciplinary and formal expression. Table 13.1, which gives the frequencies for each figure normalised to a text length of 10,000 words, clearly shows that academic writing is not the

TABLE 13.1 Frequency of self-mention forms per discipline (per 10,000 words)

Discipline	Total	Citation	I	Me	My	We	Us	Our	Other
Physics	64.6	8.7	0.0	0.1	0	39.3	0.6	14.4	1.4
Marketing	61.3	6.9	1.6	0.0	0.7	31.0	1.1	18.9	0.6
Biology	56.2	22.6	0.0	0.1	0.1	24.0	1.1	7.2	0.7
Philosophy	52.7	3.1	35.6	2.5	7.7	1.4	0.2	0.6	0.0
App ling	51.8	9.1	36.1	3.0	9.7	25.4	2.8	14.5	0.2
Sociology	47.1	6.8	12.7	1.0	2.0	15.3	0.7	7.6	0.2
Electronic eng	44.4	10.7	0.0	0.0	0.0	23.3	0.4	8.6	0.5
Mechanical eng	17.8	9.6	0.0	0.0	0.0	5.5	0.0	1.4	0.4
Overall	50.5	8.3	11.2	0.8	2.4	17.8	0.8	8.3	1.1

faceless, formal prose it is often depicted to be. While research articles may well be characterised by abstraction and high informational production (Biber, 1988), human agents are integral to their meaning. There are sufficient cases of author-reference to suggest that writers have conspicuous promotional and interactional purposes, with every article containing at least one first person reference. Overall, there were roughly 28 expressions of self-mention in each paper; 81% of these were pronouns, 16% were self-citations, and 2% were other mentions to the authors of the paper. The pronouns *we* and *I* were the most commonly used devices for self-representation in these texts.

Although the overall frequencies per 10,000 words show no obvious correlation with the traditional distinction of hard-soft knowledge domains (sciences/engineering and social sciences/humanities) (Becher, 1989; Kolb, 1981), there are considerable differences between the disciplines represented here (Table 13.2). At one extreme, there was an average of 44 cases per article in marketing, and only seven in mechanical engineering. Self-mention was particularly dense per 10,000 words in physics, marketing, and biology, and while mechanical engineers referred to themselves far less often than writers in other fields, they relied quite heavily on self-citation in linking their work into

TABLE 13.2 Average frequency of self-mention per paper

Discipline	Totals	Self-citations	Pronouns	Other
Biology	26.9	10.8	15.5	0.5
Physics	21.0	2.8	17.7	0.5
Electronic eng	15.9	3.8	11.6	0.5
Mechanical eng	6.8	3.7	2.6	0.5
Average hard fields	17.6	5.3	11.9	0.5
Marketing	43.9	4.9	38.2	1.0
Philosophy	36.7	2.2	34.5	0.0
App ling	36.5	3.2	32.3	1.0
Sociology	35.3	5.1	29.4	0.8
Average soft fields	38.1	3.9	33.6	0.7
Overall	27.8	4.6	22.7	0.6

the disciplinary fabric. More generally, when we ignore text length and look at the raw scores, we find that some 69% of all cases of self-mention occurred in the humanities and social science papers, with an average of 38 per article, compared with only 17 in science and engineering. This difference was largely due to the much greater use of first person pronouns in the soft disciplines.

These broad differences in the use of first person and self-citation suggest that self mention might vary with different assumptions about the effects of authorial presence and rhetorical intrusion in different knowledge-making communities. I will outline some of the ways it is expressed and used below.

5 Self-citation, disciplinary identity, and knowledge making

Perhaps the most obvious form of self-mention is to refer to one's earlier research, either as an element within the sentence or as a superscript note. The extent of self-citation in these papers was surprising. As Table 13.3 shows, about 70% of the papers in the study contained a reference to the author and

TABLE 13.3 Self-citation by discipline

Discipline	Totals		Per paper			Cases per paper				
	Total	% of all	Max	Min	Av	0	1–3	4–6	7–9	10+
Biology	324	11.8	30	0	10.8	2	9	5	4	10
Sociology	153	5.6	35	0	5.1	7	14	5	1	4
Marketing	148	6.1	17	0	4.9	7	9	3	2	8
Mechanical eng	110	10.6	11	0	3.7	8	15	4	2	1
Electronic eng	115	9.3	15	0	3.8	11	12	5	2	0
App ling	96	5.0	32	0	3.2	13	6	8	1	2
Physics	85	11.0	5	0	2.8	7	15	8	0	0
Philosophy	65	3.9	3	0	2.2	15	11	4	0	0
Overall	1096	7.9			4.6	29%	38%	18%	5%	10%

these comprised 8% of all references. Self-citation was particularly frequent in biology, with an average of 11 citations per paper, but was also common in sociology and marketing. Overall, self-citation appears to be a prominent feature of academic writing in the sciences and engineering where it made up almost 11% of all references, compared with only 5% in the soft fields, and constituted 60% of all expressions of self-mention. The 'cases per paper' columns in Table 13.4 indicates the dispersal of items, with about a third of all papers containing four or more references to the authors.

Doubtless the circumstances which motivate writers to cite their own work are varied and complex, involving psychological, rhetorical, and social factors that contain elements of confidence, experience and self-promotion. Self-citation is obviously an important means of demonstrating one's disciplinary credentials and credibility. Quite clearly, no research occurs in a social vacuum and always has to be contextualized by grounding it in the issues which engage the discipline. Self-citation can help to emphasize these links one has to one's colleagues through an engagement in a common literature and the professional intimacy one shares with a set of current disciplinary problems. This is part of what it means to display disciplinarity (Sullivan, 1996).

TABLE 13.4 Frequency of first person pronouns (cases per paper)

Discipline	Total	I	Me	My	We	Us	Our
Physics	17.7	0.0	0.0	0.0	12.7	0.2	4.7
Biology	15.5	0.0	0.0	0.0	11.5	0.5	3.4
Electronic eng	11.6	0.0	0.0	0.0	8.4	0.1	3.1
Mechanical eng	2.6	0.0	0.0	0.0	2.1	0.0	0.5
Average hard fields	11.9	0.0	0.0	0.0	8.7	0.9	2.9
Marketing	38.2	1.1	0.0	0.5	22.2	0.8	13.5
Philosophy	34.5	26.1	1.6	5.3	0.9	0.2	0.4
App ling	32.3	12.7	1.1	3.4	8.9	1.0	5.1
Sociology	29.4	9.5	0.7	1.5	11.5	0.5	5.7
Average soft fields	33.6	12.4	0.8	2.7	10.9	0.6	6.2
Overall	22.7	6.2	0.4	1.3	9.8	0.4	4.6

Participation in published research is perhaps the strongest demonstration a writer can make to establish his or her claim to be seen as an important player in a field and to have work taken seriously. This was highlighted by a number of my informants in the interviews:

> Everyone has to fit their research into a framework to make sense. I always cite my own work because people are more likely to listen if you are part of that framework. (Soc)

> Citing yourself is an important way of showing your familiarity with an issue. It shows you know what you are talking about and have something worth saying. (Bio)

In other words, situating research within the wider narratives of their community can enhance a writer's credibility. Constructing a solid disciplinary identity increases the likelihood that their work will be read and perhaps accepted.

The broad variations in the frequency of self-citation however also indicate differences in genre conventions which suggest underlying differences in the research practices of these academic communities. Issues in the humanities and social sciences, for instance, tend to be relatively diverse, range over a wide academic territory, and are detached from immediately prior developments (Becher, 1989). This means that references are relatively diffuse and there is comparatively little opportunity for self-citation.

References in sciences and engineering, on the other hand, tend to be tightly bound to a particular research topic. References here closely define a specific context and contribute to the sense of linear progression which is often said to characterize hard knowledge (Kolb, 1981; Kuhn, 1970). This is partly a consequence of the fact that scientists tend to participate in highly discrete and specialised areas of research, largely because of the heavy investments in procedural capability and technical equipment that hard knowledge production often requires. It is also related to the sheer volume of knowledge and its rapid expansion. These factors coerce scientists into a niche of expertise from where they can follow defined paths and make precise contributions. Research on particular issues is therefore often conducted at a restricted number of sites and by a limited number of researchers, allowing writers to draw on their own work to a considerable extent:

> We aren't just blowing our own trumpets here. There just aren't that many people doing work in this particular field. (Phy)

> A paper in biology is not just a one off bit of isolated research. Projects tend to be expensive and may take a long time to set up and produce anything important. What we write up probably reports a piece of research that may be going on for years. We are continuously building on what we've done. (Bio)

Clearly these kinds of self-reports underplay the possible promotional role of self-mention, but their credence as candidate explanations are given some plausibility by the social practices of disciplinary communities. I now turn to examine the part played by first person pronouns in academic writing and the management of interpersonal rhetoric.

6 First person and authorial presence

Pronouns are clearly the most salient forms of self-mention in the corpus and, once again, there are substantial differences in how they are employed in different disciplines, both in overall frequency and in preferred patterns of use. Table 13.4 shows that 3/4 of all first person pronouns occurred in the humanities and social science papers and that they were rarely used in mechanical engineering. The subjective cases (*I* and *we*) comprise 70% of all pronouns.

Once again, this distinction in the frequency of first person use reflects the fact that writers in these broad domains of knowledge have very different ways of conducting research and persuading readers to accept their results. Generally speaking, the writer in the hard sciences is seeking to establish empirical uniformities through research activities that involve precise measurement and systematic scrutiny of a limited number of controlled variables. Hard knowledge tends to be universalistic and to be motivated by conceptual issues. Research therefore usually consists of conducting experiments to propose solutions to specific disciplinary problems and typically involves familiar procedures, broadly predictable outcomes, and relatively clear criteria of acceptability (e.g. Becher, 1989; Whitley, 1984).

As a result the writer can downplay his or her personal role in the research in order to highlight the phenomena under study, the replicability of research activities, and the generality of the findings. By electing to adopt a less intrusive or personal style, writers can strengthen the objectivity of their interpretations and subordinate their own voice to that of unmediated nature. Such a strategy subtly conveys an empiricist ideology that suggests research outcomes would be the same irrespective of the individual conducting it. One of my respondents expressed this view clearly:

I feel a paper is stronger if we are allowed to see what was done without 'we did this' and 'we think that'. Of course we know there are researchers there, making interpretations and so on, but this is just assumed. It's part of the background. I'm looking for something interesting in the study and it shouldn't really matter who did what in any case. (Bio interview)

In theory anyone should be able to follow the same procedures and get the same results. Of course reputation is important and I often look at the writer before I look at a paper, but the important thing is whether the results seem right. (Bio interview)

The high proportion of personal pronouns in the soft RAs, on the other hand, suggests a quite different rhetorical stance. Establishing an appropriately authorial persona and maintaining an effective degree of personal engagement with one's audience are valuable strategies for probing relationships and connections between entities that are generally more particular, less precisely measurable, and less clear-cut than in the hard sciences. Variables are often more heterogeneous and causal connections more tenuous. As a result, successful communication depends to a larger extent on the author's ability to invoke a real writer in the text. Self-mention can help construct an intelligent, credible, and engaging colleague, by presenting an authorial self firmly established in the norms of the discipline and reflecting an appropriate degree of confidence and authority. Examples abound in the corpus:

> (1) I argue that their treatment is superficial because, despite appearances, it relies solely on a sociological, as opposed to an ethical, orientation to develop a response. (Soc)
>
> . . . in short, we demonstrate that what consumers know about a company can influence their evaluations of products introduced by the company. (Mkt)
>
> I bring to bear on the problem my own experience. This experience contains ideas derived from reading I have done which might be relevant to my puzzlement as well as my personal contacts with teaching contexts. (AL)

The decision to employ a first person pronoun by writers in these fields appears to be closely related to authorial stance, and with the desire to both strongly identify oneself with a particular argument and to gain credit for one's individual perspective or research decisions:

> Using 'I' emphasizes what you have done. What is yours in any piece of research. I notice it in papers and use it a lot myself. (Soc interview)
>
> The personal pronoun 'I' is very important in philosophy. It not only tells people that it is your own unique point of view, but that you believe what you are saying. It shows your colleagues where you stand in relation to the issues and in relation to where they stand on them. It marks out the differences. (Phil interview)

The first person therefore assists authors to make a personal standing in their texts and to demarcate their own work from that of others. It helps them distinguish who they are and what they have to say.

While writers in the hard sciences were less explicitly present in their texts, they were not invisible. Table 13.4 above shows that first person singular pronouns were virtually absent from the science and engineering articles and were very low in marketing. Although papers in these five fields were overwhelmingly multiple-authored, use of the plural is only partly explained by patterns of authorship. The decision to use *we* by writers of single-authored articles is often said to indicate an intention to reduce personal attributions, but it is not always the self-effacing device it is sometimes thought to be. Pennycook (1994, p. 176), for example, observes that "there is an instant claiming of authority and communality in the use of *we*."

The distancing which attends the plural meaning also seems to create a temporary dominance by giving the writer the right to speak with authority. These examples from single-authored papers suggest how writers can simultaneously reduce their personal intrusion and yet emphasise the importance that should be given to their unique procedural choices or views:

(2) *We* do not agree with this statement, . . . (EE)

Some years ago *we* proposed a new evaluation method providing an automatic and . . . (EE)

In this section, *we* give detailed information about Canada's energy resources, . . . (ME)

We believe that, when used in this manner, the word "degree" is simply a numerical suffix whose value is 27r/360 (not 27r rad/360'). (Phy)

This appears to be predominantly a feature of scientific writing, as the 16 single authored papers in the hard knowledge corpora yielded only a handful of singular forms but contained 80 plural first person pronouns. In contrast, the 75 single authored papers in the humanities and social sciences included only eight plural first person forms. Commenting on his understanding of this usage, a professor of physics observed that:

I suppose we are generally encouraged to keep ourselves in the background in our writing, to give prominence to objective physical events, but of course we are involved in research and using 'we' emphasizes this. It avoids generalities and focuses on specifics without being too aggressively personal. (Phy interview)

Another interesting remark was made by an electronic engineer, who suggested that this usage was less a means of withdrawing from the claims made in the paper than a reminder of the collaborative nature of the research activity itself:

> In terms of what it refers to, I often think that 'we' is right to some extent. I am always reporting research that I've done as part of a team, even if I am writing a solo paper. It's a kind of shorthand acknowledgement of the part played by my colleagues. (EE interview)

In sum, rhetorical conventions of pronoun self-mention contribute to the kind of relationship that the writer establishes with the reader and these rhetorical preferences are at least partly influenced by a disciplinary community's epistemological beliefs and social practices. But while writers in the softer disciplines projected a more prominent identity, scientists did not entirely subordinate their discoursal selves to their research activities.

7 Theme and personal prominence

In addition to selecting a first person pronoun to construct a more engaged and committed presence in their texts, writers also have syntactic options, and in particular the decision to thematize (Halliday, 1994). The concept of theme is controversial (e.g. Huddleston, 1988), but the choice of initial position, what we are asked to attend to first, can contextualize the sentence by providing the reader with an interpretive framework for the newsworthy information that follows. The decision to front a clause with a first person pronoun gives it special focus, representing it as a significant aspect of the message and signaling the overt presence of the writer as a visible participant in the research process. The persuasive impact of such choices has been discussed by Gosden (1993) who found that this highlighting of the author's role occurs most frequently in Introduction and Discussion sections, the most rhetorically charged parts of research papers.

Thematisation of first person pronouns is therefore an additional element of self-presence in academic discourse, and writers made considerable use of this powerful strategy in my corpus. Some 45% of all exclusive first person pronouns were clause initial, and although the soft fields generally had higher percentages than the sciences, first person themes were frequent in all disciplines:

> (3) We believe that an achievable performance goal for a commercial system under optimum conditions should be 10 kg kW h. (ME)

I argue both accounts are unacceptable. (Phil)

Our method allows a more accurate identification of such defects in the assembly. (EE)

. . . my own experience suggests that it is more feasible to achieve multi-disciplinary than truly integrated, inter-disciplinary research. (AL)

But while theme choices which gave writers maximum visibility were more common in the soft papers, they also had a powerful impact in scientific contexts where author mention is more often subordinated to research processes. In the science and engineering disciplines the thematic selection of writers represents an important departure and clearly signals a more authoritative stance. In (4), for example, the writers emphasize the importance and scale of their contribution through a series of personal theme choices, while in (5) the writers simultaneously open a gap for their research and highlight their study by a theme choice which contrasts their own role with earlier activities:

(4) In this study, we demonstrate that the SIE DNA-binding site of the c-fos promoter, which is a target for STAT1/ STAT3 in EGF-treated cells (17, 65), is specifically bound by a STAT3-containing protein in ES cells maintained in the presence of LIF. *We examine* the expression and phosphorylation of the STAT3 protein in these cells in response to LIF treatment or withdrawal. *We show* that the c-fos promoter is LIF responsive in ES cells, and that the multimerized SIE element confers LIF-dependent transcription to the minimal TK promoter. *We demonstrate* that STAT3 mutants, which behave as dominant negative factors in the IL-6 pathway (38), can repress LIF-dependent transcription. *We also show* that stable expression of one of these mutants leads to morphological differentiation of the ES cells. (Bio)

(5) A problem that has not been studied in any detail is the nonuniformity of the longitudinal magnetic field of a *z*-gradient coil in the transverse (*x* and *y*) directions, which is associated with the elliptical geometry. *We address this issue* in the present paper by expanding, in ascending powers of *x* and *y,* the magnetic field associated with an azimuthal current. (Phy)

8 Self-mention and discourse purposes

So far in this paper I have mainly focused on the issue of author visibility and the role of self-mention in constructing a credible authorial identity. In this section I want to briefly examine some of the main ways that writers use self-mention in their texts, combining collocational analysis and the comments of informants to identify the points at which writers choose to intrude most explicitly.

In all disciplines writers' principal use of first person was to explain the work that they had carried out; representing their unique role in constructing a plausible interpretation for a phenomenon. In the hard knowledge corpus, and in the more practical and quantitatively focused papers in the soft fields, this mainly involved setting out the procedures they had performed. Author prominence here serves to reassure us of the writer's professional credentials through a demonstrable familiarity with disciplinary research practices. But in addition, it also acts to highlight the part the writer has played in a process that is often represented as having no agents at all. By inserting themselves into their research activities then, writers inject an element of qualitative judgement that reminds us that, in other hands, things could have been done differently and that personal choices have been made:

(6) *I also examined* the linguistic process of nominalization. (AL)

We acid-shocked cells in the presence of the nonspecific $Ca2^+$ channel inhibitor $La3^+$. (Bio)

We used a measure based on interlaced batches to compute an unbiased estimate S2 of the variance of the performance characteristic within a single run, . . . (EE)

I reviewed the case material and found solid grounds for these activities. (Mkt)

In 1995 *I went to Mexico and Chiapas* to better understand the Zapatista movement. *I visited* many people, men and women, from and around the movement, and discussed with them the questions treated in this article. *I am formulating* them as "questions to Ramona". (Soc)

In more theoretically-oriented articles writers sought less to figure as practical agents than as builders of coherent theories of reality. Explicit self-mention here does not collapse personal credibility into a demonstration of specific procedural competencies, but establishes a more personal authority based on confidence and command of one's arguments:

I'm very much aware that I'm building a façade of authority when I write, I really like to get behind my work and get it out there. Strong. Committed. That's the voice I'm trying to promote, even when I'm uncertain I want to be behind what I say. (Soc interview)

You have to be seen to believe what you say. That they are your arguments. It's what gives you credibility. It's the whole point. (Phil interview)

In the soft fields, and in philosophy in particular, a writer's style is a significant element of both his or her immediate credibility in the paper and wider reputation in the discipline. An effective argument here depends in no small way on writers' success in convincing readers of their reasonableness, seriousness, and sincerity, and this is, in part, achieved by balancing caution with commitment. Writers must display appropriate respect for alternatives, but back their views with a personal warrant where necessary. The personal voice here works to address readers directly through a firm alignment with their views, pledging certainty and an interpersonal assurance of conviction. Inevitably, this often shades into explicit claim making:

(7) However, *I believe that* this is a one dimensional view, in that sensations may well be necessary components of emotion while not being the defining feature. *I also think that* sensations and perceptions are not simply natural phenomena that are closed off from cultural conditioning, . . . (Soc)

I suggest then that beyond reporting what research has taught us about the researched situation, we might also address what research has taught us about research. (AL)

Much as *I hate to admit it,* mathematicians may have good grounds for rejecting PDP. (Phil)

Self-mention of this kind seems to help bind the writer and the reader together as co-participants in an ongoing debate; portraying them as interactants engaged in a disciplinary dialogue.

One function of the first person in argument is as an exemplification device. Largely confined to philosophy, in this usage the *I* pronoun becomes not only an instrument of self-mention for the writer as scholar, but a device for inserting him or her as the main protagonist in illustrative examples:

(8) For instance, suppose that I think up a whole bunch of possible paths through the graph and then verify that none of them is an HP.

E.g., if the amount in my envelope is $10, then by swapping I stand to gain $10 whereas I only stand to lose $5.

What I believe, for example, or what I feel, makes a difference to what I do.

For, Forbes maintains, the skeptic can evade Putnam's argument and achieve all she has ever wanted by switching to the hypothesis that I am 'relevantly like a brain in a vat'.

More commonly, writers in the discursive soft knowledge disciplines used self-mention to provide an overt structure for their discourse. Often however, such framing did more than clarify the schematic structure of the argument, it explicitly stated the goal or purpose of the paper, providing an opportunity for the writer to promote both themselves and their stance. Interestingly, this personal claim-staking also occurred in the physical sciences, often considered to be free from such personal agendas:

(9) *We shall prove,* however, that this is not the case. (Mkt)

We argue that a 'zero-sum' formulation of power dynamics is simplistic and misleading. (Soc)

In this paper, *we clearly demonstrate that* Tax can activate transcription of the CQB promoter through the NF-Y element. (Bio)

First and foremost, one could claim that the model is too oversimplified to produce physically meaningful results. *We argue against this claim* on the basis of the model's ability to produce results consistent with experiment. (Phy)

This explicitly persuasive use of self-mention is perhaps clearer, and more widely used, where writers employ *I/we* to summarise a viewpoint or make a knowledge claim. This use not only serves to metadiscursively guide the reader through the discussion, but once again explicitly foregrounds the writer's distinctive contribution and commitment to his or her position. With this use the writer and the claim are strongly coupled, soliciting recognition for both.

(10) *We have demonstrated that* MCP can be used to form single- and multiple-helical microcoils by printing lines on all objects. (ME)

We have now discovered that the Byr2 kinase catalytic domain can also bind to the regulatory domain of Byr2. *We have determined* the minimum binding domain for each of these interactions by characterizing the binding profile of a series of Byr2 deletion mutants. (Bio)

Likewise, *I have offered evidence that* some critical thinking practices may marginalize subcultural groups, such as women, within US society itself. (AL)

Perhaps in more than any other function, this use suggests the conscious exploitation of a strategy to manage the reader's awareness of the writer's role; his or her attempt to take a position in relation to the community and to seek credit for that position.

> Of course you have to recognise what other people have said and take care not to tread on toes, but it is important to leave your readers in no doubt about your own view. (Soc interview)

> If there are good reasons for a particular interpretation, all the data point the same way to the same conclusion, then I'm happy to pin my colours to the mast. You have to make sure that what you've done gets noticed so that you get recognised for it. (Mkt interview)

> It's conventional to use these formulas to keep yourself out of the picture. They are just conventional ways of expressing inference. Sometimes though you need to be explicit about what you think, that the contribution is your own. (ME interview)

In addition to the subjective first person, the possessive forms are also used to promote the writer's contribution by associating them closely with their work. The most common collocations were *analysis, approach, research, argument, results, study,* etc. (cf. Kuo, 1999, p. 135). These constructions worked to highlight the author's close involvement in research outcomes or activities

> (11) As demonstrated through *my analysis,* a significant rhetorical task in this process . . . (AL)

> *Our method* allows a more accurate identification of such defects in the assembly. (EE)

> The benefit of *my account* is that the application of Davidson's paratactic machinery enables us to circumvent this problem. (Phil)

> Lastly, *our model* offers insights into consumer behavior and clientele effects. (Mkt)

> It is in this spirit that I offer *my own contribution* to the debate. I want to set out a slightly different approach to those taken in the above articles . . . (Soc)

> *Our results* demonstrate that flagellar RNA accumulations are . . . (Bio)

Once again then, this is not a simple reporting of results or procedures, but an expression of the participants' custody and personal ownership of what they report. It is a rhetorical strategy of promotion.

9 Some conclusions and teaching implications

I have argued that first person pronouns and self-citation are not just stylistic optional extras but significant ingredients for promoting a competent scholarly identity and gaining accreditation for research claims. Self-mention is important because it plays a crucial role in mediating the relationship between writers' arguments and their discourse communities, allowing writers to create an identity as both disciplinary servant and persuasive originator. The points at which writers choose to announce their presence in the discourse are those where they are best able to promote themselves and their individual contributions. Their intrusion helps to strengthen both their credibility and their role in the research, and to help them gain acceptance and credit for their claims.

The distribution of these features shows that not all disciplines sanction the same degree of authorial presence. Writers' decisions are closely related to the social and epistemological practices of their disciplines and represent an important way of signaling membership and honoring what is accepted as professional engagement, appropriate intrusion, and persuasive conviction. But disciplinary conventions are enabling rather than deterministic, and typical patterns of self-mention only provide broad perimeters of choice. In constructing their texts writers also construct themselves, and self-reference represents the confidence to speak authoritatively rather than concealing authorship behind the impersonality options the genre provides. Issues of seniority, experience, relationship to the community, and general sense of self are also likely to influence these decisions. What is essentially involved however is an author's desire to be seen in his or her text and to affirm a commitment to his or her work. This expression of self is an important way of creating one's own voice, of speaking with authority, and of securing reader support.

Academic writers have rhetorical options then, and the effects of manipulating these options suggest that there are considerable advantages for our students in being aware of them. The whole issue of consciousness raising is crucial in EAP and is central to learning to write confidently. For teachers this means helping students to move beyond the conservative prescriptions of the style guides and into the rhetorical contexts of their disciplines, investigating the preferred patterns of expression in different communities. This kind of inquiry can take a variety of forms, but most simply it involves training students to read rhetorically and to reflect, perhaps through diaries, on the practices they observe and use themselves (e.g. Johns, 1990, 1997). What, for example, is an author's purpose in using a personal pronoun here? Why has she chosen to cite herself at this point? What is achieved by him using *we* in this paper? When do writers typically move to self-mention?

We might also allow data to drive learning more directly by guiding students' exploration of authentic models, asking small groups to collect frequency counts of these forms and discuss typical collocations in a computer corpus, perhaps comparing the practices found in articles and theses. Students can also interview faculty experts on their own writing practices or their reactions to the practices of others in the discipline. The findings from these are likely to provide a useful basis for group feedback discussions and further consideration on the decisions behind certain forms and the impressions one can make in employing them. Finally, students should be encouraged to experiment with their academic writing. By employing a personal voice in their texts they will be able to get feedback on their practices and perhaps be able to evaluate the impact of their decisions more clearly.

To close, it has to be said that the relationships between knowledge, the linguistic conventions of different disciplines, and personal identity, are fuzzy and complex. Yet it is equally true that these are issues worth addressing and exploring further with our students. Only by developing a rhetorical consciousness of the kinds of features I have discussed in this paper can we help them gain control over their writing and meet the challenges of academic writing in either a first or second language.

14

Is there an "academic vocabulary"?

Ken Hyland and Polly Tse

The concept of an *academic vocabulary*

The idea of an academic vocabulary has a long history in teaching English for academic or specific purposes (EAP and ESP). Variously known as *subtechnical vocabulary* (Anderson, 1980; Yang, 1986), *semitechnical vocabulary* (Farrell, 1990), or *specialized nontechnical lexis* (Cohen, Glasman, Rosenbaum-Cohen, Ferrara, & Fine, 1988), the term is used to refer to items which are reasonably frequent in a wide range of academic genres but are relatively uncommon in other kinds of texts (Coxhead & Nation, 2001). This vocabulary is seen as a key element of *essayist literacy* (Lillis, 2001) and an *academic style* of writing and is considered to be more advanced (Jordan, 1997) than the core 2,000–3,000 words that typically make up around 80% of the words students are likely to encounter in reading English at university (Carter, 1998).

Many teachers regard helping undergraduates develop control over such a specialist vocabulary as an important part of their role, and attempts have been made to assemble lists of key terms to guide materials writers and help students plan their learning more efficiently. Early ESP materials, for example, sought to identify and present forms with a high frequency in scientific and technical writing (e.g., Barber, 1988; Herbert, 1965) and considerable effort has been devoted to investigating the vocabulary needed for academic study (e.g., Campion & Elley, 1971; Coxhead, 2000; Nation, 1990). Such research is usually based on the assumption that learners are seeking to build a repertoire of specialized academic words in addition to their existing basic or general service vocabulary, and this repertoire building is often seen as the purpose of developing university vocabulary. Consequently, vocabulary is typically seen as falling into three main groups (Nation, 2001):

1 *High frequency words* such as those included in West's (1953) General Service List (GSL) of the most widely useful 2,000-word families in English, covering about 80% of most texts.

2 An *academic vocabulary* of words which are reasonably frequent in academic writing and comprise some 8%–10% of running words in academic texts.

3 A *technical vocabulary* which differs by subject area and covers up to 5% of texts.

First year undergraduate students are said to find an academic vocabulary (2) a particularly challenging aspect of their learning (Li & Pemberton, 1994). This aspect of their learning is challenging because, although technical vocabulary is central to students' specialized areas, general academic vocabulary serves a largely supportive role and the words are "not likely to be glossed by the content teacher" (Flowerdew, 1993, p. 236). Many of these words also occur too infrequently to allow incidental learning (Worthington & Nation, 1996), encouraging researchers and teachers to develop vocabulary lists for directly teaching these terms.

The notion that some words occur more frequently in academic texts than in other domains is generally accepted. It also appears to correspond with EAP's distinctive approach to language teaching, based on the identification of the specific language features and communicative skills of target groups, and devoted to learners' particular subject-matter needs. However, whether it is useful for learners to possess a general academic vocabulary is more contentious because it may involve considerable learning effort with little return. It is by no means certain that there is a single literacy which university students need to acquire to participate in academic environments, and we believe that a perspective which seeks to identify and teach such a vocabulary fails to engage with current conceptions of literacy and EAP, ignores important differences in the collocational and semantic behavior of words, and does not correspond with the ways language is actually used in academic writing. It is, in other words, an assumption which could seriously mislead students.

In this article we therefore set out to question some of the assumptions underlying the idea of a general academic vocabulary by analyzing the distribution of items in the widely used *Academic Word List* (AWL) (Coxhead, 2000) in a corpus of 3.3 million words from a variety of genres and disciplines.

Lists of academic vocabulary

To identify the most valuable words in academic contexts, a variety of vocabulary lists have been compiled from *corpora,* or collections, of academic texts. Corpus studies have shown that the most frequent words in English cover a large percentage of word occurrences in any text. The top three words (*the, of, to*) make up some 10% of uses in the 400-million-word Bank of English corpus, for instance, and the first 100 comprise about one half of all written and spoken texts (e.g., Hunston, 2002; Kennedy, 1998). Extending the value of such frequency counts to academic uses, researchers have sought to compile lists of frequently occurring words found in different disciplines and different kinds of texts (e.g., Coxhead, 2000; Farrell, 1990; Praninskas, 1972; Xue & Nation, 1984).

Common to these lists is the focus on *word families,* that is, the base word plus its inflected forms and transparent derivations (Bauer & Nation, 1993). This approach overcomes the difficulty of specifying what counts as a word by including all closely related affixed forms as well as the stem's most frequent, productive, and regular prefixes and affixes. It is also supported by the view that knowledge of a base word can facilitate the understanding of its derived forms and evidence which suggests that members of the same word family are stored together in the mental lexicon (Nagy, Anderson, Schommer, Scott, & Stallman, 1989, p. 262; Nation, 2001).

The most recent compilation is the AWL (Coxhead, 2000), which contains 570 word families believed to be essential for students pursuing higher education irrespective of their chosen field of specialization.[1] The 3,112 individual items in this inventory do not occur in West's (1953) GSL and were included in the AWL if they met certain frequency and range criteria. Words were selected on the basis of occurring at least 100 times in an academic corpus of 3.5 million words of varied genres and in at least 15 of the 28 disciplines within the four broad subject groupings of the corpus: arts, commerce, law, and science (Coxhead, p. 221). Cox-head found that the AWL covered 10% of the words in her corpus and only 1.4% of a similar-sized corpus of fiction, suggesting that the items are more relevant for learners with academic purposes. Taken together with items on West's GSL, the AWL accounted for 86% of the academic corpus.

There is no doubt that the AWL is an impressive undertaking, representing the most extensive investigation to date into core academic vocabulary. By stressing students' target goals and the need to prioritize items in their lexical development, the AWL helps distinguish EAP from general English and

---

[1] The headwords of families in the AWL together with the items grouped by sublist are available from the Massey University (2004) Web site.

sets an agenda for focused language learning. But although it has become a benchmark for developing teaching materials for EAP (e.g., Schmitt & Schmitt, 2005), the list has not been independently evaluated. It remains unclear how well the AWL can be said to represent the lexical composition of academic writing in English, and we have little idea of its coverage in particular disciplines and genres. In fact, a major difficulty of such lists, and not just the AWL, is the assumption that a single inventory can represent the vocabulary of academic discourse and so be valuable to all students irrespective of their field of study. We explore this hypothesis by examining the frequency, range, preferred meanings and forms, and the collocational patterns of items in the AWL.

Corpus and methods

Our study employs both qualitative and quantitative analyses of a medium-sized academic corpus organized by discipline, genre, and writer expertise. Our academic corpus offers a broad cross section of writing in the disciplines and includes a range of professional and learner texts representing key academic genres across a broad span of disciplines (Table 14.1). The disciplines are biology, physics, and computer science (*sciences*); mechanical

TABLE 14.1 Academic Corpus

	Sciences	Engineering	Social Sciences	Totals
Research articles	189,800	178,900	633,400	1,002,100
Textbooks	106,100	108,000	176,800	390,900
Book reviews	31,600	15,900	77,000	124,500
Scientific letters	122,000	0	0	122,000
Expert writers (subtotal)	449,500	302,800	887,200	1,639,500
Master's theses	139,300	96,200	205,700	441,200
doctoral dissertations	191,800	93,900	457,000	742,700
Final year project theses	87,100	77,000	305,100	469,200
Student writers (subtotal)	418,200	267,100	967,800	1,653,100
Overall	867,700	569,900	1,855,000	3,292,600

and electronic engineering (*engineering*), and sociology, business studies, and applied linguistics (*social sciences*). Within each of these fields we collected 30 research articles, seven textbook chapters, and 20 academic book reviews in each of seven disciplines; and 45 scientific letters in physics and biology. This corpus represents the range of sources students are often asked to read at university and so include the kinds of lexical items they will frequently encounter. In addition to these professionally written texts we added eight master's theses, six doctoral dissertations, and eight final-year undergraduate project theses by L2 students in each of six disciplines to represent students' productive uses of vocabulary.

The academic corpus thus comprises contemporary examples of the main academic discourse genres and includes both long and short texts, published peer-refereed articles, pedagogic texts, and student writing. Unlike Coxhead, whose corpus was opportunistic, including unequal numbers of texts in each field, a number of 2,000 word text fragments from the now dated Lancaster-Oslo/Bergen (LOB) Corpus and Brown Corpus as well as the Wellington Corpus, and no examples of student writing, we attempted to more systematically represent a range of key genres in several fields, with equal numbers of entire texts in each genre. Although our procedure resulted in subcorpora of different sizes, we compensated by comparing item frequencies per 10,000 words.

Having compiled our corpus, we then combined the texts into disciplines and fields using Corpus Builder (Cobb, 2004) and explored these using RANGE, a program developed by Nation (2002) and used to create the AWL (Coxhead, 2002). This software is preloaded with West's (1953) GSL of the most frequent 2,000 English words and the AWL, and it shows the frequency of items from each list in any corpus together with its range, or the number of different subcorpora they occurred in. To determine the frequency and spread of AWL items, we examined these figures for the entire academic corpus, for the three subcorpora of engineering, science, and social science, and then for the disciplines within these fields. Finally, we ran concordances on selected items to see if they were used in the same ways and with the same meanings.

Results: in search of an academic vocabulary

Overall frequencies and distributions

Coxhead and Nation (2001, p. 254) claim that a word should be included in a general academic vocabulary if it is common to a range of academic texts and accounts for a substantial number of words in an academic corpus. In our study, all 570 of the AWL word families occurred in the academic corpus, with 541 occurring in all three subcorpora. As Table 14.2 shows, the AWL covers an impressive 10.6% of the words in the corpus, and together with the 2,000 words of the GSL it provides an accumulative coverage of 85%, representing roughly one unknown word in every seven words of text.

Although the list offers a good overall coverage of our academic corpus, we can see that this coverage is not evenly distributed. Students in the sciences are far less well served because the combined AWL and GSL failed to account for 22% of the words in our science corpus, meaning that students would stumble over an unknown item about every five words, making the text incomprehensible. This variation may suggest that writing in the sciences demands a more specialized and technical vocabulary, but as we discuss later, the fact that all disciplines shape words for their own uses seriously undermines attempts to describe a core academic vocabulary.

To explore the issue of range further, we examined the frequency of individual items in the three subcorpora themselves. Coxhead's criteria for uniformity of frequency was 100 occurrences overall with at least 10 in each of four fields (Coxhead, 2000, p. 221), but this seemed a remarkably low threshold given the size of her corpus and the fact that the word list contains more than 3,000 individual words. We therefore used a more rigorous and systematic standard, identifying items as *frequent* if they occurred above the mean for all AWL items in the corpus (i.e., 597). Using this measure, only 192 families, or about a third of the AWL items, could be regarded as frequent, with the research terms *process, analyze, research, data,* and *method* being the most common.

TABLE 14.2 Coverage of Academic Word List (AWL) in Academic Corpus

	Frequency words	AWL items	Mean	Coverage % AWL	GSL	Overall
Engineering	551,891	61,408	108	11.1	73.3	84.4
Social Sciences	1,822,660	200,393	352	11.0	77.0	88.0
Sciences	838,926	78,234	137	9.3	69.0	78.3
Overall	3,213,477	340,035	597	10.6	74.0	84.7

Note. GSL = General Service List (West, 1953).

Interestingly, the top 60 items in our corpus provided a similar coverage (3.9%) to Coxhead's most frequent items, but only 35 items were common to both lists. The most frequent 60 families form an important sublist in Coxhead's compilation because they represent a coverage more than twice as numerous as the next most frequent 60 items, about one occurrence every four pages. At the other end of the scale, *commence, concurrent, levy,* and *forthcoming* were among 23 families we judged to be extremely infrequent because they occurred less than 60 times in the corpus (below 10% of the overall mean).

Distributions across fields

Looking more closely at the distributions across subcorpora, some items are frequent overall because of their concentration in one or two subcorpora, 15 of our top 50 items, for example, had more than 70% of their occurrences in one field. Taking the means of individual fields as a benchmark, we found that of the 192 families which were frequent overall, only 82 were frequent in all three subcorpora and 50 in just one. Nor were the same items the most frequent in all subcorpora. Table 14.3 shows that only *analyze* and *process* in the overall most frequent list also occurred in the top 10 most frequent families in each subcorpora. We can also see here that the AWL items differed considerably in their spread in each subcorpora, with the top 10 comprising nearly 16% of families in engineering and only 11.4% in social sciences.

Distributions are also unequal when we consider the least frequent words. Using 10% of the mean in each subcorpora as a reference, we found 78 families to be extremely infrequent in one subcorpora, 63 in two subcorpora and 6 in all three. In other words, 27% of all the AWL families have a very low occurrence in at least one subcorpora and so have an extremely low chance of being encountered by students.

Although comparing the words' frequency of occurrence relative to the mean helps to determine the relative significance of particular words in different subcorpora, variations in the sizes of the subcorpora mean that this procedure gives us only a general impression of distributions. A more accurate picture is obtained by norming frequencies to give occurrences per 10,000 words and then looking at the percentage spread of each item across the subcorpora. Because we considered three fields, an even distribution would be roughly 33% of an item in each subcorpora, but no family met this criteria and more than half of all items occurred mainly in one subcorpora only. Table 14.4 shows that of the 570 AWL families, 534 (94%) have irregular distributions across the three subcorpora. Of these, 227 (40% of items) have

TABLE 14.3 Most Frequent Items by Field With Percentages of Families in That Field

Overall (all three fields)				Engineering			
Family	Freq	Item %	Cum %	Family	Freq	Item %	Cum %
Process	4,501	1.3	1.3	Equate	1,418	2.3	2.3
Analyze	4,498	1.3	2.6	*Process*	1,143	1.9	4.2
Research	3,841	1.1	3.8	Design	999	1.6	5.8
Data	3,789	1.1	4.9	Method	920	1.5	7.3
Method	3,214	0.9	5.8	Data	913	1.5	8.8
Vary	3,156	0.9	6.8	*Analyze*	895	1.5	10.2
Strategy	3001	0.9	7.6	Function	847	1.4	11.6
Culture	2962	0.9	8.5	Require	844	1.4	13.0
Function	2909	0.9	9.4	Output	839	1.4	14.4
Significant	2742	0.8	10.2	Input	818	1.3	15.7
Sciences				Social Sciences			
Family	Freq	Item %	Cum %	Family	Freq	Item %	Cum %
Data	1395	1.8	1.8	Research	3261	1.6	1.6
Method	1271	1.6	3.4	Strategy	2795	1.4	3.0
Process	1118	1.4	4.8	Culture	2583	1.3	4.3
Analyze	1029	1.3	6.2	*Analyze*	2574	1.3	5.6
Concentrate	865	1.1	7.3	*Process*	2240	1.1	6.7
Require	848	1.1	8.3	Consume	1947	1.0	7.7
Function	759	1.0	9.3	Response	1910	1.0	8.6
Obtain	750	1.0	10.3	Individual	1894	0.9	9.6
Extract	739	0.9	11.2	Participate	1800	0.9	10.5
Similar	726	0.9	12.1	Significant	1762	0.9	11.4

Note. Freq = frequency. Cum = cumulative.

TABLE 14.4 Concentration of Items in One Field (Adjusting for Text Size)

Concentration of items	Number of families	
	All items	Most frequent items (above mean)
40%–59% of occurrences in one field	307 (53.9%)	103 (53.6%)
60%–79% of occurrences in one field	154 (27.0%)	48 (25.0%)
Over 80% of occurrences in one field	73 (12.8%)	19 (9.9%)
Total families with uneven distribution	534 (93.7%)	170 (88.5%)

at least 60% of all instances concentrated in just one subcorpora, and 13% have at least 80% in a single subcorpora. Among the most frequent items, more than 90% of all cases of *participate, communicate, output, attitude, conflict, authority, perspective,* and *simulate* occurred in one subcorpora.

Overall, only 36 word families were relatively evenly distributed across the science, engineering, and social science subcorpora, and so might therefore be regarded as candidates for an academic word list, albeit a very limited one. Of these, however, only 22 were frequent by our criterion of being above the overall mean, and only seven were in the top 60 items. Just six families appeared in the top 60 of both Coxhead's list and our own: *analyze, consist, factor, indicate, period,* and *structure,* which together comprised about 0.5% of the corpus. In other words, although the AWL may describe certain high frequency words in the register, the distributions in our data suggest that most items have a limited range across subcorpora.

The concentration of items is no less pronounced when we move to a finer level of delicacy and examine how words are distributed within subcorpora. Table 14.5 shows the extent of these concentrations, with 283 items in engineering (52% of all families) having more than 65% of all cases in just one discipline, 244 items in the sciences (43%) with more than 65% in just one discipline, and 128 (22.5%) of items in the social sciences with more than 65% in one discipline. Overall, only one family occurred roughly equally across the three disciplines in the sciences, seven families spread evenly across the social sciences, and 47 across the two engineering fields.

Once again then, the patterns suggest a more complex picture of language use in the disciplines than notions of a general academic vocabulary allow, pointing to more specialized language uses.

Word meanings and uses

There is a further difficulty with compiling a so-called common core of academic vocabulary in that not only should it include items that meet frequency and range criteria, but also items which behave in roughly similar ways across disciplines. When reading academic texts, students need to be confident that they are understanding words in the right way, which means a vocabulary list must either avoid items with clearly different meanings and dissimilar co-occurrence patterns, or these items must be taught separately rather than as parts of families. We need, then, to be cautious about claiming generality for families whose meanings and collocational environments may differ across each inflected and derived word form (Oakey, 2003).

Most words have more than one sense and Wang and Nation (2004) have recently examined the potential *monosemic bias* of the AWL to discover if cases of *homography,* or unrelated meanings of the same written form, misrepresent the composition of word families and so affect the learning burden for students. Although Wang and Nation found only a small number of families containing homographs (e.g., *major* which can mean both *important* and *military rank*), they further suggest that words have essentially similar meanings across fields. In our corpus, however, different disciplines showed clear preferences for particular meanings and collocations. This finding has implications for the notion of an academic vocabulary and attempts to base word lists on it. As brief examples, we might take the two most frequent AWL items in the corpus, *process* and *analyze,* both of which occur far more often in academic discourse than in other registers.

Despite its high frequency in all three subcorpora, the word *process* is far more likely to be encountered as a noun by science and engineering students

TABLE 14.5 Concentration of Items in Disciplines (% Adjusted for Corpus Size)

Disciplines	Total Families	Total of all items occurring in one discipline		
		40%–64%	65%–79%	Over 80%
Engineering	542	259 (47.8%)	133 (24.5%)	150 (27.7%)
Sciences	568	322 (56.7%)	116 (20.5%)	128 (22.5%)
Social Sciences	570	409 (71.8%)	74 (13.0%)	54 (9.5%)
Overall	570	336 (59.0%)	110 (19.4%)	114 (20.0%)

346 THE ESSENTIAL HYLAND

346 THE ESSENTIAL HYLAND

than by social scientists. This likelihood is the result of the well known practice of nominalization, or *grammatical metaphor* (Halliday, 1998), which refers to the way that writers in the sciences regularly transform experiences into abstractions to create new conceptual objects. Embedding an item such as *process* into complex abstract nominal groups produces terms such as

- *A constant volume combustion process . . .*
- *the trouble call handling process . . .*
- *processing dependent saturation junction factors . . .*
- *the graphical process configuration editor . . .*

Though nominalization enables writers in the sciences to give new objects stable names and discuss their properties without further investigation, it does not help novice users unpack specialized meanings from the individual lexical item. We believe this practice is therefore likely to present difficulties to both native and nonnative-English-speaking students.

Similarly, *analyze* appears to be used differently across fields. In the social sciences, it tends to occur more regularly as a noun, while in engineering, students are six times more likely to come across the form *analytical.* There are also semantic differences. The word *analysis,* for instance, is often associated with particular types of approach, so that it appears in discipline-specific compound nouns such as *genre analysis* or *neutron activation analysis.* The verb form also has field-specific meanings. In engineering, it tends to refer to methods of determining the constituent parts or composition of a substance (see Examples 1 and 2), and in the social sciences it tends to mean simply considering something carefully (see Examples 3 and 4):

> *Example 1.* The proteins were separated by electrophoresis on 10% polyacrylamide SDS gels and *analyzed* by autoradiography. (Physics doctoral dissertation)
>
> *Example 2.* We used a variety of methods to *analyse* fungal spore load, volatiles and toxins. (Biology article)
>
> *Example 3.* The major objective of this report is to *analyze* developments in political sociology over the last half century. (Sociology article)
>
> *Example 4.* Following this, the results were *analyzed* to determine buyer behaviour. (Business master's thesis)

A random analysis of AWL families with potential homographs reveals a considerable amount of semantic variation across fields. Table 14.6 shows the main meanings for selected words with different overall frequencies in

TABLE 14.6 Distribution of Meanings of Selected Academic Word List Word Families Across Fields (%)

Family	Meaning	Science	Engineering	Social Science
Consist	stay the same	34	15	55
(rank 41)	made up of	66	75	45
Issue	flow out	7	6	18
(46)	topic	93	94	82
Attribute	feature	83	35	60
(93)	ascribe to	17	65	40
Volume	book	1	7	50
(148)	quantity	99	93	50
Generation	growth stage	2	2	36
(245)	create	98	98	64
Credit	acknowledge	0	60	52
(320)	payment	100	40	48
Abstract	précis/extract	76	100	13
(461)	theoretical	14	0	87
offset	counter	0	14	100
(547)	out of line	100	86	0

the AWL together with their distributions. The first four are from our high frequency list, with occurrences above the overall mean, and show that even where uses are very frequent, preferred uses still vary widely, with social science students far more likely to meet *consist* as meaning "to stay the same" and science and engineering students very unlikely to come across *volume* meaning "a book or journal series" unless they are reading book reviews. With less frequent words the preferred meanings differ dramatically, and we have to question the value of teaching the secondary meanings at all.

These preferred uses become more apparent when we consider patterns at the disciplinary level. In Table 14.7, for instance, we can see minimal uses of *issue* to mean "flow out" in all disciplines but business, an overwhelming preference for *attribute* to mean a "feature" in business, sociology, and computer science and as the verb "to accredit" in applied linguistics, biology, and physics. Similar disciplinary reversals arise in the preferred uses of *volume, credit, abstract,* and *offset* among these examples, and no doubt there are more in the corpus that we have not analyzed.

In addition, all disciplines adapt words to their own ends, displaying considerable creativity in both shaping words and combining them with others to convey specific, theory-laden meanings associated with disciplinary models

TABLE 14.7 Distribution of Meanings of Selected Academic Word List Word Families Across Disciplines (%)

Families	Meanings	AL	BS	Soc	EE	ME	Bio	CS	Phy
Consist	stay the same	54	74	63	30	20	38	64	35
(41)	made up of	46	26	37	70	80	62	36	65
Issue	flow out	3	28	4	14	0	0	0	0
(46)	topic	97	72	96	86	100	100	100	100
Attribute	feature	21	90	71	36	56	12	99	0
(93)	ascribe to	79	10	29	64	44	88	1	100
Volume	book	96	1	80	5	41	4	0	14
(148)	quantity	4	99	20	95	59	96	100	86
Generation	growth stage	38	18	71	1	6	3	8	1
(245)	create	62	82	29	99	94	97	92	99
Credit	acknowledge	65	9	71	100	44	0	0	0
(320)	payment	35	91	29	0	66	0	100	0
Abstract	précis/extract	12	28	20	75	80	100	92	0
(461)	theoretical	88	72	80	25	20	0	8	100
offset	counter	100	100	100	4	0	25	0	0
(547)	out of line	0	0	0	96	100	75	100	100

Note. AL = applied linguistics, BS = business sciences, Soc = sociology, EE = electronic engineering, ME = mechanical engineering, Bio = biology, CS = computer science, Phy = physics.

and concepts. Thus, only texts in business studies contain the form *issuer,* for example, and 56% of instances of *convert* are related to "convertible security" (42% of all cases) and "convertible bond" (14% of all cases).

Also creating difficulties for learners is the fact that words take on additional meanings as a result of their regular co-occurrence with other items (e.g., Arnaud & Bejoint, 1992), so that the term *value* in computer science, for instance, is often found as *value stream* (21% of all cases) and *multiple-value attribute mapping* (7% of all cases). Even high frequency items such as *strategy* have preferred associations with *marketing strategy* forming 11% of all cases in business, *learning strategy* making up 9% of cases in applied linguistics, and *coping strategy* constituting 31% of cases in sociology. Nor are word choices themselves simply arbitrary selections from an equally appropriate pool of semantically equivalent candidates. Frequency patterns reveal clear disciplinary preferences as routine uses take on the constancy of convention. The word *phase,* for example, is favored in biology, where it co-occurs with a limited range of items such as *stationary* and *mobile;* in mechanical engineering, *stage* is preferred, and in applied linguistics, the preferred term is *period.*

In sum, these different word choices, collocates, and fixed phrases colour the everyday uses of words with more particular discipline-specific meanings, reflecting how writers need to represent themselves and their ideas through a locally appropriate theoretical and methodological framework. These patterns reinforce the view that particular *lexical bundles,* or strings of words which "commonly go together in natural discourse" (Biber, Johansson, Leech, Conrad, & Finegan, 1999, p. 990), also contribute to meaning-making in academic contexts. Such stable word combinations are an important part of a discipline's discoursal resources but enormously complicate the business of constructing general word lists. By breaking into single words items which may be better learnt as wholes, vocabulary lists simultaneously misrepresent discipline-specific meanings and mislead students.

Discussion

Despite the pedagogic attraction of a universal academic vocabulary, there appear to be good reasons to approach the concept with caution. A growing body of research suggests that the discourses of the academy do not form an undifferentiated, unitary mass, as might be inferred from such general lists as the AWL, but constitute a variety of subject-specific literacies (e.g., Hyland, 2000a, 2002b). Such ideas as co*mmunicative competence* in applied linguistics, *situated learning* in education, and *social constructivism* in the social sciences have contributed to a view which places the notion of *community* at the heart of writing. Each subject discipline represents a way of making sense of human experience that has evolved over generations, and each is dependent on its own particular rhetorical practices.

Words are often associated with different meanings and uses across registers (e.g., Biber et al., 1999) and we tend to find similar variations in the practices in different disciplines. As Trimble (1985) cautions, academic vocabulary can take on extended meanings in technical contexts, and in different disciplinary environments words may have quite different meanings. We noted earlier, for instance, that the AWL families were underrepresented in the sciences, suggesting the need to recognise a distinct scientific vocabulary. The resources which have developed in the sciences for construing reality as a world of logical relations and abstract entities are far removed from our routine ways of describing the world and so represent a more precise disciplinary lexical arsenal. But all fields, not only the sciences, draw on a specialized lexis.

All academic representations shape and manipulate language for disciplinary purposes, often refashioning everyday terms so that words take on more specific meanings. In fact, it would be surprising if specialised meanings had not evolved to discuss field-specific topics more precisely. Persuasion in academic writing involves using language to relate individual beliefs to shared experience, fitting observations and data into patterns which are meaningful to disciplinary insiders (Bazerman, 1994; Geertz, 1988). Writers must encode ideas and frame arguments in ways that their particular audience will find most convincing, drawing on conventional ways of producing agreement between members and frequently moulding everyday words to the distinctive meanings of the disciplines (e.g., Hyland, 2000a; Myers, 1990). In other words, different views of knowledge, different research practices, and different ways of seeing the world are associated with different forms of argument, preferred forms of expression, and, most relevantly, specialised uses of lexis.

Corpus evidence shows that words do not occur randomly in language use and that choices are governed by both rule-based systems of categories (e.g., Sinclair, 1991) and community-based conventional practices. The findings of this study show that these practices do not just operate at the level of register, as assumed in the notion of *academic vocabulary,* but indicate greater specificity which undermine claims that there is a general vocabulary of value to all students preparing for, or engaged in, university study. We have found that various parameters of variation support this assertion, and analysis of our corpus using the most recent list reveals uneven word frequencies, restricted item range, disciplinary preferences for particular items over semantic equivalents, and additional meanings lent to items by disciplinary convention and associations in lexical bundles.

Many items are considerably underrepresented in particular fields or disciplines, and this clustering in one field suggests a considerable degree of disciplinary specificity in their use. The AWL seems to be most useful to students in computer sciences, where 16% of the words are covered by the list, and least useful to students in biology, with only 6.2% coverage. Coxhead acknowledged a bias toward the commerce subcorpus in her study but investigated no further. It seems likely, however, that this bias resulted from the inclusion of disciplines which shared greater similarities in her commerce corpus (e.g., accounting, economics, and finance), and dissimilar disciplines in arts (e.g., psychology, history, and linguistics) and sciences (e.g., geography, mathematics, and biology). This procedure seems to have produced remarkably high frequencies of words in the AWL common to finance-oriented disciplines, such as *consume, corporate, invest, partner, purchase,* and *finance.* These words all occur overwhelmingly in business studies and, because it tends to develop business-related applications, in the computer studies texts in our corpus.

Similarly, a law subcorpus comprises a full quarter of Coxhead's database, and a glance at the AWL shows a considerable number of law-related headwords (*legal, legislate, regulate, compensate,* etc.) which are unlikely to figure significantly in the texts of most undergraduates. We are also concerned that the AWL corpus is partly made up of 2,000-word text fragments rather than whole texts and that some of these, namely the Brown and LOB Corpora, date back almost 30 years. In short, we believe that the selection of Coxhead's corpus may have skewed the AWL and squeezed other potential candidates out of the corpus altogether.

These findings therefore serve to undermine the value of relying on decontextualized lists of vocabulary as a source of generally available and equally valid items for student writers across the disciplines. Within each discipline or course, students need to acquire the specialized discourse competencies that will allow them to succeed in their studies and participate as group members. This means seeing literacy as something we do, an activity "located in the interactions between people" (Barton and Hamilton, 1998, p. 3), rather than as a thing "distanced from both teacher and learner and imposing on them external rules and requirements" (Street, 1995, p. 114). Because literacy is a practice integral to particular social and institutional contexts, we are forced to acknowledge that lists which claim to represent an academic vocabulary are likely to have a limited usefulness. The words they contain are unlikely to be of equal value to all students, and many words will be of almost no use to them at all. As teachers, we have to recognize that students in different fields will require different ways of using language and so we cannot depend on a list of academic vocabulary.

We are therefore unable to support the division between *academic* and *technical* vocabulary assumed by list compilers. The fact is that although some words appear to be more generally used in the register, ranging across several disciplines, these items are not used in the same way and do not mean exactly the same thing in different disciplinary contexts. Instead, it might be more accurate to regard academic vocabulary as a cline of technically loaded or specialized words ranging from terms which are only used in a particular discipline to those which share some features of meaning and use with words in other fields. The general point is, however, that we need to identify students' target language needs as clearly as possible and to address these needs as well as we can, and part of this work will involve introducing, making salient, and practicing the specialized vocabulary of their fields or disciplines.

Nor do we find reasons to be sanguine about the unexamined view that students first acquire a multipurpose general service vocabulary which they only then top up with an academic repertoire. Clearly, many students learn English for purposes other than higher education and may not encounter large numbers of academic words in their studies. But often students do not follow this kind of trajectory and, instead, build their vocabularies less systematically, as they come across words from a variety of sources. Second language acquisition research indicates that students are likely to acquire items as they need them rather than in a taught sequence, and students will meet many of these academic words before gaining control of a general service vocabulary. This sort of acquisition is particularly likely given the fact that the AWL assumes knowledge of the words in the rather archaic GSL, which is now more than 50 years old and does not reflect current usage. It is not hard to imagine, for example, a student encountering so-called academic items (from the AWL) such as *available, percent, achieve, similar,* and *assist* before GSL items such as *headdress, shilling, redden, cultivator,* and *beak.*

In sum, because academic knowledge is embedded in processes of argument and consensus-making, it will always be particular to specific disciplines and the socially agreed on ways of discussing problems in those disciplines. The fact that writing actually helps to create disciplines, rather than being just another aspect of what goes on in them, is a serious challenge to identifying uniformities in academic language use that apply in the same ways to all disciplines. If this view is correct, then it is also a serious challenge for students crossing disciplinary boundaries in modular and joint degree courses. As Bhatia (2002) observes: "Students interacting with different disciplines need to develop communication skills that may not be an extension of general literacy to handle academic discourse, but a range of literacies to handle disciplinary variation in academic discourse" (p. 27). It is difficult to see how students equipped with a general academic vocabulary are in a position to make these transitions.

Although this article may tend toward the negative, we have not set out to discredit the AWL. On the contrary, we feel it offers a useful characterization of register-level vocabulary choices which may provide learners with a basis for challenging stereotypes or examining specific practices in their own fields. We also agree with the pedagogical principles that lay behind it: that teachers should seek to teach the most relevant and useful vocabulary to their students and that corpus analyses are the best way of ascertaining which vocabulary to teach (Coxhead, 2002). However, the evidence from this study urges caution in seeing the surfaces of texts as assemblies of discrete words with similar uses and meanings. By considering context, cotext, and use, *academic vocabulary* becomes a chimera. It gives a misleading impression of uniform practices and offers an inadequate foundation for understanding disciplinary conventions or developing academic writing skills.

Conclusions and implications

Designing EAP materials is a complex task and, by arguing for more specific approaches to vocabulary teaching, it may seem that we are making it even harder. There are, however, important theoretical and pedagogic issues at stake which make it important to examine the assumptions that inform the idea of a general academic vocabulary. Modern conceptions of EAP, and language teaching more generally, stress the role of communication rather than language and the processes by which texts are created and used as much as they stress the texts themselves. Thus, teaching looks beyond lists of common core features and the autonomous views of literacy that such lists assume, recognizing that contextual factors are crucial to language choices because we communicate as members of social groups.

Vocabulary lists, such as the AWL, may provide some guidelines for teaching purposes, but if we find that individual items occur and behave in dissimilar ways in different disciplines then we are forced to acknowledge the importance of contextual environments which reflect different disciplinary practices and norms. Obviously vocabulary is more than individual words acting separately in a discourse, and compilers of vocabulary lists rightly insist that items should not be learnt out of context (e.g., Coxhead, 2000, 2002; Coxhead & Nation, 2001; Nation, 2001). Acquisition clearly needs to be part of a well-planned and sequenced program, with a mix of explicit teaching and incidental learning, a range of activities which focus on elaboration and consolidation, and sufficient information about contexts and definitions. We would argue, however, that the most appropriate starting point for such a program, offering the best return for learning effort, is the student's specific target context.

Like the AWL itself, corpus-informed lists and concordances can be used to help establish vocabulary learning goals for EAP courses, design relevant teaching materials, and generally target instruction more carefully. We believe it is important, however, that these lists and concordances are derived from the genres students will need to write and the texts they will need to read. Learners should be encouraged, for example, to notice these high frequency items and multiword units through repeated exposure and through temporary decontextualisation activities such as matching and item identification activities. Consciousness-raising tasks which offer opportunities to retrieve, use, and manipulate items can be productive, as can activities which require learners to produce the items in their extended writing. Students can, in fact, be encouraged to explore specialised corpora themselves and to try to identify meanings inductively from repeated examples or to familiarise themselves with common collocates.

In sum, although the generic label *academic vocabulary* may be a convenient shorthand for describing a general variety, it conceals a wealth of discursive variability which can misrepresent academic literacy as a uniform practice and mislead learners into believing that there is a single collection of words which they can learn and transfer across fields. As we have learnt more about the different contexts in which students find themselves at university, the complexity of the particular communicative demands their studies placed on them has become increasingly clear. We have began to see that many language features, including vocabulary, are specific to particular disciplines, and that the best way to prepare students for their studies is not to search for overarching, universally appropriate teaching items, but to provide them with an understanding of the features of the discourses they will encounter in their particular courses.

15

As can be seen: Lexical bundles and disciplinary variation

1 Introduction

Multi-word expressions are an important component of fluent linguistic production and a key factor in successful language learning. While students might struggle to master phrasal verbs such as *look after* and *sell out* or idioms like *in a nutshell* and *beat about the bush,* these are relatively rare compared with the frequently occurring word sequences which Biber, Johansson, Leech, Conrad, and Finegan (1999) call *lexical bundles* and Scott (1996) refers to as *clusters.* Essentially, these are words which follow each other more frequently than expected by chance, helping to shape text meanings and contributing to our sense of distinctiveness in a register. Thus the presence of extended collocations like *as a result of, it should be noted that,* and *as can be seen* help identify a text as belonging to an academic register while *with regard to, in pursuance of,* and *in accordance with* are likely to mark out a legal text.

These bundles are familiar to writers and readers who regularly participate in a particular discourse, their very 'naturalness' signalling competent participation in a given community. Conversely, the absence of such clusters might reveal the lack of fluency of a novice or newcomer to that community. Haswell (1991, p. 236), for example, suggests that

> there can be little doubt that as writers mature they rely more and more on collocations and that the lesser use of them accounts for some characteristic behaviour of apprentice writers.

In other words, gaining control of a new language or register requires a sensitivity to expert users' preferences for certain sequences of words over others that might seem equally possible. So, if learning to use the more frequent fixed phrases of a discipline can contribute to gaining a communicative competence in a field of study, there are advantages to identifying these clusters to better help learners acquire the specific rhetorical practices of their communities.

Yet while studies point to the considerable variation of bundles in different genres (e.g. Biber, 2006; Biber, Conrad, & Cortes, 2004; Hyland, 2008a; Scott & Tribble, 2006), how far they differ by discipline remains uncertain. This is the issue I address in this paper, examining a 3.5 million word corpus to identify the forms and functions of 4-word bundles across four contrasting disciplines.

2　Bundles, collocations and communities

The study of formulaic patterns has a long and distinguished history in applied linguistics, dating back to Jespersen (1924) and to Firth (1951), who popularised the term 'collocation' along with the famous slogan that 'you shall judge a word by the company it keeps'. More recently, Nattinger and DeCarrico (1992) have emphasised the importance of frequent multi-word combinations as a way of assisting communication by making language more predictable to the hearer. Wray and Perkins (2000), for instance, argue that such sequences function as processing short-cuts by being stored and retrieved whole from memory at the time of use rather than generated anew on each occasion. The extensive use of such pre-fabricated sequences as *it has been noted that* in academic written genres, for instance, helps to signal the text register to readers and reduce processing time by using familiar patterns to link elements of new information.

Text receivers are therefore able to sort out what is natural from what is merely grammatical and judge whether a particular collocation 'sounds right' in that context. Thus, *as can be seen* is a frequent and unremarkable collocation in academic writing while the equally possible *as you can see* or *as can be observed* are rarely encountered. What I shall call 'bundles', or frequently recurrent strings of uninterrupted word-forms, thus appear to represent a psychological association between words and reflect a very real part of users' communicative experiences. The key idea here is that of *collocation* or "the relationship that a lexical item has with items that appear with greater than random probability in its textual context" (Hoey, 1991:6). This extension of formulaic phrases to regular collocations such as *Have a nice day* and *I want to make three points* therefore hints at the extent of formulaicity in language use, with Altenberg (1998) suggesting that as much as 80% of natural language could be patterned in this way.

This pervasiveness has, in fact, led writers such as Sinclair (1991) and Hoey (2005) to propose radical new theories of language to replace our traditional conceptions of grammar. Instead of seeing lexical choices as constrained by the slots which grammar make available for them, they regard lexis as systematically structured through repeated patterns of use. As Sinclair (1991, p. 108) observes:

> By far the majority of text is made of the occurrence of common words in common patterns, or in slight variants of those common patterns. Most everyday words do not have an independent meaning, or meanings, but are components of a rich repertoire of multi-word patterns that make up a text. This is totally obscured by the procedures of conventional grammar.

In other words, grammar is the output of repeated collocational groupings. Sentences are typically made up of interlocking bundles as words are mentally 'primed' for use with other words through our experience of them in frequent associations (Hoey, 2005). Everything we know about a word is a result of our encounters with it, so that when we formulate what we want to say, the wordings we choose are shaped by the way we regularly encounter them in similar texts.

A range of corpus studies have shown how ubiquitous these bundles are in academic genres. Defining lexical bundles as combinations that recur at least 10 times per million words and across five or more texts, Biber et al. (1999, p. 994) suggest that 3-word bundles occur over 60,000 times and 4-word bundles over 5000 times per million words in academic prose. While the majority of words in any text do not occur in recurrent combinations, about 21% of the 5.3 million words of the academic component of the Longman Spoken and Written English corpus make up these common bundles, with the most frequent strings featuring over 200 times per million words.

To illustrate some of the features of these forms, Table 15.1 shows the most frequent 3-, 4-and 5-word bundles in my 3.5 million word corpus of academic writing in articles, PhD dissertations, and Master's theses. The lists highlight the fact that many of the most frequent bundles in academic writing are extremely common indeed, and that these frequencies drop dramatically as strings are extended to five words and beyond. It is also clear that many 3-word bundles such as *on the other* and *it can be* frequently expand into the 5-word bundles *on the other hand the* and *it can be seen that,* supporting Cortes (2004) observation that many four and five word strings 'hold 3-word bundles in their structure'.

TABLE 15.1 Most frequent 3-word, 4-word and 5-word bundles in the 3.5 million word academic corpus

3-Word	Freq.	4-Word	Freq.	5-Word	Freq.
in order to	1629	on the other hand	726	on the other hand the	153
in terms of	1203	at the same time	337	at the end of the	138
one of the	1092	in the case of	334	it should be noted that	109
the use of	1081	the end of the	258	it can be seen that	102
as well as	1044	as well as the	253	due to the fact that	99
the number of	992	at the end of	252	at the beginning of the	98
due to the	886	in terms of the	251	may be due to the	64
on the other	810	on the basis of	247	it was found that the	57
based on the	801	in the present study	225	to the fact that the	52
the other hand	730	is one of the	209	there are a number of	51
in this study	712	in the form of	191	in the case of the	50
a number of	690	the nature of the	191	as a result of the	48
the fact that	630	the results of the	189	at the same time the	41

Bundle	Freq	Bundle	Freq	Bundle	Freq
most of the	605	the fact that the	177	is one of the most	37
there is a	575	as a result of	175	it is possible that the	36
according to the	562	in relation to the	163	one of the most important	36
the present study	549	at the beginning of	158	play an important role in	36
part of the	514	with respect to the	156	can be seen as a	35
the end of	501	the other hand the	154	the results of this study	35
the relationship between	487	the relationship between the	152	from the point of view	34
in the following	478	in the context of	150	the point of view of	34
the role of	478	can be used to	148	it can be observed that	33
some of the	474	to the fact that	143	this may be due to	32
as a result	472	as shown in figure	136	an important role in the	31
it can be	468	it was found that	133	in the form of a	31

Additionally, the table indicates that most bundles, unlike idiomatic phrases, are semantically transparent and formally regular, providing the building blocks of coherent discourse. They are, in other words, identified empirically purely on the basis of their frequency rather than their structure, as they typically span structural units. In particular, we might note the considerable use of what Biber et al. (1999, p. 995) call noun phrase + post modifier fragments (*the number of, the relationship between the, one of the most important*), preposition + *of* phrase fragments (*in terms of, on the basis of, at the beginning of the*), as well as anticipatory *it* fragments (*it can be, it was found that, it should be noted that*) (Hyland & Tse, 2005).

An important feature of bundles however is their variability across different genres. Biber (2006), for instance, discovered that the spoken genre of classroom teaching uses about twice as many different bundles as conversations and about four times as many as textbooks. He suggests that this extremely high density can be explained by the fact that teaching draws heavily on both oral and written genres. He also found that the bundles are required to do very different jobs in the two genres, with classroom talk comprising much higher proportions of discourse organisers (*going to talk about, it has to do with*) and stance bundles (*I don't know if, I want you to*) than textbooks. Similarly, Cortes (2004), Scott and Tribble (2006) and Hyland (2008a), the latter using the corpus discussed here, also found systematic differences between genres, with bundles typical of published academic prose being far less common in writing by second language students.

In fact, it is often a failure to use native-like formulaic sequences which identifies students as outsiders and there is a general consensus that formulaic sequences are difficult for L2 learners to acquire (e.g. Yorio, 1989). Control of a language involves a sensitivity to the preferences of expert users for certain sequences of words over others, but students can have enormous difficulty distinguishing the idiomatic from the merely grammatical. But while there seem to be potentially enormous benefits in identifying the most frequent forms for teaching, we need to be cautious in making assumptions about the generality of academic bundles. Both Sinclair and Hoey point out, for instance, that because we all have different textual experiences, we all have a different mental concordance to draw on so that particular patterns are cumulatively loaded with the contexts we participate in. So just as individual lexical items occur and behave in different ways across disciplines (Hyland & Tse, 2007), we need to be sure we are assisting learners towards an appropriate disciplinary-sensitive repertoire of bundles.

Applied linguists and language teachers have therefore increasingly come to see bundles as important building blocks of coherent discourse and characteristic features of language use in particular settings. But despite their importance to language production, questions remain concerning their disciplinary specific use. Analysis of specialist corpora can therefore help us to understand the kinds of language data which particular communities of users might encounter and which will inform their use. I turn to this issue now, examining bundles in the principal research genres of four contrasting disciplines.

3 Corpora and methods

Data for the study consist of three electronic corpora of written texts comprising research articles, PhD dissertations and MA/MSc theses from four disciplines (Table 15.2). The disciplines were chosen to represent a cross-section of academic practice: electrical engineering (EE) and microbiology (Bio) from the applied and pure sciences, and business studies (BS) and applied linguistics (AL) from the social sciences. The research article (RA) corpus consists of 120 published papers comprising 30 in the leading journals of each of the four disciplines. The PhD and Master's corpora were written by Cantonese L1 speakers studying at five Hong Kong universities and contained 20 texts in each discipline.

I decided to focus on 4-word bundles because they are far more common than 5-word strings and offer a clearer range of structures and functions than 3-word bundles. Bundles are essentially extended collocations defined by their frequency of occurrence and breadth of use, but the actual frequency cut offs are somewhat arbitrary. This study takes a conservative approach by setting a minimum frequency of 20 times per million words and an occurrence in at least 10% of texts. Many of the higher frequency items, of course, figure far more often than this. WordSmith Tools 4 (Scott, 1996) was used to generate

TABLE 15.2 Corpora word counts

Discipline	Articles	Doctoral	Masters	Totals
Electrical engineering	107,700	334,800	190,000	632,500
Biology	143,500	458,000	192,600	794,100
Business studies	214,900	437,200	192,300	844,400
Applied linguistics	211,400	670,000	248,000	1,129,400
Totals	677,500	1,900,000	822,900	3,400,400

4-word bundle lists for the texts in each discipline, then to concordance examples to determine their functions. I then compared the frequencies and patterns across the disciplinary corpora.

The bundles were categorised both structurally, in terms of their grammatical types, and functionally, according to their meanings in the texts. While few bundles represent complete structural units in academic writing, it is possible to group them and Biber et al.'s classification was used for this purpose (Table 15.3).

It is also possible to identify general meanings and purposes of bundles (e.g. Biber, 2006; Cortes, 2004). Here I modified these earlier taxonomies to group bundles inductively, in ways which seemed to best represent their functions in my corpus. The framework, discussed further below, is organised around three broad functional categories (research-oriented, text-oriented, and participant oriented), with sub-categories grouping more specific roles. Now I turn to explore these general observations in more detail by comparing the preferences of the different groups.

TABLE 15.3 Most common patterns of 4-word bundles in academic writing (Biber et al., 1999, pp. 997–1025)

Structure	Examples
Noun phrase + of	the end of the, the nature of the, the beginning of the, a large number of
Other noun phrases	the fact that the, one of the most, the extent to which
Prepositional phrase + of	at the end of, as a result of, on the basis of, in the context of
Other prepositional phrases	on the other hand, at the same time, in the present study, with respect to the
Passive + prep phrase fragment	is shown in figure, is based on the, is defined as the, can be found in
Anticipatory it + verb/adj	it is important to, it is possible that, it was found that, it should be noted
Be + noun/adjectival phrase	is the same as, is a matter of, is due to the, be the result of
Others	as shown in figure, should be noted that, is likely to be, as well as the

4 Frequencies and structures of disciplinary bundles

There were 240 different 4-word bundles altogether in the 3.5 million word corpus, totalling nearly 16,000 individual cases or just over 2% of the total words. Table 15.1 above shows that *on the other hand* was by far the most frequent of these, that it occurred about 200 times per million words, and was over twice as common as the next placed bundles, *at the same time* and *in the case of.* The top ten bundles all occurred over 60 times per million words and the analysis suggests that most bundles in academic writing are parts of noun or prepositional phrases. There are, however, some interesting disciplinary differences. The electrical engineering texts contained the greatest range of bundles with 213 different 4-word strings meeting the 20 per million words threshold (across 10% of texts), and also the highest proportion of words in the texts occurring in 4-word bundles. Biology, on the other hand, had the smallest range of bundles, the fewest examples, and the lowest proportion of texts comprised of words in bundles. Table 15.4 summarises this frequency information.

Many bundles used by engineers are not found in the other disciplines and there is considerably greater reliance on pre-fabricated structures than in the other fields. It is difficult to say why this might be, but speculatively it could be a consequence of the relatively abstract and graphical nature of technical communication. The density of bundles in this corpus perhaps reflects the dependence of Engineering rhetoric on visual and numeric representation, so while arguments are based on plausible interpretations of data, they ultimately rest on findings which are often presented in visual form. The job of language is to fashion interpretations of course, but in technical subjects it also weaves an argument by linking data or findings in routinely patterned, almost formulaic ways.

TABLE 15.4 Bundle frequency information

Discipline	Different bundles	Total cases	% of total words in bundles
Electrical engineering	213	4562	3.5
Business studies	144	3728	2.2
Applied linguistics	141	4631	1.9
Biology	131	2909	1.7

At the other end of the table, Biology employs the smallest range of different bundles and the fewest bundles overall, although the actual proportion is similar to those in the two social science disciplines. Again, the reasons for these differences are unclear, but they are related to the distinctive ways that Biology pursues and argues problems. Although Biology is like electrical engineering in that it employs visuals to buttress its arguments, this is an altogether more discursive and descriptive discipline, with a less active and applied agenda. It is also a discipline more concerned with naming and coding than the other fields, with a more specialised readership, speaking to a relatively narrow group of scientist end users with specific interests in findings which inform their own research.

In addition to different frequencies, the corpora show that the principal structures of bundles also differ across fields. Table 15.5 gives the percentages of the main structures in each discipline in the corpus using the Biber et al. (1999, pp. 1014-1024) classification.

As can be seen, the noun phrase with *of*-phrase fragment is the most common structure overall, comprising about a quarter of all forms in the corpus. This covers a range of meanings in academic discourse and in particular is widely used to identify quantity, place or size (*the temperature of the, the base of the*), to mark existence (*a wide range of the presence of the*), or highlight qualities (*the nature of the, a function of the*). More interesting is

TABLE 15.5 Main structures of bundles across disciplines (%)

Structure	Biology	Electrical engineering	Applied linguistics	Business studies	Totals
Noun phrase + of	23.7	22.3	22.9	28.5	24.4
Passive + prepositional phrase	31.3	29.8	6.9	9.0	19.3
Other prepositional phrase	13.7	11.6	24.4	19.7	17.5
Prepositional phrase + of	9.2	7.9	19.9	16.0	13.5
Noun phrase + other modification	9.4	10.8	9.6	12.4	10.6
Others	6.4	9.2	10.7	9.9	9.5
Anticipatory it structure	6.3	8.4	5.6	4.5	2.5
Totals	100	100	100	100	

the difference between disciplines, with the social science corpora making far greater use of bundles beginning with a prepositional phrase. The majority of these have an embedded *of*-phrase, typically indicating logical relations between propositional elements:

(1) We generated multi-item scales *on the basis of* previous measures, a review of the relevant literature, and interviews with marketing and purchasing personnel. (BS)

. . .such transformations should be studied *in terms of the* semantic and ideological transformations they entail. (AL)

Alternatively, *in the case of* up-front financing, the VC is required to provide the amount of k in a lump-sum way up front. (BS)

Here then we see the emphasis of the soft knowledge fields on the discursive exploration of possibilities and limiting conditions, identifying and elaborating relationships in argument.

The Science and Engineering texts, on the other hand, employed significantly more passive bundles, normally followed by a prepositional phrase fragment typically marking a locative or logical relation. Generally, writers here either seek to guide readers through the text (2) or to identify the basis for an assertion in an argument (3):

(2) The experiment setup *is shown in Fig. 4.13.* (EE)

The value of Rs *is given by Eq. (3.11).* (EE)

All important events for pot trials *are summarised in Table 4.11.* (Bio)

(3) This apparent stability might *be due to the* complexing of plasma/serum DNA with proteins in the circulation. (Bio)

The measurement *is based on the* evaluation of infrared images produced by thermal waves. (EE)

Therefore, an antisense approach *can be used to* block the expression of endogenous agrin in PC12 cells. (Bio)

Identifying tabular or graphic displays of data and the bases of an assertion are typically constructed through formulaic passive constructions in the hard sciences. This both highlights the research or text feature being discussed and can help downplay the personal role of the scientist in the interpretation of data to suggest that the results would be the same whoever conducted the research.

Interestingly, the science writers also tended to employ more examples of the anticipatory-it pattern, which is another means of disguising authorial interpretations. These bundles introduce extraposed structures and function to foreground the writer's evaluation without explicitly identifying its source:

(4) *It is possible that* an increase in ethylene production in these fruits is mediated by CABA.   (Bio)

It is found that the optimal number of processing elements is application-dependent.   (Bio)

Referring to Fig. 1, *it can be seen that* each stage of the early bucket brigade circuit consists of an NPN bipolar transistor and a storage capacitor.   (EE)

I now turn to look at the patterns themselves and their distributions across the disciplines.

5 Patterns and variations

There were also considerable differences in the 4-word bundles themselves across disciplines. Table 15.6 shows the fifty most commonly used bundles in the four fields in frequency order, with items occurring in all four disciplines marked in bold and those occurring in three disciplines italicized.

The table may make depressing reading for commercial materials writers seeking to identify universals of academic writing and compile word lists for general academic purposes. Over half the items in each list do not occur at all in any other discipline and only 30% of the strings in each discipline are found in two other fields. Applied linguistics has 29 items in the top 50 which do not occur in any of the other lists and electrical engineering has 28. The discipline-specificity of these preferences for 4-word bundles is illustrated by the bold and italicized items, with only five bundles shared across all four disciplines and just 14 bundles occurring in three disciplines. Electronic engineering and applied linguistics shared just nine bundles, for example. The best candidate bundles for a general EAP course are *on the other hand, in the case of, as well as the,* and *the end of the,* all of which occur in the top band of bundles in at least three disciplines and so comprise bundles with high frequencies across fields.

The greatest affinity is between broadly cognate fields, as business studies and applied linguistics share 18 items, although only *on the basis of, in the context of, the relationship between the,* and *it is important to* are exclusive

TABLE 15.6 Most frequent 50 4-word bundles in four disciplines (Bold = item occurs in 4 disciplines, italic = items occurs in 3 disciplines)

Biology	Electrical engineering	Applied linguistics	Business studies
in the presence of	**on the other hand**	**on the other hand**	**on the other hand**
in the present study	as shown in figure	**at the same time**	**in the case of**
on the other hand	**in the case of**	*in terms of the*	**at the same time**
the end of the	is shown in figure	on the basis of	*at the end of*
is one of the	it can be seen	in relation to the	on the basis of
at the end of	as shown in fig	**in the case of**	**as well as the**
it was found that	is shown in fig	in the present study	the extent to which
at the beginning of	can be seen that	*the end of the*	*the end of the*
as well as the	*can be used to*	*the nature of the*	significantly different from zero
as a result of	the performance of the	*in the form of*	are more likely to
it is possible that	as a function of	**as well as the**	the relationship between the
are shown in figure	is based on the	*at the end of*	**the results of the**

(Continued)

TABLE 15.6 (Continued)

Biology	Electrical engineering	Applied linguistics	Business studies
was found to be	*with respect to the*	*the fact that the*	the hang seng index
be due to the	is given by equation	in the context of	*the other hand the*
in the case of	*the effect of the*	*is one of the*	in the context of
is shown in figure	the magnitude of the	in the process of	*as a result of*
the beginning of the	**at the same time**	**the results of the**	the performance of the
the nature of the	in this case the	in terms of their	hong kong stock market
the fact that the	it is found that	to the fact that	is positively related to
may be due to	the size of the	in the sense that	are significantly different from
are summarised in table	be seen that the	the relationship between the	*in terms of the*
has been shown to	the accuracy of the	of the hong kong	the degree to which
an important role in	**as well as the**	at the beginning of	in the long run
at room temperature for	the same as the	the role of the	in the united states
at the same time	*is one of the*	of the present study	*the nature of the*

can be used to	a function of the	as a result of	the total number of
in the absence of	*as a result the*	one of the most	the size of the
as shown in figure	**the results of the**	can be seen as	in the number of
with respect to the	*in the form of*	it is important to	it is important to
used in this study	is assumed to be	it should be noted	the standard deviation of
was added to the	of the power system	on the one hand	in the hong kong
a result of the	it is necessary to	can be found in	*with respect to the*
in addition to the	it is possible to	the ways in which	of the number of
the quality of the	the length of the	in other words the	*in the form of*
are listed in table	are shown in fig	*the other hand the*	the difference between the
is due to the	can be obtained by	the starting point of	by the end of
the presence of a	*in terms of the*	be seen as a	*the effect of the*
the results of the	are shown in figure	in the eyes of	is consistent with the
was found in the	is due to the	the beginning of the	the quality of the

(Continued)

TABLE 15.6 (Continued)

Biology	Electrical engineering	Applied linguistics	Business studies
were found to be	the structure of the	should be noted that	*as a result the*
a wide range of	is defined as the	that there is a	*can be used to*
the effect of the	it was found that	at the level of	in addition to the
the presence of the	*the other hand the*	for the purpose of	standard deviation of the
to the presence of	*the presence of the*	in hong kong and	*the fact that the*
was used as a	with the use of	are more likely to	in the presence of
as a result the	is the same as	the meaning of the	we assume that the
have been shown to	it can be observed	on the part of	is more likely to
in this study the	it is because the	the purpose of the	the efficiency of the
is possible that the	than that of the	*a wide range of*	the price of the
the base of the	will be discussed in	the use of the	*a wide range of*

to these two fields. Biology and electrical engineering have 16 bundles in common, with *it was found that, is shown in figure, as shown in figure, is due to the,* and *the presence of the* not found in the social science lists. The contrasts between these two short lists reflect something of the argument patterns in the two domains, with those in the first group largely connecting aspects of argument and those in the second group avoiding authorial presence while pointing to graphs and findings. It is worth noting that while there were no bundles referring to tables or figures in the applied linguistics corpus and only two in the business texts, both science lists included these as among their most frequent strings.

While consideration of the lexical composition of these formulaic strings is useful, we are better able to understand the roles they play in academic discourse by examining their discourse function and I turn to this in the next section.

6 Functions of bundles

A framework for analysing the bundles found in this corpus was developed from Biber's (Biber, 2006; Biber et al., 2004) classification. While my main categories are similar, differences in the two corpora necessitated modifications. Biber's taxonomy emerged from a much broader corpus of spoken and written registers which included casual conversation, textbooks, course packs, service encounters, institutional texts, and so on, and this seems to have yielded far more personal, referential, and directive bundles than my more research-focused genres. Biber, for instance employs stance as a *super-ordinate* category while I have folded it into a grouping in which bundles refer to either the writer or reader. In addition, the realisations in different categories are so different that use of the same sub-groups would invite unproductive comparisons. This classification therefore collects bundles into the three broad foci of research, text and participants, and introduces sub-categories which specifically reflect the concerns of research writing. These are:

Research-oriented – help writers to structure their activities and experiences of the real world includes:

- **Location** – indicating time/place (at the beginning of, at the same time, in the present study).

- **Procedure** (the use of the, the role of the, the purpose of the, the operation of the).

- **Quantification** (the magnitude of the, a wide range of, one of the most).
- **Description** (the structure of the, the size of the, the surface of the).
- **Topic** – related to the field of research (in the Hong Kong, the currency board system).

Text-oriented – concerned with the organisation of the text and its meaning as a message or argument includes:

- **Transition signals** – establishing additive or contrastive links between elements (*on the other hand, in addition to the, in contrast to the*).
- **Resultative signals** – mark inferential or causative relations between elements (*as a result of, it was found that, these results suggest that*).
- **Structuring signals** – text-reflexive markers which organise stretches of discourse or direct reader elsewhere in text (*in the present study, in the next section, as shown in figure*).
- **Framing signals** – situate arguments by specifying limiting conditions (*in the case of, with respect to the, on the basis of, in the presence of, with the exception of*).

Participant-oriented – these are focused on the writer or reader of the text (Hyland, 2005b) includes:

- **Stance features** – convey the writer's attitudes and evaluations (*are likely to be, may be due to, it is possible that*).
- **Engagement features** – address readers directly (*it should be noted that, as can be seen*).

Using this classification scheme, we find some functional categories are strongly connected to the structural patterns discussed earlier, with noun phrases + of structures prominent in research-oriented functions, prepositional phrase patterns in text-oriented functions, and *anticipatory it* largely occurring in participant functions. We can also note a roughly even split between research- and text-oriented bundles overall, with participant strings being far less frequent. Table 15.7, however, shows that once again there are differences in disciplinary distributions, pointing to variations in what writers are attempting to achieve through their linguistic choices.

6.1 Research-oriented bundles

One clear difference is the greater concentration of research-oriented bundles in the Science and Engineering texts, a preference which amounted to almost half of all bundles in the science/technology corpora. The scale of this use functions to impart a greater real-world, laboratory-focused sense to writing in the hard sciences. Many of these bundles contributed to the description of research objects or contexts, specifying aspects of models, equipment, materials or aspects of the research environment, and were typically realised by noun phrase + of structures:

(5) *The structure of the* coasting-point identification model (see Fig. 5.6) can be divided into the following areas for description. (EE)

. . . *the performance of the* coder is less affected by neither improper voiced-unvoiced classification nor voiced-unvoiced speech transitions of different durations. (EE)

The size of the perforations becomes progressively smaller towards *the base of the* apparatus. (Bio)

Over half of all cases, however, were used to depict research procedures, showing the ways that experiments and research were conducted:

(6) The DNA was precipitated *in the presence of* 2.5 volumes of ethanol and 0.1 volume of 3.0 M sodium acetate pH. (Bio)

Transmission phase angle modulation *can be used to* increase the stability of the system, by maintaining the angle at a low value. (EE)

All of the precipitate *was added to the* cells in a 100 mm culture plate or 300 mm of the precipitate to a 60 mm culture plate. (Bio)

TABLE 15.7 Distribution of bundle functions by discipline (%)

Discipline	Research-oriented	Text-oriented	Participant-oriented	Totals
Biology	48.1	43.5	8.4	100
Electrical engineering	49.4	40.4	9.2	100
Applied linguistics	31.2	49.5	18.6	100
Business studies	36.0	48.4	16.6	100
Overall	41.2	45.5	13.2	100

This emphasis on the ways the research was conducted plays an important role in conveying the grounded, experimental basis of research in the hard sciences. The physical practicalities of scientific study played a far greater part in the student discourses than in the articles, however, perhaps reflecting the ways that the Master's level students conceptualised their studies and approached the writing task. The Master's thesis is a pedagogic genre with a display and assessment purpose which clearly puts students under some pressure to demonstrate their ability to handle research methods appropriately and stake a claim to being comfortable with the subject content of the discipline. Consequently, bundles which set out procedures comprise a far higher proportion in the theses and dissertations, as do those which refer to the specific topic or context of the research:

(7) Thus by studying this type of faults, the transient stability *of the power system* under the most adverse condition can be determined.   (EE MSc)

This can improve the *signal-to-noise ratio* of the reconstruction and the reconstructed signal will become more natural.   (EE MSc)

Forty percent of the total land area *in Hong Kong is* designated as country parks which together covers an area of over 40,000 ha.   (Bio MSc)

These patterns may, therefore, reveal the preoccupations of the apprentice, and perhaps specifically the *second language* apprentice, demonstrating competence through the control of physical resources and disciplinary research practices. But the significantly greater use of research-oriented bundles in the hard knowledge fields also expresses something of a scientific ideology which emphasises the empirical over the interpretive, minimising the presence of researchers and contributing to the "strong" claims of the sciences. Highlighting research rather than its presentation places greater burden on research practices and the methods, procedures and equipment used, and this allows scientists to emphasise demonstrable generalisations rather than interpreting individuals. New knowledge, then, is accepted on the basis of empirical demonstration and experimental results designed to test hypotheses related to gaps in knowledge. The rhetorical conventions of the field, including the preferred patterns of 4-word bundles, help contribute to this epistemological framework.

6.2 *Text-oriented bundles*

In contrast, the applied linguistics and business studies corpora were dominated by text-oriented strings, which were particularly marked in the research articles where they comprised almost two-thirds of all bundles. This reflects the more discursive and evaluative patterns of argument in the soft knowledge fields, where persuasion is more explicitly interpretative and less empiricist, producing discourses which often recast knowledge as sympathetic understanding, promoting tolerance in readers through an ethical rather than cognitive progression (Hyland, 2004). So while claims are often based on observations of real-world phenomena, knowledge is typically constructed as plausible reasoning rather than as nature speaking directly through experimental findings. The presentation of research is therefore altogether more discursively elaborate, and text-oriented bundles are heavily used to provide familiar and shorthand ways of engaging with a literature, providing warrants, connecting ideas, directing readers around the text, and specifying limitations.

Perhaps not surprisingly, about 50% of text-oriented bundles in the social science texts worked to frame arguments by highlighting connections, specifying cases and pointing to limitations:

(8) The term 'linguistics' might be too narrow *in terms of the* diverse knowledge-base and expertise that is required in the applied linguist's job.   (AL)

However, *in the case of* Kodak's KIOO, which is an intricate piece of film, words are kept minimum to keep the viewer's attention.   (BS)

The levels are connected *in the sense that* it is impossible to appreciate the functioning at any one level without taking account of the other levels.
                                                                          (AL)

These bundles tend to be preposition + of structures and are used to focus readers on a particular instance or to specify the conditions under which a statement can be accepted, working to elaborate, compare and emphasise aspects of an argument.

While framing devices also comprised a high proportion of text-oriented bundles in the hard science corpora, there were far more in the applied linguistics and business texts. Here readers are often drawn from a wider knowledge base and include both those from other specialisms and disciplines and practitioners looking to apply the research in different areas. This readership is not only less cohesive than in the sciences but the research often has to be contextualised far more carefully and the connections between components explained in greater detail for readers unfamiliar with the thread of prior research.

The next most frequent group of bundles in the text-oriented category were structuring signals. A substantial portion of these help organise the text by providing a frame within which new arguments can be both anchored and projected, announcing discourse goals and referring to text stages:

(9) The *purpose of this paper* is to investigate the perceptions of consumers in the Hong Kong market toward a foreign service offering, specifically fast food.   (BS)

In this chapter we introduce a forecasting technique utilizing the notion of, global optimization to define the input-output membership functions with respect to. . . .   (EE)

In this section we offer evidence on the effect of corporate investment decisions on the market value of the firm.   (BS)

These bundles help frame, scaffold, and present arguments as a coherently managed and organised arrangement, reflecting writers' awareness of the discursive conventions of a sustained discussion and in consideration of the discoursal expectations and processing needs of a particular disciplinary audience. They are especially widespread in the much longer doctoral texts, where they help to structure arguments over a greater span of text. As Bunton (1999, p. S41) observes:

. . .it is the very length of the research thesis which makes it all the more important for the writer to continue to orient the reader throughout the thesis as to how the current subject matter relates to the overall thesis, i.e. to maintain cohesion and coherence.

Equally, however, these bundles represent an awareness of both argument and audience, and their use suggests writers' attempts to position themselves as competent academics able to control the rhetorical conventions of their fields.

Another group of structuring signals point to other parts of the texts to make additional material salient and available to the reader in recovering the writer's intentions. As mentioned earlier, the electrical engineers were particularly heavy users of these signals, reflecting their dependence on graphical and numerical information and the need to refer to these in their arguments:

(10) Their styles of being a facilitator *will be discussed in* the next chapter, indicating the favourable student factors that contributed to being a facilitator.   (AL)

As shown in Fig. 2, VDSATH is approximately equal to VDS when the transistor operates in the triode region.   (EE)

As shown in the example, process steps can be parameterised with materials object names. (EE)

Biologists, on the other hand, made considerable use of resultative markers, bundles which introduce writer's interpretations and understandings of research processes and outcomes. This is a key function in the rhetorical presentation of research as these bundles signal the main conclusions to be drawn from the study and highlight the inferences the writer wants readers to draw from the discussion:

(11) *The results of the* mating experiments clearly indicate the existence of two ISGs in C. subnuda. (Bio)

This *is due to the* precipitation of solid state CdS in the anoxic paddy soil. (Bio)

These results suggest that the observed variability is largely statistical, but that spatial variations cannot be entirely neglected. (Bio)

Resultative markers can frame an assertive construal of events, boosting the writer's position and directing readers to a categorical understanding, but more often they preceded a more conciliatory stance, as the last example here, downplaying any confidence the writer might have in his or her interpretation and opening a discursive space in which the reader might feel free to dispute it. Such considerations are at the heart of participant-oriented selections.

6.3 *Participant-oriented bundles*

Participant bundles provide a structure for interpreting a following proposition, conveying two main kinds of meaning: stance and engagement. These labels refer to writer-and reader-focused features of the discourse respectively, representing key aspects of interaction in texts (Hyland, 2005). While *stance* concerns the ways writers explicitly intrude into the discourse to convey epistemic and affective judgements, evaluations and degrees of commitment to what they say, *engagement* refers to the ways writers intervene to actively address readers as participants in the unfolding discourse.

Some two thirds of all participant-oriented bundles indicated the writer's stance, and the overwhelming majority of these were in the social science texts. Here, writers have to establish their claims through more explicit evaluation and engagement: personal credibility, and explicitly getting behind arguments, plays a far greater part in creating a convincing discourse for these writers:

(12) Such a dilemma *may be due to* the fact that they generally are unable to get support on English difficulties.   (AL)

Ventures with superior performance *are more likely to* keep the original designs or even develop towards separate entities.   (BS)

Nevertheless, *it is possible that* greater social interaction between marketing and Engineering managers would be beneficial to organizational interests.   (BS)

These few examples not only illustrate the use of stance bundles in the social science texts, but also the fact that they largely convey a reluctance to express complete commitment to a proposition, allowing writers to present information as an opinion rather than accredited fact. Hedges figure prominently here as do the anticipatory-*it* structures discussed above. These realisations help to protect the writer from possible false interpretations and indicate the degree of confidence that it may be prudent to attribute to the accompanying statement.

Not only are these stance bundles largely used to communicate uncertainty or caution, but they are also almost entirely expressed impersonally; in fact there is only one personal stance structure in the entire corpus, found in the applied linguistics collection:

(13) In concluding this chapter, *I would like to* emphasize that this study does not reject any theories proposed in previous studies on code-switching.   (AL)

Finally, *I would like to* suggest that the teaching of LSP should re-assess its current emphasis on the differences between professional groupings. (AL)

More usually, however, stance is expressed impersonally through bundles which employ models, epistemic adverbs and anticipatory-it patterns, as in example (12) above.

While stance bundles occurred principally in the social science corpora, and here overwhelmingly in the research articles, writers in the hard sciences largely employed strings which sought to engage readers. These were almost all directives (Hyland, 2002a), bundles which explicitly mark the presence of the 'reader-in-the-text' (Thompson, 2001) and instruct readers to perform an action or to see things in a way determined by the writer. Here the writer pulls the audience into the discourse at critical points to guide them to particular interpretations, typically by the use of a modal of obligation or a predicative adjective expressing the writer's judgement of necessity/importance:

(14) Intuitively, *we can see that* if the income levels of two economies become more similar over time, it must be the case that the poor economy is growing faster. (BS)

It should be noted that the extracted MAPs are associated with the polymerized tubulin. (Bio)

Second, *it is important to* recognize that the current state of knowledge in this area is still in its infancy. (AL)

In other words, although mixtures of zero al exists, *it is necessary to* carefully optimize the material parameters associated with the rotational viscosity. (EE)

Here the writer acknowledges the dialogic dimension of research writing, intervening to direct the reader to some action or understanding. These bundles therefore act to position readers, requiring them to notice something in the text and thereby leading them to a particular interpretation.

The relatively substantial presence of these items in the hard science corpora reflects the fact that these disciplines place considerable emphasis on precision, particularly to ensure the accurate understanding of procedures and results. The more linear and problem-oriented approach to knowledge construction found in the sciences allows arguments to be formulated in highly standardised, almost shorthand, ways which presuppose a degree of theoretical knowledge and routine practices not possible in the soft fields. As a result, directives offer writers an economical and precise form of expression which cuts more immediately to the heart of technical arguments. This high proportion of engagement bundles, however, also represents a reluctance to adopt a more intrusive personal voice through stance options, a rhetorical choice which reduces the writer's role as agent and interpreter and allows research to be presented as independent of any particular scientist.

I should point out that participant bundles were predominantly a feature of the research articles and that virtually all cases in the two student genres were examples of engagement. This avoidance of participant-oriented bundles may be a result of my student corpus and perhaps reflect the influence of a second language factor on these patterns. It certainly underlines a preference for impersonality by Hong Kong students found in other studies, which seems to result from both educational experiences and cultural preferences for a conciliatory, non-interventionist stance (Scollon & Scollon, 1995). While it is worth mentioning that stance and engagement are often expressed in other ways than 4-word bundles (e.g. Biber, 2006; Hyland, 2005), the relative absence of their use in the student corpus suggests that these writers may be uncomfortable in explicitly aligning themselves with a particular evaluation or personally attesting to the weight they want to attribute to their claims. Such investment clearly carries a certain risk in this extremely high stakes genre, and it appears to be one they do not wish to take.

7 Conclusions

My main purpose in this study has been to explore the extent to which phraseology contributes to academic writing by identifying the most frequent 4-word bundles in the key genres of four disciplines. The findings support studies by Cortes (2004) and Biber (2006) which show considerable variations in the frequency of forms, structures and functions across types of academic writing, but extend these studies by examining several disciplines and relating variations in the social and rhetorical practices of academic communities. The study indicates that writers in different fields draw on different resources to develop their arguments, establish their credibility and persuade their readers, with less than half of the top 50 bundles in each list occurring in any other list.

The results need to be treated with some caution, of course. I have not discussed the possible influence of first language on the findings in any detail and a corpus of first language students might well suggest different preferences, although at this level of proficiency I would be surprised if this were the case. I am also aware of the limitations of the size of my sample, as 3.5 million words is not a large corpus in terms of work being conducted today. More work with different disciplines, genres, and first language groups is likely to yield a fuller picture of community-specific practices. I hope, however, that I have done enough here to suggest that 4-word bundles should be regarded as a basic linguistic construct and that their distributions can help characterise disciplinary discourses.

While there is little space remaining for elaboration, these findings have clear implications for EAP practitioners. Not only do they reinforce the calls by Nattinger and DeCar-rico (1992), Lewis (1997), Willis (2003) and others for an increased pedagogical focus on bundles, but they also help undermine the widely held assumption that there is a single core vocabulary needed for academic study. Bundles occur and behave in dissimilar ways in different disciplinary environments and it is important that EAP course designers recognise this, with the most appropriate starting point for instruction being the student's specific target context. Corpus-informed lists and concordances can be used to help establish frequently occurring and otherwise productive bundles for EAP courses and the design of relevant teaching materials. It is important, however, that these lists and concordances are derived from the genres students will need to write and read. This means, for example, encouraging learners to *notice* these multi-word units through repeated exposure and through activities such as matching and item identification. Consciousness raising tasks which offer opportunities to retrieve, use and manipulate items can be productive, as can activities which require learners to produce the items in their extended writing.

Numerous studies now show the extent to which language features are specific to particular disciplines, and that the best way to prepare students for their studies is not to search for universally appropriate teaching items, but to provide them with an understanding of the features of the discourses they will encounter in their particular courses. The further study of bundles, I suggest, can offer insights into a crucial, and often overlooked, dimension of genre analysis and help provide us with a better understanding of the ways writers employ the resources of English in different academic contexts.

Commentary on Part IV

Diane Belcher

Applied linguistics prides itself, with good reason, on being descriptive, rather than prescriptive, of language use. This authentic-usage-oriented view, is, of course, the opposite of what we see in academic writing style manuals and many popular writing textbooks, which give us prescriptions for 'good writing' without the apparent backing of research on written discourse. Applied linguists' evidence-based approach to understanding academic discourse thus, arguably, has been a leap forward. This empirical advance, however, has not come without growing pains, as descriptiveness has frequently focused on surface, very often micro rather than macro, features of text, with texts taken as representative of context, but not necessarily examined in context, and scrutinized as end products rather than as turns taken in an ongoing conversation. The empirical data have also often been analysed by researchers working in isolation, outside the originating context, expecting us, their audience, to trust their reading of the data. Too much pedagogical eagerness to trust in the implications of such data can lead, as Canagarajah (2013), invoking Street (1984), has warned, to a largely autonomous view of literacy as a decontextualized set of skills, or a neo-normativeness, rather than acknowledgement of literacy as boundary-crossing, fluid and open to negotiation.

Some applied linguists have been eloquent advocates of more multiperspectival, contextualized empirical research on academic discourse. More than a decade ago, Flowerdew (2005) called for integration of corpus and genre-based approaches, combining the power of computer-assisted analysis of the micro-features of authentic academic discourse with a genre-theory-based macro-level view (see also Biber, Connor and Upton 2007). Others have argued for the combinatory methodological power of text analysis and ethnography, Swales's (1998) version of which he dubbed 'textography' (see also Paltridge and Wang 2011).

For Ken Hyland, theoretically and methodologically complex, contextualized and technologically cutting-edge approaches to exploring the features of academic discourse have been his modus operandi for several decades.

Hyland has employed corpus tools to construct specialized academic corpora focused on specific genres in specific disciplines as produced by writers of varying degrees of expertise, from students to professional academics. He has examined linguistic features of texts as lenses not only on the strategies used to realize the routine moves of genres and part-genres, for example, research articles and their sections, but also on the epistemological and ontological uniqueness of disciplines and the politics of their knowledge construction. Hyland has relied not just on his own analysis but enlisted the observations of disciplinary informants as well. In doing so, he has brought depth and breadth to our understanding of many of the features of various academic genres: on what motivates their use and how remarkably varied that usage is across disciplines. His findings, in turn, have offered a wealth of possible pedagogical implications to EAP materials developers, curriculum designers, classroom practitioners and teacher-trainers. Hyland's findings, however, will not, as he himself has cautioned more than once, make EAP professionals' lives easier, and they definitely will not support the autonomous discrete-skills view of literacy that Canagarajah and others have alerted us to. The dream of discovering a readily teachable fixed set of features of an all-encompassing academic discourse has little connection with the nuanced, variegated, dynamic reality of authentic academic usage that Hyland's research uncovers. But if the window Hyland opens complicates the lives of EAP professionals, it also offers keys to awareness of the inner workings of disciplinary discourse communities and to how newcomers might gain entry and exercise agency in them.

Just how varied discourse is across disciplines and how revelatory textual features can be of a field's 'epistemological and social convictions' (this book, p. 310) was made abundantly clear when Hyland published his 1999 article (Chapter 12, this book, p. 310) 'Academic attribution: Citation and the construction of disciplinary knowledge'. While citations, as Hyland noted, had already been identified as having rhetorical force (e.g. Swales 1990), citation usage in the context of specific disciplines was not well established. Adopting the mixed-methods approach that can now be seen as characteristic of much of Hyland's research, he employed corpus tools to build and analyse a database of eighty research articles from major journals in eight very different disciplines, ranging from philosophy to physics, and sought the emic insights of expert informants into the discourse practices of their fields. Although unique citation usage was found in each discipline, especially revealing were the differences between the hard and soft disciplines, that is, natural and applied sciences versus the humanities and social sciences, with the soft fields exhibiting decidedly more citations, more agentive integral citations (with cited authors foregrounded), more stance-attributing citations and more discourse-oriented (rather than research-action) reporting verbs. This citation-usage hard/soft divide more than suggests, as further confirmed by the disciplinary informants, the wider contextualization and more overt position-taking of the discursive soft disciplines as compared with the incremental, seemingly agentless consensus-building knowledge construction of the positivist hard disciplines. Although Hyland does not offer pedagogical implications in this study, the message for EAP practitioners is clear: that citation use should be taught as far more than obligatory knowledge display, and certainly more than mainly a way of avoiding plagiarism, but, instead, as a means of understanding and participating in the knowledge-making activities of individual disciplines.

At the end of his 1999 article on academic attribution (Chapter 12), Hyland observed that a productive next step would be to investigate self-citation and its interaction with both other-citation and writer purposes. Hyland did exactly this in his 2001 study 'Humble servants of the discipline? Self-mention in research articles' (Chapter 13). In this study, Hyland once again challenges our view of a generic entity called 'academic discourse', addressing, in particular, characterizations of it as an 'author-evacuated' type of prose (Geertz 1988; see also Elbow 1991) and the conflicting advice about authorial voice often given to academic writers. In this 2001 study, Hyland also again employed a mixed-methods, that is, quantitative/qualitative, approach, building a computer-analysable corpus complemented by discourse community-insider input. The corpus for this study is much like that of the 1999 work, comprising research articles from ten highly ranked journals in each of eight fields in the hard and soft disciplines, but this later corpus is three times the size of the earlier one, with 240 research articles, or 1,400,000 words, in the 2001 study. What Hyland found in this corpus was that there were no research articles, whether in mechanical engineering or sociology, that completely steered clear of some form of writer self-mention. Perhaps unsurprisingly, self-mention, especially in the guise of personal pronouns, was found to be far more prevalent in the more discursive soft fields, in which claims of uniqueness, in argument or perspective, tend to be more explicit. However, self-citation, a less common type of self-mention, was actually most frequent in one of the hard disciplines, biology, in which writers' research areas tend to be more narrowly defined than in the softer fields, and those working on exactly the same topic may be relatively few in number. So once more, disciplinary variation is evident, and textual features, in this case self-mention, index how research articles function as 'sites of disciplinary engagement' (this book, p. 317) Hyland offers a number of pedagogical suggestions based on these findings, encouraging such activities as guided rhetorical reading and student interviews of experts in their fields of interest. Perhaps the most myth-deflating revelation of this study for writing pedagogists to attend to is that there is always room for self-expression in academic writing, though the forms it takes, from more to less explicit and frequent, go hand in hand with the preferred styles of meaning-making and community interaction in each field.

The concept of a general and generalizable academic discourse is still more directly challenged in Hyland's 2007 article co-authored with Polly Tse, 'Is there an "academic vocabulary?"' (Chapter 14). Hyland and Tse were not unappreciative of what Coxhead (2000) attempted through her corpus approach to an inventory of academic vocabulary across the curriculum, her Academic Word List (AWL), but they did question its methodological robustness and pedagogical value. To test the AWL, they constructed a 3.3-million word corpus of key professional academic and pedagogical genres, including published research articles, textbook chapters and student research projects, in eight science, engineering and social science fields. Their findings pointed to only approximately a third of the AWL's 570 word families (base words with their inflections and derivations) as reasonably frequent in Hyland and Tse's own cross-disciplinary corpus. Beyond limited frequency, the AWL could also be viewed as exhibiting 'potential *monosemic bias*' (this book, p. 345, authors' italics), in other words, as suggesting that the same written forms of words have the same meanings in different disciplinary contexts, when actual usage as seen in the Hyland-Tse corpus tells a very different story. Not surprisingly, Hyland and Tse's conclusion is that the AWL is far from likely the shortcut to academic literacy sought by many EAP instructors, who would be better advised to guide their students in exploration of words and their collocations in specific disciplines. Nevertheless, despite Hyland and Tse's findings, a decade after the publication of their empirically based cautionary tale, it would appear not to have been as persuasive as their data would have led us to expect. Coxhead (2011, 2016) has amply documented the continued, if not increasing, influence of the AWL. The lure of a seemingly learnable and widely usable general academic vocabulary, no doubt especially for those teaching students not yet committed to a discipline, appears as strong as ever. In other words, the academic-core, autonomous-literacy pedagogical approach remains ripe for additional challenges.

In their 2007 article, Hyland and Tse noted that 'obviously vocabulary is more than individual words acting separately in a discourse' (this book, p. 353, Chapter 14). In an article published the following year, 'As can be seen: Lexical bundles and disciplinary variation' (Chapter 15), Hyland indeed turned his attention to more than individual words. Bundles, or strings of words that co-occur more often than by chance, had, at that point, the mid-2000s, been examined in the context of specific genres, but not, Hyland observed, in specific disciplines. Again taking a corpus-informed approach, Hyland compiled a corpus, this time 3.5 million words, of published research articles and student dissertations and theses in four hard and soft disciplines. As in his other corpus-based studies, Hyland found discipline-specific variation pointing to substantial epistemological and social-interactional differences, in this case, with the lexical composition of bundles more likely to mask authorial presence in the hard sciences, and the functions of bundles more likely to underscore the interpretive roles of research writers in the soft sciences. Echoing observations in Hyland and Tse (2007), Hyland (2008) remarks in reference to his bundles findings that the extent of disciplinary specificity revealed, with more than half of the fifty most common bundles in each field found in no other field in the corpus, is probably less than heartening for commercial materials developers in search of academic discourse universals. In other words, the findings provide still further evidence of the need to build writer awareness of the 'features of the discourses they will encounter' (this book, p. 381, Chapter 15) in their fields of study.

Taken together, what emerges from the four studies just discussed, as in much of Hyland's other work, is an 'academic discourse' whose features are far more varied and indicative of profoundly different ways of seeing and understanding the world than the term itself, especially facile use of it, would seem to suggest. While the appeal of an unproblematized construct called 'academic discourse' appears to persist, as implied by continued interest in an academic core, a prime example of which is the popular AWL, there is also evidence of growing interest in enabling learners in Anglophone academia to do just as Hyland's pedagogical implications encourage, to explore its diverse and far-from-static discourse features on their own (Lee and Swales 2006; Charles 2012, 2014) and with instructor guidance (Charles 2007). Armed with more user-friendly corpus tools and capable of compiling their own customized corpora, or given access to authentic corpus-derived materials (Poole 2016), learners may no longer need, as Lee and Swales (2006) have noted, to depend on the intuitions or the overgeneralized conceptualizations of academic discourse of others. Is this wishful thinking? Perhaps, but it is a wish fast becoming a reality in the classrooms of increasingly corpus-savvy EAP and writing studies professionals (e.g. Aull 2015), some number of whom will probably have read the work of Ken Hyland.

PART FIVE

Pedagogy and EAP

Introduction

The four chapters in this final section represent the area of my work which most explicitly seeks to intervene in practice and to directly engage with teachers. The papers here also amount to some of the work that I am most satisfied with. Each of them takes a theme which has preoccupied me for most of my career and which I've spent some effort in researching over the years: genre, hedging, teacher feedback and specificity, and each, I hope, expresses a clear point of view on the topic.

Chapters 16 and 19 here were written to engage with perennial debates in writing instruction and are the most polemical chapters in this book. Chapter 16 appeared in a special issue of *Journal of Second Language Writing* on 'L2 writing in the post-process era' which emerged from a colloquium at the annual conference of the American Association of Applied Linguistics. I was in the audience at that session and astonished that half a dozen presentations by luminaries in L2 writing could talk for several hours without mentioning the G-word. I think this showed the lack of interest, or perhaps awareness, of genre in US writing scholarship at the time and my question to the panel must have jogged the organizer, Dwight Atkinson, to recognize this absence and invite me to contribute to the issue. This is probably why the paper is something of a genre-based attack on process, and it probably takes a more rabidly anti-process stance than I would today, even though I still believe what I wrote back then.

Chapter 19 that concludes this book with the subtitle 'how far should we go now?' deliberately echoes Ruth Spack's rhetorical question concerning the role that English teachers can play in teaching specialist genres of writing. Coming from an American liberal arts tradition of undergraduate education in which all first-year students are required to pass courses in essay writing, Spack is largely hostile to 'vocational' or disciplinary-oriented writing. Her conclusion was that teachers' lack of subject knowledge expertise limits their effectiveness and she argues that teachers of English should not be asked to teach genres of academic disciplines to which they do not belong. Instead, she maintains that they should teach the academic writing process, focusing on 'appropriate inquiry strategies, planning, drafting, consulting, revising, and editing'. This chapter is, like the genre paper earlier, therefore a critique of this position, but it goes further to try and establish the value of EAP in higher education by challenging EAP teachers to take a stance on how they view language and learning and to examine their courses in the light of this stance. It raises the issue of whether there are skills and features of language that are transferable across different disciplines or whether we should focus on the texts, skills and forms needed by learners in distinct disciplines.

While I am less fervent about this position now (e.g. Hyland 2016), research consistently shows us that the discourses of the academy do not form an undifferentiated, unitary mass but a variety of subject-specific literacies. I have always argued that disciplines have different views of knowledge, different research practices and different ways of seeing the world, and investigating the practices of those disciplines will take us to greater specificity. I also recognize, however, that not all contexts are the same and that circumstances often require teachers to identify more register-level skills and to appreciate students' own subject-specific knowledge in order to hand over control of subject content to them, providing them with the tools to explore texts in their subject contexts. Having led a language centre that offers both an English for General Academic Purposes course (to help first-year undergraduates bridge the gap between school and disciplinary English) as well as 'English in the Discipline' courses, I see the issue as less polarized now and better recognize the complexities of instructional contexts. It may be more useful for practitioners to see the two positions as ends of a continuum rather than a dichotomy: a dilemma rather than a conflict, but the chapter here still raises key issues of EAP practice and theory.

The other two chapters in this section are very different, both from the other two chapters and from each other. They both deal with aspects of writing instruction, one with teaching hedges in academic writing and the other with feedback, but one is a practical paper offering suggestions for practice and the other is a research paper which explores practice rather than models it.

Chapter 17 was cannibalized from my doctoral dissertation and is one of the most practical papers I've ever written. Hedging has been a central interest of mine since the early 1990s when I hit upon the topic as a way to explore the knowledge-creating practices of the sciences for my PhD at the University of Queensland. Back then, hedging was buried under the concept of modality and the preserve of serious, larger-than-life linguists like Geoffrey Leech and Frank Palmer, who saw it as an abstract grammatical category. Alternatively it attracted epistemic philosophers, who were often more concerned about *how* we can know about the world rather than *how we make claims* about what we know. I thought it would be interesting to look at how scientists talked about how they knew something, or actually how they tried to persuade others of what they felt they knew. The 'author evacuated' prose of the sciences allowed writers to construct facts and a stance as knowledgeable insiders and hedging seemed to be a way of managing these goals. For a while, to me at least, hedging seemed the most important concept in rhetoric and fortunately others have agreed, probably with suitably reduced enthusiasm for the subject.

Hedging, then, is now a taken-for-granted instrument in the academic's toolbox. It is a way of moderating the certainty writers invest in a statement because they are uncertain about how accurate it is, how long it might survive the scrutiny of further research or how it might affect the response of the reader. It is also now a well-researched element of academic discourse and routinely introduced to undergraduate writers as a way to demonstrate a mature sensitivity to the things they are asked to discuss. What is different now is that claims about the functions hedging might perform and the patterns in which it routinely occurs require much stronger support. A photocopied collection of sixteen research articles was my corpus and regarded as perfectly adequate to discuss hedging – even for a PhD. There is also something rather quaint about the tasks I have suggested in the paper. But while flipped classrooms, out-of-class learning and online pedagogies have encroached into many EAP courses, tasks of this kind still offer useful and usable approaches.

Chapter 18, written with Fiona Hyland, remains one of my favourite papers. It was clear to both of us that teachers give feedback to students rather than texts and one reason for believing this was the way feedback is often constructed interpersonally. The teacher's consideration of the affective dimensions of feedback – how the student writer will receive the comments – is often overlooked as it does not trouble participants engaged in the low-stakes tasks given to them in the experimental contexts which tend to dominate research in this area. However, when one looks at teacher-written feedback created in real classrooms, this interpersonal element is striking. Combining the feedback and interview transcripts Fiona had collected for her PhD and my interest in writer mitigation, we decided to collaborate to see how teachers managed these interpersonal relations in what is, after all, a very threatening genre. As the chapter shows, there is hardly any comment which does not carry some respect for the student's face and consideration for how he or she will react to the critique of the work. Feedback remains a central area of research in L2 writing instruction and while this paper has had some influence, there is still what seems to be an overemphasis on the correction of errors rather than engagement with students.

The section concludes with a response by Professor Ann Johns of San Diego State University, USA, one of the leading figures in L2 student writing and a pioneer in developing research-informed pedagogy in this area.

16

Genre-based pedagogies: A social response to process

Introduction

Process approaches have had a major impact on the ways writing is both understood and taught, transforming narrowly-conceived product models and raising awareness of how complex writing actually is. Few teachers now see writing as an exercise in formal accuracy, and most set pre-writing activities, require multiple drafts, give extensive feedback, encourage peer review, and delay surface correction. But while process approaches have served to instil greater respect for individual writers and for the writing process itself, there is little hard evidence that they actually lead to significantly better writing in L2 contexts. The main reason for this is that their rich amalgam of methods collect around a discovery-oriented, ego-centred core which lacks a well-formulated theory of how language works in human interaction. Because process approaches have little to say about the ways meanings are socially constructed, they fail to consider the forces outside the individual which help guide purposes, establish relationships, and ultimately shape writing.

Genre-based pedagogies address this deficit by offering students explicit and systematic explanations of the ways language functions in social contexts. As such they represent the most theoretically developed and fruitful response to process orthodoxies. In this brief overview I will seek to elaborate this point. I will sketch out some of the ways that genre approaches have influenced second language pedagogies by moving away from a highly restricted view of human activity over-reliant on psychological factors, to a socially informed theory of language and an authoritative pedagogy grounded in research of texts and contexts.

A social take on process

It is hazardous to speak of process as a single approach to teaching since, like genre, it is a term which embraces a range of orientations and practices. At the heart of this model, however, is the view that writing is a "non-linear, exploratory, and generative process whereby writers discover and reformulate their ideas as they attempt to approximate meaning" (Zamel, 1983, p. 165). Following Emig's (1983) description of composing as 'recursive', rather than as an uninterrupted, left-to-right *Pre-writing → Writing → Post-writing* activity, this paradigm sees writing as essentially individual problem-solving. It thus seeks to construct cognitive models of what writers do when they write, emphasising the complexity of planning, the influence of task, and the value of guiding novices to greater competence by awareness of expert strategies. Writing in this view is essentially learnt, not taught, and the teacher's role is to be non-directive and facilitating, assisting writers to express their own meanings through an encouraging and co-operative environment with minimal interference. In this section I want to consider some limitations of this model from a social perspective before offering a genre response to them.

First, process represents writing as a decontextualised skill by foregrounding the writer as an isolated individual struggling to express personal meanings. Process approaches are what Bizzell (1992) calls "inner-directed," where language use is the outcome of individual capacities and writing processes which are "so fundamental as to be universal." Basically, the writer needs to draw on general principles of thinking and composing to formulate and express his or her ideas. But while this view directs us to acknowledge the cognitive dimensions of writing and to see the learner as an active processor of information, it neglects the actual processes of language use. Put simply, there is little systematic understanding of the ways language is patterned in particular domains. From a genre perspective, on the other hand, people don't just write, they write to accomplish different purposes in different contexts and this involves variation in the ways they use language, not universal rules (Halliday, 1994). So while process models can perhaps expose *how* some writers write, they do not reveal *why* they make certain linguistic and rhetorical choices. As a result, such models do not allow teachers to confidently advise students on their writing.

Second, process models disempower teachers and cast them in the role of well-meaning bystanders (e.g., Cope & Kalantzis, 1993). This is a model of learning based on individual motivation, personal freedom, self-expression and learner responsibility, all of which might be stifled by too much teacher intervention. Methods require little of the teacher because they rely on an intuitive understanding of language use, so his or her involvement is reduced to developing students' metacognitive awareness of their writing processes and responding to writing. Response is potentially the most influential step because this is the point at which overt intervention and explicit language teaching are most likely to occur. Unfortunately, however, in learner-centred classrooms this is necessarily a reactive and extemporised solution to learners' writing difficulties. Because language and rhetorical organisation tend to be things tacked on to the end of the process as "editing," rather than the central resources for constructing meanings, students are offered no way of seeing how different texts are codified in distinct and recognisable ways in terms of their purpose, audience and message (Macken-Horarik, 2002).

Third, this inductive, discovery-based approach to instruction fails to make plain what is to be learnt (e.g., Feez, 2002; Hasan, 1996). In process classrooms students are not typically given explicit teaching in the structure of target text types. Instead they are expected to discover appropriate forms in the process of writing itself, gleaning this knowledge from unanalysed samples of expert writing, from the growing experience of repetition, and from suggestions in the margins of their drafts. This deflects attention from language and presupposes a knowledge of genre outcomes. While well-intentioned, this is a procedure which principally advantages middle class L1 students who, immersed in the values of the cultural mainstream, share the teacher's familiarity with key genres (Christie, 1996; Martin, 1993). L2 learners commonly do not have access to this cultural resource and so lack knowledge of the typical patterns and possibilities of variation within the texts that possess cultural capital (Cope & Kalantzis, 1993; Hasan, 1996).

Delpit (1988, p. 287), writing from the context of an African American teacher's experience, makes a similar argument:

[A]dherents to process approaches to writing create situations in which students ultimately find themselves held accountable for knowing a set of rules about which no one has ever directly informed them. Teachers do students no service to suggest, even implicitly, that 'product' is not important. In this country students will be judged on their product regardless of the process they utilized to achieve it. And that product, based as it is on the specific codes of a particular culture, is more readily produced when the directives of how to produce it are made explicit.

Students outside the mainstream, therefore, find themselves in an invisible curriculum, denied access to the sources of understanding they need to succeed. Thrown back on their own resources, they are forced to draw on the discourse conventions of their own cultures and may fail to produce texts that are either contextually adequate or educationally valued.

A related difficulty is that process pedagogies also draw heavily on inaccessible cultural knowledge in their instructional practices and in the concepts which inform judgements of good writing. Ramanathan and Atkinson (1999), for instance, point to the role that hidden mainstream US values play in process methods. Key principles which originated in L1 classrooms such as personal voice, peer review, critical thinking, and textual ownership tacitly incorporate an ideology of individualism which L2 learners may have serious trouble accessing. So, once again, while such crucial culturally specific norms of thought and expression in process classrooms may be unreflectively transparent for mainstream American undergraduates, they may not always be recognised or accepted by students from cultures less entrenched in the ideology of individualism.

A final point I want to make about process models of learning concerns their lack of engagement with the socio-political realities of students' everyday lives and target situations. In process methodologies personal growth and self-actualisation are core learning principles, as writers develop confidence and self-awareness in the process of reflecting on their ideas and their writing. But while this approach responds to the individual needs and personalities of learners, it offers them little by way of the resources to participate in, understand, or challenge valued discourses (e.g., Hasan, 1996; Martin, 1993). It leaves students innocent of the valued ways of acting and being in society, despite the fact that they need ways to manage the appropriate linguistic and rhetorical tools to both gain access to the powerful genres of mainstream culture and the means to conduct a critical appraisal of them. Hammond and Macken-Horarik (1999) argue that an effective critical literacy in English must presuppose control of mainstream literacy practices. Importantly, however, process models fail to introduce students to the cultural and linguistic resources necessary for them to engage critically with texts.

I should hasten to point out here that I raise these issues not to condemn process approaches or to criticise the many teachers who implement learner-centeredness in their classrooms. Progressive pedagogies have done much to inform the teaching of writing by moving us away from grammar practice and authoritarian teaching roles to facilitate more equal, respectful and interactive relationships in settings that value reflection and negotiation. I have simply tried to highlight the problems posed by an approach uninformed by an explicit theory of how language works or the ways that social context affects linguistic outcomes. These are areas where genre-based models have made their strongest impact. Put simply, social theorists argue that because process approaches emphasise individual cognition at the expense of language use, they fail to offer any clear standpoint on the social nature of writing (Martin, Christie & Rothery, 1987).

From a social perspective, a writer's choices are always context-dependent, motivated by variations in social activity, in writer–reader relations, and by constraints on the progress of the interaction. As a result, teachers cannot expect weak writers to improve simply by equipping them with the strategies of good writers. Not only are such strategies only part of the process, but they too are likely to vary with context. Instead, we need to explore ways of scaffolding students' learning and using knowledge of language to guide them towards a conscious understanding of target genres and the ways language creates meanings in context. This is the goal of genre pedagogies.

A brief overview of genre

Genre refers to abstract, socially recognised ways of using language. It is based on the assumptions that the features of a similar group of texts depend on the social context of their creation and use, and that those features can be described in a way that relates a text to others like it and to the choices and constraints acting on text producers. Language is seen as embedded in (and constitutive of) social realities, since it is through recurrent use of conventionalised forms that individuals develop relationships, establish communities, and get things done. Genre theorists, therefore, locate participant relationships at the heart of language use and assume that every successful text will display the writer's awareness of its context and the readers who form part of that context. Genres, then, are "the effects of the action of individual social agents acting *both* within the bounds of their history and the constraints of particular contexts, *and* with a knowledge of existing generic types" (Kress, 1989, p. 10, Kress's emphasis).

It is customary to identify three broad, overlapping schools of genre theory (Hyon, 1996; Johns, 2002). *The New Rhetoric approach,* influenced by post-structuralism, rhetoric and first language composition, studies genre "as the motivated, functional relationship between text type and rhetorical situation" (Coe, 2002, p. 195). The focus here is mainly on the rhetorical contexts in which genres are employed rather than detailed analyses of text elements (e.g., Freedman & Medway, 1994). *The ESP approach* is more linguistic in orientation and sees genre as a class of structured communicative events employed by specific discourse communities whose members share broad social purposes (Swales, 1990, pp. 45–47). These purposes are the rationale of a genre and help to shape the ways it is structured and the choices of content and style it makes available (Johns, 1997). A third orientation is based on Halliday's (1994) *Systemic Functional Linguistics* (SFL). Known in the US as the "Sydney School" (e.g., Hyon, 1996; Johns, 2002), this model of genre stresses the purposeful, interactive, and sequential character of different genres and the ways language is systematically linked to context through patterns of lexico-grammatical and rhetorical features (Christie & Martin, 1997).

While these approaches are united by a common attempt to describe and explain regularities of purpose, form, and situated social action, they clearly differ in the emphasis they give to text or context, the research methods they employ, and the types of pedagogies they encourage (Hyland, 2002c). New Rhetoric, with its emphasis on the socially constructed nature of genre, has helped unpack some of the complex relations between text and context and the ways that one reshapes the other. But while New Rhetoric underlines that literacy is not the monolithic competence it is often perceived to be, its contribution to L2 writing instruction has been minimal. Australian and ESP genre theorists, however, have been closely engaged with issues of L2 teaching, and unswerving in their efforts to provide students with a knowledge of relevant genres so they can act effectively in their target contexts.

ESP genre approaches have perhaps had the most influence on L2 writing instruction worldwide, grounding teaching in a solid research base and drawing strength from an eclectic set of pedagogies and linguistic theories. SFL, however, perhaps offers the most theoretically sophisticated and pedagogically developed approach of the three, underpinned by a highly evolved and insightful theory of language and motivated by a commitment to language and literacy education. Basically, Halliday's theory systematically links language to its contexts of use, studying how language varies from one context to another and, within that variation, the underlying patterns which organise texts so they are culturally and socially recognised as performing particular functions. The exploration and description of these patterns and their variations has been the focus of genre theory and the resources it exploits to provide disadvantaged learners with access to the cultural capital of socially valued genres.

A genre view of language and writing

Genre theory seeks to (i) understand the ways individuals use language to orient to and interpret particular communicative situations, and (ii) employ this knowledge for literacy education. This second purpose complements research in the cross-disciplinary movement known as the New Literacy Studies, which stresses that all writing is situated and indicative of broader social practices (e.g., Barton & Hamilton, 1998).

Basically, genres are rhetorical actions that writers draw on to respond to perceived repeated situations; they are choices which represent effective ways of getting things done in familiar contexts. Some genre theorists have, therefore, sought to identify the recognisable structural identity, or "generic integrity," of particular academic and workplace genres in terms of their *stages* (or rhetorical structures) and the constraints on typical move sequences (Bhatia, 1999; Butt, Fahey, Feez, Spinks, & Yalop et al., 2000). Another research direction has looked at language variation *across* genres and the resources available for creating meanings in a culture (Hunston & Thompson, 2001). This research attempts to show how clusters of register, style, lexis, and other features reflect the different personal and institutional purposes of writers, the different assumptions they make about their audiences, and the different kinds of interactions they create with their readers. As a result, a lot more is known about the ways writers frame their ideas for particular readers, construct an appropriate authorial self, and negotiate participant relationships in writing (e.g., Bondi, 1999; Hyland, 2000a, 2002a, 2002e; Thompson, 2001).

One important assumption made by genre adherents is that writing is *dialogic* (Bakhtin, 1986), both because it presupposes and responds to an active audience, and because it involves a plurality of voices through links to other texts. Writing involves drawing on the texts we typically encounter and are familiar with. Consequently, the concepts of intertextuality and interdiscursivity (Bakhtin, 1986) have been extremely influential in genre theory. One influence has been that analysts are not simply concerned with describing text similarities, but with exploring the contextual constraints on allowable configurations. Variation is just as important as similarity because texts spread along a continuum of approximation to core genre examples, with varying options and restrictions operating in particular cases (Swales, 1990). Genre research, thus, extends beyond texts to the sites where relationships can facilitate and constrain composing and to the discourse communities in which texts will be used and judged (Hyland, 2000a).

Discourse community is a concept central to genre views of writing as it is a powerful metaphor joining writers, texts and readers in a particular discursive space (Porter, 1992; Swales, 1990, 1998). While often criticised as altogether too structuralist, static, and deterministic, the notion of discourse community foregrounds the socially situated nature of genre and helps illuminate something of what writers and readers bring to a text, implying a certain degree of intercommunity diversity and intra-community homogeneity in generic forms. Genre theory has, therefore, often relied on some sense of "discourse community" to account for this kind of variation, seeking to draw on its explanatory and predictive authority without framing communities as utopias of shared and agreed-upon values and conventions. While reservations about the concept persist, it is currently the most useful tool available to explain the situated cognition required for interpretation and engagement. Communities are where genres make sense; they are the systems where the multiple beliefs and practices of text users overlap and intersect (Swales, 1998).

It is also worth mentioning here that while process and genre are often contrasted in terms of their views of writer creativity, genres are not overbearing structures which impose uniformity on users. There is huge potential for internal heterogeneity of genres, and issues of unity and identity are frequently raised in the literature. The fact that language users routinely and unreflectively recognise similarities and differences between texts with sufficient agreement to successfully negotiate and interpret meanings is itself highly significant. Our abstract, more-or-less shared knowledge of texts, intertextuality, audience, and standard purposes makes writing and reading efficient and contributes to mutual understanding. Genres help unite the social and the cognitive because they are central to how writers understand, construct, and reproduce their social realities. But while a shared sense of genre is needed to accomplish understanding, it is not necessary to assume that these are fixed, monolithic, discrete and unchanging.

Genre and second language literacy

Genre-based pedagogies rest on the idea that literacies are community resources which are realised in social relationships, rather than the property of individual writers struggling with personal expression. This view offers writing teachers a radical new perspective on what they do, for the naïve assumptions that writing, and teaching writing, are somehow neutral, value-free activities are no longer defensible. It encourages us to acknowledge that literacies are situated and multiple — positioned in relation to the social institutions and power relations that sustain them. Expressed most simply, writing is used in many ways across many social contexts, but only some of these have institutional and cultural stature. It is not the case that all genres are created equal, because they are associated with, and are used to regulate entry into, social communities possessing more or less prestige and influence. The question of access to, and production of, valued texts is central to the notions of power and control in modern society, and underlines the genre theorist's emphasis on which genres should be taught.

What this means is that writing cannot be distilled down to a set of cognitive processes. Genre knowledge is important to students' understanding of their

L2 environments, and crucial to their life chances in those environments. The teaching of key genres is, therefore, a means of helping learners gain access to ways of communicating that have accrued cultural capital in particular professional, academic, and occupational communities. By making the genres of power visible and attainable through explicit instruction, genre pedagogies seek to demystify the kinds of writing that will enhance learners' career opportunities and provide access to a greater range of life choices. Without the resources to understand these genres, students in university and WAC contexts will continue to find their own writing practices regarded merely as failed attempts to approximate prestigious forms (Johns, 1997).

For some critics, however, providing L2 students with more effective access to the dominant genres of our culture does nothing to change the power structures that support them, or to challenge the social inequalities which are maintained through exclusion from them (e.g., Benesch, 2001). Luke (1996, p. 314), for example, writes:

> A salient criticism of the 'genre model' is that its emphasis on the direct transmission of text types does not necessarily lead on to a critical reappraisal of that disciplinary corpus, its field or its related institutions, but rather may lend itself to an uncritical reproduction of discipline.

Thus, teaching genres may only reproduce the dominant discourses of the powerful and the social relations which they construct and maintain.

A similar charge could, of course, be levelled at process and other pedagogies which simply perpetuate inequalities by failing to provide students with better access to powerful genres (e.g., Hasan, 1996). In fact, learning about genres does not preclude critical analysis but provides a necessary basis for critical engagement with cultural and textual practices. As Bakhtin (1986, p. 80) has suggested, writers must be able to control the genres they use before they can exploit them. Hammond and Macken-Horarik (1999, p. 529) make this point forcefully:

> Systematic discussion of language choices in text construction and the development of metalanguage — that is, of functional ways of talking and thinking about language — facilitates critical analysis. It helps students see written texts as constructs that can be discussed in quite precise and explicit ways and that can therefore be analysed, compared, criticised, deconstructed, and reconstructed.

In other words, to fail to provide learners with what we know about how language works as communication denies them both the means of communicating effectively in writing and of analysing texts critically.

Genre approaches seem to offer the most effective means for learners to both access and critique cultural and linguistic resources (Hasan, 1996). By providing learners with an explicit rhetorical understanding of texts and a metalanguage by which to analyse them, genre teachers can assist students to see texts as artifacts that can be explicitly questioned, compared, and deconstructed, thereby revealing their underlying assumptions and ideologies.

To sum up, from a genre perspective writing is not an abstract activity, but a social practice. What is considered good writing, appropriate engagement, convincing argument, effective persuasion, and creative expression does not depend on mastery of universal processes, but varies from one community context to the next. By focusing on the literacy practices writers encounter at school, at work, and at university, genre pedagogies help them to distinguish differences and provide them with a means of conceptualising their varied experiential frameworks. Highlighting variability thus helps undermine a deficit view which sees writing difficulties as learner weaknesses and which misrepresents writing as a universal, naturalised and non-contestable way of participating in communities.

Genre-based pedagogies

Genre not only presents teachers and students with a different view of writing, but also with a distinct set of teaching practices. In contrast to process models, genre-based pedagogies support learners within a contextual framework for writing which foregrounds the meanings and text-types at stake in a situation. At their core, these methods offer writers an explicit understanding of how texts in target genres are structured and why they are written in the ways they are. To create a well-formed and effective text, students need to know the lexico-grammatical patterns which typically occur in its different stages, and the teacher's task is to assist students towards a command of this through an awareness of target genres and an explicit grammar of linguistic choices. Providing writers with a knowledge of grammar shifts writing instruction from the implicit and exploratory to a conscious manipulation of language and choice.

Inside genre classrooms a range of methods are employed. These include investigating the texts and contexts of students' target situations, encouraging reflection on writing practices, exploiting genre sets, and creating mixed-genre portfolios (Johns, 1997; Paltridge, 2001). In SFL approaches the teaching–learning process is typically seen as a *cycle* which takes writers through modelling, joint negotiation, and independent construction, allowing students different points of entry and enabling teachers to systematically expand the meanings students can create (e.g., Feez, 2002). This model represents a "visible pedagogy" in which what is to be learned and assessed is made clear to students, as opposed to the invisible pedagogy of process approaches (e.g., Delpit, 1988).

The theoretical underpinning of this pedagogical approach is provided by Vygotsky's (1978) emphasis on the interactive collaboration between teacher and student, with the teacher taking an authoritative role to "scaffold" or support learners as they move towards their potential level of performance. This scaffolding is most evident at the early stages of learning a genre where the teacher contributes what learners cannot do alone. The teacher intervenes at this stage to model and discuss texts, deconstructing and analysing their language and structure. This support is strategically diminished as students progress, with teachers and learners sharing responsibility in the joint negotiation and construction of texts, often through several drafts and with peer assistance, until the learner has the knowledge and skills to perform independently. Here is an approach to writing instruction with a central role for both language and teachers. It is teaching which supports L2 students with an explicit pedagogy and which presupposes little prior understanding of cultural practices.

Genre pedagogies assume that writing instruction will be more successful if students are aware of what target discourses look like, but it is this reproductive element which process adherents have been most critical. The argument is that the explicit teaching of genres imposes restrictive formulae which can straightjacket creativity through conformity and prescriptivism; that genres might be taught as moulds into which content is poured, rather than as ways of making meanings (e.g., Dixon, 1987; Raimes, 1991). There is always some danger of reifying genres with a text-intensive focus, as inexperienced or unimaginative teachers may fail to acknowledge variation and choice, applying what Freedman and Medway (1994, p. 46) calls "a recipe theory of genre" so that students see genres as 'how-to-do' lists. Obviously the dangers of a static, decontextualised pedagogy exist and must be guarded against, but there is nothing *inherently* prescriptive in a genre approach. I can see no reason why providing students with an understanding of discourse should be any more prescriptive than, say, providing them with a description of a clause, or even of stages in a writing process.

In sum, genre is a socially informed theory of language offering an authoritative pedagogy grounded in research on texts and contexts, strongly committed to empowering students to participate effectively in target situations. Genre pedagogy is buttressed by the belief that learning is best accomplished through explicit awareness of language, rather than through experiment and exploration, but this does not mean replacing communicative practices with teacher-centred ones. There is nothing here that excludes the familiar tools of the process teacher's trade. Genre simply requires that they be used in the transparent, language-rich, and supportive contexts which will most effectively help students to mean.

Conclusion

Genre is, in part, a social response to process. It suggests that because writing is a means of connecting people with each other in ways that carry particular social meanings, it cannot be only a set of cognitive abilities. The process of writing is a rich collection of elements of which cognition is only one, and to understand it fully and to teach it effectively we need to include in this mix the writer's experiences together with a sense of self, of others, of situation, of purpose and — above all — of the linguistic resources to address these effectively in social action. Writing is a basic resource for constructing our relationships with others and for understanding our experience of the world, and as such genre is centrally involved in the ways we negotiate, construct, and change our understanding of our societies and ourselves. As Christie (1987, p. 30) has observed, "Learning the genres of one's culture is both part of entering into it with understanding, and part of developing the necessary ability to change it."

17

Nurturing hedges in the ESP curriculum

Introduction

The term hedging was introduced to linguistics by Lakoff (1972) to describe "words whose job it is to make things more or less fuzzy". It has subsequently been used by sociologists to describe a means to avoid face-threatening behaviour and by applied linguists to discuss devices such as *I think, perhaps, might* and *maybe* which qualify the speaker's confidence in the truth of a proposition. In scientific writing these effective and propositional functions work in rhetorical partnership to persuade readers to accept knowledge claims (e.g. Myers, 1985). Hedges express tentativeness and possibility in communication and their appropriate use is a critical, although largely neglected, area of scientific discourse.

The purpose of this paper is to discuss how ESP teachers can help develop L2 learners' understanding of the principles and mechanics of the appropriate use of this critical pragmatic feature. First however, I will give a brief overview of hedging in academic writing, sketching its importance and principal means of realisation.

The importance of scientific hedging

Definitions of hedging are surprisingly rare in the literature. Zuck and Zuck (1986) refer to hedging as the process whereby the author reduces the strength of a statement, while for Markkanen and Schröder (1989), it is any manipulative, non-direct sentence strategy of saying less than one means. My use of the term is closer to the first of these definitions as it includes statements that express exactly what the author means, of saying no more than is warranted by the available evidence; in science, hedging is not an obfuscation strategy any more than it is merely a convention of academic style. A "hedge" is any linguistic means used to indicate either (a) a lack of complete commitment to the truth of an accompanying proposition or (b) a desire not to express that commitment categorically. Hedges are therefore the means by which a writer can present a proposition as an opinion rather than a fact.

Hedges therefore have an important role in a form of discourse characterised by uncertainty and frequent reinterpretation of how natural phenomena are understood. Despite a widely held belief that scientific writing is a series of impersonal statements of fact which add up to the truth, hedges are abundant in science and play a critical role in academic writing. Hedges are important to scientists because even the most assured propositions have an inherently limited period of acceptance and in these circumstances categorical assertions of truth are decidedly hazardous. Science involves weighing up the evidence and stating the extent to which conclusions can be accepted:

> The writer . . . . currently evaluates and criticises the information and the propositions he or she tries to set down as fully, accurately and objectively as possible. For centuries this dialectical processing of objective fact and subjective evaluation has been the goal of academic writing and of the training that leads to academic writing (Nash, 1990: p. 10).

Academic discourse involves interpretive statements because cognition is invariably hedged, writers offering an assessment of the referential information they provide. Rather than being factual and impersonal, effective scientific writing actually depends on augmenting propositional information in order to alert readers to the writer's opinion.

Functions of hedging in science

A contextual analysis of hedging in a 75,000 word corpus of 26 research articles in cell and molecular biology has identified three principal reasons for hedging in this genre.

Firstly hedges allow writers to express propositions with greater precision, recognising the impossibility of exactly quantifying the world. Scientific writers seek to balance fact and interpretation as they try to present information as accurately as possible, always recognising that only a segment of reality can be described and that any account will eventually be overturned. Hedges are therefore an important means of stating uncertain scientific claims with appropriate caution. So writers often say "X may cause Y" rather than "X causes Y" to specify the actual state of knowledge on the subject. This has been noted by a number of writers (e.g. Prince et al., 1982; Rounds, 1982; Skelton, 1988; Salager-Meyer, 1994; Thompson, 1993).

Hedges here imply that a proposition is based on the writer's plausible reasoning rather than certain knowledge and readers are expected to understand that the proposition is true as far as can be determined. The function of hedges to convey purposive vagueness in certain registers noted by some writers (e.g. Powell, 1985; Channell, 1994) is simply not a viable option in most scientific writing. This is because precision is central to persuasive argument and the clear and full expression of information is the only control a writer has in influencing the reader's views.

Secondly hedges allow writers to anticipate possible negative consequences of being proved wrong. Academics seek agreement for the strongest claims they can for their evidence, as this is how they gain their academic credibility, but they also need to cover themselves against the embarrassment of categorical commitment to statements that later may be shown to be wildly inaccurate. Hedges here allow writers to refer to speculative possibilities while at the same time avoiding direct personal responsibility for their statements. This function has also been recognised in the literature (Rounds, 1982; Powell, 1985; Nash, 1990; Swales, 1990) and Prince et al. (1982) call this kind of insurance "Shields". Hedges here tone down, not the claims that are made for the research, but the language used to express them, and it is typically realised by reducing the author's linguistic role through use of the passive, existential subjects or by attributing claims to the text or data. In science, writers may hedge in this way because of small samples, preliminary results, uncertain evidence or imperfect measuring techniques.

Finally hedges help writers to develop a relationship with the reader, addressing affective expectations in gaining acceptance for claims. This interpersonal function has been noted in conversational settings (e.g. Holmes, 1984; Coates, 1987) and has been attributed to deference (Fraser, 1980; Fraser and Nolan, 1981) or politeness (Brown and Levinson, 1987) and Myers (1989) has suggested that mitigation performs similar politeness functions in science. But scientists are concerned less with judgements of power and status than with acknowledging the reader's role in ratifying knowledge and expressing claims with appropriate modesty. Although academic papers try to persuade readers of a claim, this can always be rejected and so hedges help writers show deference and respect for colleagues while at the same time conforming to expected limits on self-assurance in presenting findings.

Essentially categorical assertions allow no room for negotiation and indicate that the arguments need no feedback; they force readers into a passive role. Hedged statements, on the other hand, appeal to readers as intelligent colleagues who are capable of deciding about the issues. Hedges therefore mark statements as provisional while pending acceptance by the discourse community.

The repertoire of hedging devices however does not provide separate items for each type of meaning and a complex overlap of usage suggests that the precise motivation for employing a particular device may not always be clear. Hedges are polypragmatic, denying the possibility of assigning specific formal devices exclusively to particular functional categories, and the fact that one device can perform several functions simultaneously means that neither a purely formal treatment nor a detailed contextual analysis will always determine an unequivocal pragmatic function. There is often overlap between different meanings and cases assigned to one category will include meanings associated with another.

The expression of hedging in science

Given the importance of hedging, it is not surprising to find scientific research writing carefully and extensively hedged. Hedges constitute more than one word in every 50 in my corpus and Skelton (1988), Adams Smith (1984) and Hanania and Akhtar (1985) similarly found one hedge every 2 or 3 sentences in their corpora. Hedging is most commonly expressed by lexical verbs (1), epistemic adverbs (2), epistemic adjectives (3) and modal verbs (4):

(1) This would *appear* to be in significant conflict with . . .

   I *believe* that the overall orientation of . . .

(2) *Possibly,* phosphorylation of ACC synthase . .

   There is *apparently* a relationship between . . .

(3) . . is *likely* to be due primarily to a deficiency of functional . . .

   . . it appears *possible* that the mechanism causing the . . .

(4) These results *may* have relevance to . .

   . . . . it *should* be possible to test predictions . . .

In addition to lexical items, there are a number of strategies that provide a significant means of hedging scientific statements. These may be realised syntactically by such means as conditional clauses, embedding or contrast markers, or by formulaic expressions. The most numerous strategies are those which refer to experimental weaknesses (5), to the limitations of a model, theory or method (6) or mention of inadequate knowledge (7).

(5) *Under these conditions* phosphorylation of . . . .

So it is difficult to conclude whether the 100 . . .

(6) *In our hands,* there was no significant change in V_{max} . . .

. . . approx 70% *according to our method* and some b turn . .

If this scheme is correct, then the orientation of the heme . . .

(7) *Nothing is known about* the chemical constitution of . .

It is not known whether such a weak temperature response . .

The overall importance of these formal and strategic categories is shown in Fig. 17.1.

Lexical verbs are the most frequent hedges, representing 23.3% of the total, then adverbials and adjectives followed by a restricted range of modal verbs. Lexical verbs constitute the greatest range of items with 38 different forms represented and *indicate, suggest, appear* and *propose* comprising 55.7% of all instances. Lexical verbs include performative verbs such as *suggest,* and *propose,* together with cognitive verbs like *assume* and *believe* and sensory verbs such as *appear* and *seem.* The most frequent modal adjectives are *likely, possible, most* and *consistent with..* Modal adverbs include "downtoners" (Quirk *et al.,* 1972) like *quite, almost* and *usually,* which lower the effect of the force of the verb, and disjuncts that convey an attitude to the truth (*probably, generally, evidently*). *Would, may* and *could* account for over 75% of the total number of modal verbs expressing hedging. Modal

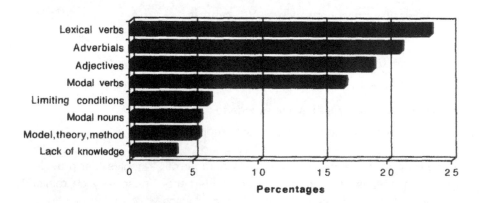

FIGURE 17.1 *Relative frequency of hedging expressions in science articles.*

nouns, such as *possibility, assumption* and *estimate,* occur less frequently than the other grammatical forms.

The most common hedging items in the corpus were *indicate, would, may* and *suggest* which occur far more often than many items that receive far more attention in ESP courses. The extent to which science research articles differ from the same hedging forms in general academic English can be seen in Table 17.1. The academic "J" sections of the combined Brown and LOB corpora represent over 160 different samples each of 2,000 words, covering published writings in diverse fields, its 350,000 words representing about 1,000 textbook pages. The Table shows that *indicate, may, suggest, could, propose* and *should* are far more likely to be found in science articles.

The significance of these figures can be judged by the fact that some 100 very frequent items account for about half the words in any text and the remaining items occur only once. The dominance of a few items is confirmed by the 160 million-word Cobuild corpus which shows that 700 words make up 70% of all English texts with *the* occurring 40 times in every 1000 words and *of,* 20 times. Some hedges are therefore among the highest frequency words in scientific writing.

L2 students and hedging

The appropriate use of hedges helps writers develop academic arguments and establish a relationship with their readers, therefore control over this feature is an important communicative resource for L2 students in university science courses. However, foreign students find the expression of commitment and detachment to propositions highly problematic and a failure to hedge statements adequately is a common feature of even formally proficient L2 writers (e.g. Skelton, 1988; Dudley-Evans, 1992).

The main reason for this is that there are clear cultural differences in the degree of indirectness permitted in academic writing (Bloor and Bloor, 1991) and proficiency in this pragmatic area is difficult to achieve in a foreign language (e.g. Clyne, 1987; Holmes, 1982). This results in what Thomas (1983) calls "cross-cultural pragmatic failure", or the inability to say what one means. This failure may be due to either inadequate linguistic knowledge (e.g. Scarcella and Brunak, 1981) or to culturally different perceptions of what constitutes appropriate linguistic behaviour, as L2 students learn to think and write differently in their own cultures (e.g. Ventola, 1992; Choi, 1988). Such differences may be related to the idea of reader involvement and whether responsibility for communication rests with the reader or the writer (Hinds, 1983; Clyne, 1987).

But while it represents a major "rhetorical gap" in many L2 students' communicative competence, hedging is generally seen as a problem to be

TABLE 17.1 Most frequently occurring hedging items compared with general academic corpora (per 10,000 words)

Item	Journal articles	Brown/ LOB (J)	Item	Journal articles	Brown/ LOB(J)
indicate	10.8	4.3	propose	2.8	0.9
would (not)	10.4	16.0	probably	2.7	2.8
may (not)	9.2	4.4	apparently	2.7	2.8
suggest	9.1	3.9	should	2.4	0.4
could	6.4	0.4	seem	2.3	7.7
about	4.0	*	possible	2.3	1.3
appear	4.0	3.7	essentially	2.3	1.1
might (not)	3.6	4.0	relatively	2.1	2.4
likely	2.8	2.7	approximate(ly)	2.1	2.1

eliminated from a form of discourse often thought to represent a factual and objective search for truth that transcends personality.

This view is most clearly represented in the style guide literature where hedges have been described as "unnecessary words" (Yarber, 1985: p. 188), "padded expressions" (Muller, 1985: p. 328) "wasteful signposting" (Smith, 1985: p. 92) and "clutter words" (Lannon, 1986: p. 135). The most eloquent position is taken by *The Elements of Style* which refers to hedges as "the leeches that infest the pond of prose, sucking the blood of words" (Strunk and White, 1959: p. 59). Some writers are more positive, advising students to hedge material that is inconclusive (Booth, 1985: p. 12) or which expresses "doubt, wishes, probability, conditions contrary to fact, or conditional statements" (Howell and Memering, 1986: p. 107). But these are rare in this literature where hedging is seen as undermining judgements (Rawlins, 1980) or robbing writing of its certainty and power (Zinsser, 1976: p. 96). Lindsay, for example, advises that

> If you have no conclusive evidence don't dither around with expressions such as 'it may be possible that.' or (worse) 'the possibility exists that . . .', which immediately suggests that you do not believe your own data. (Lindsay, 1984: p. 21)

The dominant view students receive from style guides then is to eliminate tentativeness as it adds little to the propositional information provided: "If your assertions are not made with precision, the value of your paper will be inevitably lowered" (Winkler and McCuen, 1989: p. 97). Similarly, ESP writing textbooks tend to ignore or under-represent the significance of hedging and most explanations of epistemic devices are generally ill-informed and inadequate (Hyland, 1994). Moreover, while undergraduates do a great deal of reading, they rarely take notice of the style and rhetorical techniques employed in the texts they read (Campbell, 1990: p. 226; Johns, 1991). Consequently learning to express and interpret epistemic features presents serious problems to L2 learners and their writing can often appear brusque, arrogant and overconfident.

Nurturing hedges in ESP

A major responsibility of ESP teachers is therefore, I believe, to assist students in developing an understanding of the correct use of hedging. There are two broad pedagogic approaches to this: first, to make expert writers' hedging strategies salient to students, and second, to develop the appropriate use of these forms in students' written work.

Hedging awareness: Analysing products

Concordanced material: Recognition of the most useful forms must be based on an analysis of the target language repertoire and so one approach is to introduce students to complete journal articles as good models which can be analysed and emulated. A fundamental tool in developing students' awareness of academic hedging however is a concordancer. A concordance programme is a tool for searching large amounts of computer-readable text for particular words and combinations, but while it can isolate, sort and count data, the actual analysis is performed by humans. Thus such programmes can turn students into researchers and help them search for patterns which are not readily apparent. Wu summarises the advantages of a concordance approach in this way:

> Only when words are in their habitual environments, presented in their most frequent forms and their relational patterns and structures, can they be learnt effectively, interpreted properly and used appropriately. Wu (1992: p. 32)

This both stimulates inquiry and encourages independent engagement with the language, what Johns and King (1991) describe as "data-driven learning".

Concordanced output also provides authentic data for materials which focus attention on hedging forms used in particular sciences. Gapped concordance printouts may, for example, encourage students to use contextual clues to predict missing items. This approach is especially useful if resources allow students to also analyse and compare the use of hedging in texts familiar to them from their subject courses or examples from corpora of ESL writing (e.g. Granger, 1994). Computer data also allow students to classify hedges according to their surface forms or functions while collocations enable comparisons to be made between the use of hedges in different genres or determine 'rules' for the appropriate use of confusable pairs such as *infer* and *imply* or *suggest* and *propose*. Comparisons can also be made between the strength of different forms and the importance of voice/tense choices with lexical verbs examined.

Finally, corpus materials can stimulate small projects among advanced learners which focus on hedging in their own disciplines. Thus each student might take an international journal from his or her field and analyse a number of articles in terms of the hedges it employs. The class can then pool their expertise and discuss similarities and differences, compiling the results to draft an essay or research paper on the subject (e.g. Swales, 1987). Alternatively, such data allows a deductive approach, encouraging groups to form hypotheses and examine them using empirical evidence. Relating tasks to students in this way means students can activate their acquired linguistic competence and discipline knowledge in a context of immediate discipline relevance.

Text fragments: An awareness of hedges can also be approached through the use of text fragments. To discover the effect of hedges on statements students can:

- examine a text and distinguish statements which report facts and those which are unproven;

- identify all hedges in a text, circling the forms used, and assign a meaning to them or account for their presence;

- locate and remove all hedges and discuss the effect on the meaning of the text;

- identify hedged propositions in a text and substitute them with statements of certainty;

- identify hedging forms and compile a scale ranking the amount of certainty they express, using this data to assess the accuracy of existing scales, such as that proposed by Jordon (1990: p. 66);

- consider a series of reformulations which vary the levels of certainty of a text and respond in terms of whether they accurately report the original statements;

- compare the use of different realisations of hedges in terms of meaning, syntactic pattern, lexical environment, pragmatic function, etc;

- determine whether the forms vary cross-culturally by translating them into the L1.

Longer texts: In addition to raising consciousness of forms, students have to see the importance of hedges in larger social context in order to determine why authors choose to hedge at particular points and why they use particular forms. Students need an understanding of hedges not only as text-based items but as discourse-based strategies and so hedges must be seen as relating to the writer's overall text plan. Explicit links between reading and writing have to be established and here complete texts are required to provide sufficient context to allow interpretation of the author's meanings in using hedges by drawing on particular reader assumptions.

Such understandings can be developed by exploiting contrasts where choices are related to criteria of acceptability and informativity, drawing on the functions outlined above to determine whether uses involve accuracy, uncertainty or interactive purposes. One powerful means of illustrating the effect of hedges on a text is to examine the same content in different evidential contexts, focusing on the different degrees of certainty expressed. A text with all hedges removed clearly becomes a different type of text, altogether more certain. Thus students can be asked to compare two versions of the same material, either in extracts from different drafts submitted for publication (e.g. Myers, 1985), or in treatments of the same topic in an RA and a textbook or newspaper (e.g. Fahnestock, 1986). This awareness provides students with a simple grammatical technique for distinguishing facts from opinions and identifying substantive assumptions made about a topic.

Additional techniques can require students to locate places where hedges have been removed or extraneous ones inserted, to comment on the consequences of different reporting verbs in citing prior work, to compare the meanings of the same forms in different contexts and to examine the functional effects of alternative forms. A task focusing on clarifying language use is suggested by Swales (1987) in connection with citation studies and students can similarly interview lecturers in their own disciplines concerning their published hedging behaviour. In all these ways, the goal must be to encourage students to analyse representative textual products of the genre, respond to surface features of the texts and reflect on the strength of the knowledge claims made.

Hedging use: Writing process

In addition to identifying hedging items and gaining an appreciation of their appropriate use, students require guidance to competently hedge their own writing. This requires a programme emphasising the discoursal aspects of academic writing, geared to helping the writer solve the problems of using language to both generate ideas and to shape those ideas into a form that meets both readers' needs and the demands of effective persuasion. Here hedges must be seen as part of a wider process of creating and crafting complete texts, developing both a sense of audience and a sense of purpose in using hedges correctly to produce a complete, fully contextualised piece of work.

This requires a writing environment which is as authentic as possible in terms of objectives, genre and audience. Writing tasks must specify a clear communicative context, either through classroom roles for students or by writing for experts. Equally important are supportive conditions for the development of the processes of planning, organising, composing and revising student work. Collaborative discussions concerning the levels of claims in drafting sessions, for example, or the addition or refinement of hedges by peer "reviewers" at the revision stage, provide a helpful focus. First, however, students need opportunities to practise the various forms and functions of hedges common in the genres of their disciplines and the following suggestions address this issue.

Focus on high frequency items: Despite the variety of means to convey hedging, there is some justification for concentrating initially on lexical devices. Not only are these the most numerically significant signals of hedging in science, they are also likely to be the easiest to acquire, thus offering the greatest benefit for the least learning effort. Such a strategy will provide learners with a means of expressing themselves with subtlety in situations where they might lack the confidence to use an equivalent syntactic pattern. Knowledge of lexical items will also foster familiarity with the concept and the circumstances when it can be employed, allowing more complex means of expression to be developed later. Acquiring the vocabulary to express this aspect of meaning is made easier by the fact that a comparatively limited range of devices have an extremely high frequency in science.

Rather than encouraging students to master the semantic complexities of the modal verb system, for example, attention could focus on use of the high frequency modal hedges *would, may* and *could*. Moreover, the findings reported above show the predominance of alternative forms in scientific writing and these are generally easier for L2 students to acquire. Lexical verbs and adverbials in particular have an extremely high frequency and present students with far less semantic confusion. They are also sufficiently common in casual conversation to permit their transfer to other contexts (cf. Holmes, 1988). The fact that lexical verbs are often used in combination with other hedges also allows devices to be taught in patterned phrases and can assist students to develop confidence in both using different devices and the degrees of certainty expressed by them.

Early stages of ESP courses might concentrate on verb phrases with *indicate, suggest, appear* and *propose,* which represent 56% of all mitigating lexical verbs and constitute a spread of both evidential and judgemental items. Students should also study a range of high frequency disjunct adverbs such as *probably, possibly, apparently* and *essentially,* and downtoners of varying force, such as *quite, slightly, rarely* and *relatively.* Instruction can then proceed to related adjectival and nominal forms with common strategies and syntactic realisations being introduced at later stages. This procedure may help students acquire an understanding of the importance of hedging, the resources to express it, and the assurance to employ forms appropriately.

Pedagogic tasks: Tasks which require an explicit focus on the productive use of hedges include the following:

- link statements by including or omitting hedges to convey an accurate assessment of the truth relations between them;

- complete sentence frames, in an appropriate context, which employ common hedges such as *in spite of .., if . . .then, under these conditions . . . , the model implies . . .,* etc;

- paraphrase a description of an experiment using hedges to refer to uncertain claims;

- rework RA titles from journals in appropriate fields by the addition or subtraction of hedges so their implied certainity and "research space" is affected (Swales, 1990: p. 222);

- use interpersonal hedges to argue against the views of a politician or journalist;

- undertake free writing activities which explore personal possibilities, such as alternative directions at certain points in the past or options for the future;

- select the most appropriate supports for a particular claim from a list and link them to the controlling idea with appropriate hedges or certainty markers.

Many of the activities suggested here can provide the basis of "reformulation" activities (e.g. Allwright *et al.*, 1988) whereby a native speaker, perhaps a subject specialist, rewrites a NNS's draft for class discussion. Using the framework provided in this thesis, such a discussion could focus on the level of claims made and the form and function of the hedges employed. This would serve to both encourage reflection on the effectiveness of the suggested alterations and assist learners in drafting their own claims by offering an expert reader's perspective.

Writing for an audience: Reader response is critical to effective writing and in addition to making appropriate cognitive choices in hedging use, students also need to make decisions about correct register, style and the likely background knowledge and expectations of their audiences. Genuinely communicative activities involve creating contexts in which students can write appropriately for particular readers. Learners generally have little experience in writing for real audiences and therefore need to be sensitive to genre-specific and community-specific issues like hedges. They have to ask who their audience is, what they know and expect and what they will use the information for. Tasks must therefore incorporate the students' particular socio-rhetorical situation, encouraging awareness of the kind of knowledge claims it is appropriate to make and the level of conviction to invest in them.

To develop a consideration of audience and purpose in writing students can:

- reformulate a textbook passage from the perspective of a scientist presenting the ideas for the first time in a research report, making the claims more tentative;

- write a speculative paper explaining possible cause and effect relationships in arguments about, say, the reasons for inflation, drug abuse, Ph.D. dropout rates, juvenile crime, etc;

- write a Discussion section of an RA which hypothesises about future possibilities from photographs, graphs, statistical trends, experimental results, etc;

- contribute a short essay for a class journal about the state of knowledge on a topic or the implications of a new development in the student's discipline;

- describe an experiment in the form of a Results or Discussion section of a research article for a subject teacher using hedges to convey the appropriate level of claim;

- conduct a group investigation of the use of hedging in students' various disciplines and write a research paper or essay on their similarities and differences.

The literature suggests various ways to teach audience awareness in argumentation (e.g. Brandt, 1990). Schriver (1992), for example, proposes the use of audience analysis heuristics to help writers anticipate reader needs. Based on this, subject experts can provide taped think-aloud reading protocols on student texts for feedback which sensitises students to the importance of hedges in gaining audience acceptance of claims. Alternatively, Johns (1993) suggests researching the expectations of real audiences to phrase claims suitably on issues of direct relevance to students. By relating writing tasks to real purposes and readers, students can learn to use hedges appropriately in argumentation. An idea of Swales (1990) for teaching "suggestions for further research" can also be adapted to hedging. This involves students providing critical comments following a seminar presentation given by an instructor. Following a review of the presentation and these comments on tape, they then switch roles to write a conclusion to the related research paper using both sources of information. Clearly the activity sets up authentic situations for the use of hedges in both the need to be tentative in the oral discussion and in cautiously making claims for the findings of the research in writing.

The tasks outlined here move the learner from the role of discussant to presenter to interactant in achieving an awareness of appropriate hedging use in academic writing. In each of these ways, students are encouraged to develop their understanding of hedges and the part they play in the achievement of rhetorical purpose, the interests and values of the audience and the given genre.

Conclusions

A competence in hedging involves both a rhetorical consciousness, or perceived rationale for its communicative use, and an ability to appropriately employ it for maximum interpersonal and persuasive effect. By exposing students to the hedging choices frequently made by expert writers an understanding of these devices can be developed and incorporated into their own work. Concordance work is useful as it can make these choices explicit to both students and lecturers. It also offers a sufficiently eclectic means of presentation to cater for discipline in heterogeneous students working in the same ESP class and for the possibility of independent learning. This competence can contribute to actively empowering students to become members of the scientific discourse community.

The use of authentic examples clearly addresses a number of other criticisms often levelled at published textbooks (e.g. Kuo, 1993; Richards, 1993). The use of concordanced printouts or research articles meets requirements of learner independence, communicative authenticity, face validity, motivation, relevance, flexibility and disciplinary appropriacy. The absence of adequate textbook coverage can thereby provide an incentive to the use of more reliable and learner-focused materials which can subsequently form the basis of resource banks which relate language practice to a particular topic. The limitations of current published materials clearly demonstrate that the lexical, syntactic and pragmatic features of writing in particular disciplines must be organised around more thorough research than is currently the case. Thus the more we can learn about the use of hedges in different contexts the more confident teachers can be in guiding the selection of material for practice in interpreting and expressing hedges.

The pedagogic tasks sketched here simply suggest ways in which students might be led to appropriate means of rhetorical expression by drawing on their linguistic resources, authentic models of language use and an explicit focus on hedging devices. By emphasising textual products and an intended readership, this approach encourages students to consider the relationship between writers, claims and readers, and to introspect on the reasons for employing hedging in scientific writing. In this process the teacher and learner can each draw on their respective specialist knowledge and contribute as equals.

18

Sugaring the pill: Praise and criticism in written feedback

Fiona Hyland and Ken Hyland

Introduction

Providing written feedback to students is one of the ESL writing teacher's most important tasks, offering the kind of individualised attention that is otherwise rarely possible under normal classroom conditions. However, while generally acknowledged as pedagogically useful (e.g., Cohen & Cavalcanti, 1990; Hedgcock & Lefkowitz, 1994), the role of written feedback has largely been seen as informational, a means of channelling reactions and advice to facilitate improvements. Because of this, its important interpersonal aspects, the part it plays in expressing a teacher's stance and beliefs about writing and in negotiating a relationship with learners, is often overlooked.

As teachers, we are usually conscious of the potential feedback has for helping to create a supportive teaching environment. In addition, we are aware of the need for care when constructing our comments. We know that writing is very personal and that students' motivation and self-confidence as writers may be damaged if they receive too much criticism (e.g., Connors & Lunsford, 1993). We may also believe that praising what a student does well is important, particularly for less able writers, and we may use praise to help reinforce appropriate language behaviours and foster students' self-esteem. However, while teachers may recognise that the use of praise and criticism in feedback is important, the use of these features as central resources for negotiating judgements and evaluations of student writing and their contribution to the pedagogic role of feedback in EFL contexts has not been systematically studied.

In this paper, we offer a detailed analysis of the ways praise and criticism are used in the written feedback given by two teachers to six ESL students in a university language enhancement course. Following Holmes' (1988) characterisation of compliments, we view *praise* as an act which attributes credit to another for some characteristic, attribute, skill, etc., which is positively valued by the person giving feedback. It, therefore, suggests a more intense or detailed response than simple agreement. *Criticism*, on the other hand, we define as "an expression of dissatisfaction or negative comment" on a text (Hyland, 2000a, p. 44a). This definition thus emphasises commentary which finds fault in aspects of a text, and we felt the need to distinguish this from a third category, *suggestion*, which we regard as coming from the more positive end of a continuum. Suggestions differ from criticisms in containing an explicit recommendation for remediation, a relatively clear and accomplishable action for improvement, which is sometimes referred to as "constructive criticism." Our definitions thus draw heavily on common sense understandings of these terms, but also embody distinctions that were recognised by both student and teacher participants in the study.

In what follows, we examine the forms and patterns of these acts and draw on classroom observations and interviews to uncover teacher and student perspectives on the ways they are used, before discussing some implications for teachers. Importantly, we see praise and criticism as not only a crucial feature of the teaching and learning context, but also as helping to constitute this context. The teachers in this study were well aware that the type of comments they gave acted to both directly contribute to learning and to create the interpersonal conditions in which learning might occur.

Significant prior research

Most of the research on praise and criticism has occurred within a framework based on politeness, has examined speech, and has focused on complimenting behaviour (e.g., Herbert, 1990; Holmes, 1995). Our knowledge of how criticism is expressed is similarly very limited and largely restricted to (often elicited) conversational routines among intimates (e.g., Beebe & Takahashi, 1989). Politeness models have also been influential in writing in explaining how particular features are used to maintain rapport and mitigate criticism in student peer reviews (Johnson, 1992) and, more recently, in published academic book reviews (Hyland, 2000a).

The use of praise and criticism in written feedback is, however, more complex than in book reviews. Teachers are usually not simply appraising writing, but are often hoping to use the opportunity for teaching and reinforcing writing behaviours. In fact, they may be fulfilling several different and possibly conflicting roles as they give feedback: sometimes acting as teacher, proofreader, facilitator, gatekeeper, evaluator, and reader at the same time (Leki, 1990; Reid, 1994). In addition, their personal knowledge of the writer is usually greater than it would be between a book reviewer and an author, and they probably have more interest in creating and maintaining a good face-to-face relationship with the student. In other words, teachers often have to weigh their choice of comments to accomplish a range of informational, pedagogic, and interpersonal goals simultaneously.

Research on praise and criticism in feedback is fairly sparse. Several L1 studies suggest that teachers attend to error more than excellence and tend to focus their feedback on the negative aspects of the writing. Dragga (1986, cited in Daiker, 1989), for instance, analysed 40 student essays and found that 94% of comments focused on what students had done poorly or incorrectly. Experimental studies have often gone further to examine the different effects of focusing on positive and negative aspects of texts. Taylor and Hoedt (1966), for example, failed to find any difference in the quality of writing produced by students receiving either positive or negative feedback, although they did show that negative feedback had a detrimental effect on writer confidence and motivation. Gee (1972) also reported no significant differences in quality of writing, but more positive attitudes from those whose writing had been praised.

One problem with these studies is that praise and criticism were contextually disembodied, simply given mechanically according to the group writers were assigned to, with no relationship to the quality of the writing, or teachers' perceptions of students' needs. Other work has recognised that to be effective praise needs to be credible and informative and that insincere praise is unlikely to encourage good writing (Brophy, 1981). This is particularly the case at early stages of the L2 writing process where it is suggested that premature praise may actually confuse students and discourage revisions (Cardelle & Corno, 1981). Studies of L2 students' reactions to teacher feedback show that learners remember and value encouraging remarks but expect to receive constructive criticism rather than simple platitudes (Ferris, 1995).

One feature which may influence patterns of praise and criticism in written feedback is teacher response style. Anson (1989), for instance, has argued that the ways teachers judge writing and define their role when giving feedback are influenced by their belief systems. Such beliefs are partly the result of personal constructs but also originate in the social context in which teachers work. He suggests that teachers typically respond to student writing in one of three ways. Dualistic responders focus largely on surface features and take the tone of a critical judge of standards. Relativistic responders attend almost exclusively to ideational aspects of the writing, often ignoring significant linguistic and rhetorical problems. Finally, reflective responders respond to both ideas and structure and attempt not to be dictatorial in their approach. Developments in teaching theory and research have perhaps moved many teachers away from a dualistic response style, seeing it as prescriptive and potentially damaging to the student's writing development. Severino (1993) adds a socio-political dimension to this, pointing out that the nature of a teacher's response can suggest a stance towards both linguistic and cultural assimilation in L1 and L2 contexts.

Teacher response style may also be influenced by other factors, which can include the language ability of students, task type, and the stage at which feedback is given. Feedback offered at a draft stage will often be different from feedback on a final product, intended to perform a different function. Many teachers view feedback on drafts as more developmental and so offer more critical comments on specific aspects of the text, while feedback on a final product is likely to give a holistic assessment of the writing, praising and criticising more general features. Thus, any study of teacher written feedback must take into account the interplay between teachers, students, texts, and writing purposes and so consider written comments as "multidimensional social acts in their own right" (Sperling, 1994, p. 202).

Participants and data

Our written data consists of all the teacher written feedback given to six ESL writers from various language backgrounds on a 14-week full-time English proficiency course at a New Zealand university. Writing was an important aspect of this course, with about 2 hours each week given over to teaching academic writing and another 2 hours devoted to writing workshops. Feedback was collected for all students in two classes: one class preparing students for undergraduate admission (Class A) and the other preparing them for postgraduate studies (Class B). The teachers of the two classes were both experienced ESL writing instructors who had taught the course several times before.

Three students from each class agreed to participate as case studies. In Class A, there were 15 students. The composition of the class was quite varied and included students from Russia, Somalia, Vietnam, Korea, Indonesia, Japan, China, Taiwan, and Hong Kong. The case studies (all pseudonyms) were Maho, a 19-year-old female from Japan, Keith a 26-year-old male Taiwanese student, and Seng Hee, a 20-year-old female Korean student. Like the rest of the class, these students were low to mid intermediate level and came to New Zealand with the aim of studying for a first degree. Class B was more homogenous and consisted of 16 students, including 7 students from Thailand and 6 from China. They had all completed a first degree in their own language and were planning to enroll for postgraduate studies after this course. The case studies here comprised Liang from Taiwan and Samorn from Thailand, both female students in their 30s, and Zhang, a 27-year-old male from China. These students were all at a high intermediate to advanced proficiency level.

All the feedback on all the case studies' writing assignments was carefully documented and categorised. This amounted to 10 pieces of work for Class A and 7 pieces of writing for Class B. Three pieces of writing in both classes involved a feedback/revision cycle, consisting of the writing of a draft, followed by written feedback, and then a revised version in response to the feedback.

Regular observations of the classes by one of the researchers, with particular attention given to the writing workshops, provided information on the context within which feedback was given. Interviews with teachers offered another perspective on this context, as well as another source of information on feedback practices. The teacher interview prompt sheet (see Appendix A) consisted of questions covering the teachers' approaches to teaching writing and giving written feedback and their expectations of student behaviour after feedback. In addition, they were asked to describe an occasion where they felt that they had given very successful feedback and to pass one piece of advice about giving effective feedback to a new teacher of ESL writing.

Teachers were also asked to conduct think-aloud protocols as they gave written feedback to the draft of one piece of writing for each case study participant. A retrospective interview was then carried out with the students within a day of revising the draft. Photocopies of the draft with feedback and the final version of the writing were used as a visual prompt during the interview (see Appendix B).

To ensure content reliability and combat researcher bias, triangulation and respondent validation were included in the research design. Triangulation involved obtaining as many different perspectives on the data as possible. These different perspectives came first from the different sources of data: the teachers, the students, and the researcher and from a triangulation of methods, as data were collected through interviews, questionnaires, analysis of texts, observation of classes, and verbal reports. Respondent validation, or "member checking" (Lincoln & Guba, 1985), involved allowing participants in the research access to data and seeking their input and their evaluation of its authenticity to correct researcher bias. All observation notes were commented on by the two class teachers to help validate the interpretations and minimise misrepresentations.

Analysis and categorisation scheme

While there have been a number of models suggested for classifying teacher comments, many of these have often focused on contrasting large-scale areas such as "content" vs. "form" (Searle & Dillon, 1980), "local vs. global" issues (Zamel, 1985), and "high order vs. low order" concerns (Keh, 1990) and have not addressed teachers' aims.

Important exceptions are Ferris' models (Ferris, 1997; Ferris, Pezone, Tade, & Tinti, 1997). However, while these are perhaps descriptively more useful, they contain rather overcomplex lists of text variables, which may be too detailed to be used by teachers wanting to examine their own feedback. More problematically, the schema tends to confuse pragmatic and formal criteria and contains considerable overlaps. Thus, coders are asked to categorically distinguish questions which give information and those which ask for it (Ferris, 1997), for example, and to identify comments where teachers are asking for "known information" and where they are rhetorically seeking to "spur the student to further thought" (Ferris et al., 1997, p. 164). We feel that the constraints acting on teachers when providing feedback make it unlikely that they deliberate between, say, "asking for unknown information" or "making a request" in their end comments. Nor is it entirely clear how far either the analyst or the reader can determine which meaning is intended. We argue instead that teachers constantly seek to engage with both the student and the assignment through a limited number of overarching pedagogically and interactionally effective comment functions.

We developed our coding categories inductively through our reading of the essays and drafts. First, we identified individual "feedback points" as single written interventions that focused on a particular aspect of the text. The two researchers, working independently, had little difficulty in determining where one point ended and another began. Even long comments could usually either be clearly identified as carrying a single pragmatic force or as breaking down into a series of related speech acts. The few occasions where we could not agree were resolved by discussion. While we acknowledge that all feedback, including both "in-text" and "end" comments, have the potential to affect students powerfully, for this study we decided early on to ignore in-text points which occurred in the margin or the body of the text. In fact, many of these were simply symbols or codes which focused on language inaccuracies and corrections. We chose to examine only end comments, those written either at the end of the essay or on a separate feedback sheet, or on both, since these were longer, more substantive, and more discursive remarks on the student's writing overall. In total, 495 feedback points were examined, totaling 4700 words from 51 student essays.

Having identified the feedback points, we then worked independently to code each one according to its main evaluative or pedagogic purpose of praise, criticism, or suggestion. All comments were, thus, double-coded for reliability, and while we sometimes disagreed on whether a feedback point was principally a suggestion or a criticism, the few discrepancies were quickly resolved. The coding system we evolved seemed potentially useful as it effectively characterised the need for teachers to both retrospectively respond to the completed draft and to proactively address the students' future work. The meanings expressed in the end comments not only centre on the key functions of feedback, but also provide a more straightforward and readily identifiable pragmatic and content focus than Ferris' categories. They indicate, for instance, that "making a request," "giving information," "making a grammar comment," etc., are essentially means of praising, criticising, and suggesting.

This is not to say that there are no overlaps between our three categories, particularly between criticisms and suggestions. Embedded in every suggestion is an assumption that the original text requires improvement, and this criticism can be more or less explicit in the way the feedback is expressed. Thus, in (1) we have a fairly clear suggestion for revision, while in (2) the same teacher has chosen to express her comment more forcefully as a criticism:

1. Try to express your ideas as simply as possible and give extra information.
2. There is no statement of intention in the essay — what is the purpose of your essay and how are you going to deal with it? You are not giving me any direction.

The recipient of the feedback in (2) clearly has to do more interpretive work to recover the implications of the comment, and presumably the teacher hopes that the careful student reader will be able to unpack the intended advice it carries. However, this may not always occur, particularly if the comment is couched in general terms:

3. Repetition of sentence beginnings does not always work when too many start the same way.

In other words, we see criticism and suggestion as opposite ends of a cline of expression ranging from a focus on what was done poorly to measures for its improvement. We distinguished suggestions then, partly through the fact that they contained a retrievable plan for action, a do-able revision of some kind, which either addressed the current text (4) or extended this to future writing behaviour (5):

4. You need to decide on your main point at some stage in the process and connect everything to that.
5. For your next essay, I suggest you use written references (books, etc.).

While suggestions were more or less directive in this way, a second way of identifying them was through their surface structure. Over 3/4 of all suggestions were expressed by a limited number of explicit formulae, principally including the modals of assumption and obligation *need to, could,* and *should,* hypothetical *would,* and the verb *try.* Comments which lacked any of these features were generally criticisms.

The important point to make here is that teachers face choices when responding to student written work, selecting from available options which carry very different pragmatic force. We will argue in what follows that these decisions are often based on a desire to negotiate interactions which recognise both the learner's struggle to make meaning in a foreign language and the fragile intimacy of the teacher–student relationship.

Teacher acts in end comments

Our results show that 44% of the almost 500 feedback points were related to praise and 31% were critical. Only a quarter were comments offering explicit suggestions (Table 18.1). These findings are immediately interesting as they contradict the work which claims that positive comments are rare in feedback (e.g., Connors & Lunsford, 1993). Our experience is that teachers frequently use praise and regard it as important in developing writers (see also Bates, Lane, & Lange, 1993).

In interviews, both teachers showed that they were very aware of the potential effects of both positive and negative feedback. Joan (Teacher A) said that her general principles for giving written feedback were to find something positive to say and to look for the most important and most generalisable problems to comment on. Nadia (Teacher B) also reported that she tried to guard against the overuse of negative comments. She felt that it was necessary to show both the positive and negative parts of the writing, but stressed that the positive was "incredibly important" since ESL writers were

TABLE 18.1 Teachers' use of feedback acts

	Praise	Criticism	Suggestion	Overall
Joan	160 (42%)	114 (30%)	109 (28%)	383
Nadia	58 (52%)	39 (35%)	15 (13%)	112
Totals	218 (44%)	153 (31%)	124 (25%)	495

very insecure. She tried to be more positive in her feedback on a final piece of writing, whereas on a draft she felt that she could be harder because the students were aware that to get a good grade later, they would need to "fix up" aspects of their writing which were weak.

One aspect of Table 18.1 which is immediately striking is the difference in the total number of comments given by the two teachers, and especially in the greater number of suggestions made by Joan. Although a similar number of pieces of writing were commented on by both teachers, Joan made 109 suggestions, whereas Nadia made only 15. This can be partly explained by the different stances of the two teachers. In her interview, Joan stressed her dual role when giving feedback "as facilitator and provider of knowledge," pointing out "which aspects are good and which can be improved." In contrast, Nadia reported that she preferred to respond globally rather than deal with discrete points. In addition, the way they gave feedback possibly affected the number of suggestions. Joan used a feedback sheet and tended to give feedback on each of the points it specified (organisation, introduction, conclusion, language accuracy, etc.). Nadia, on the other hand, wrote her feedback directly on the student paper as end comments which may have led to her making more holistic and less specific observations.

We have already noted that feedback on developmental writing may have a different purpose to that given on final essays, and so we also decided to look at comments according to whether they were given in a draft or final version. Three assignments involved students writing two drafts, the second in response to teacher written feedback. The distribution of praise, criticism, and suggestions across drafts is given in Table 18.2.

It is interesting to note that although criticism and suggestion were fairly evenly distributed between drafts and final versions, nearly three quarters of all praise was reserved for final drafts. Interviews with the two teachers

TABLE 18.2 Distribution of teacher feedback acts in drafts and final copies

	Draft	Final	Total
Praise	33 (26.2%)	93 (73.8%)	126 (100%)
Criticism	40 (54.0%)	34 (46.0%)	74 (100%)
Suggestion	34 (46.6%)	39 (53.4%)	73 (100%)
Totals	107 (39.2%)	166 (60.1%)	273 (100%)

revealed that Nadia felt happier offering critical comments on drafts where there was potential to improve them, and Joan was uncomfortable making critical comments on drafts without appending a positive comment. In final versions, it seems that praise was extensively used to motivate the students in their next writing. There also appears to be similar thinking behind the high use of suggestions as many of these moved beyond specific problems in the current text to provide more general advice on writing. When we consider the data in Table 18.2 vertically, looking first at the distribution of teacher acts in the draft and then in the final versions, we see more clearly that although the type of feedback the students received most often overall was praise, the feedback they received most on first drafts was criticism.

Our analysis of these two teachers' responding practices also sought to address the principal areas of feedback focus by categorising the general aspects of the writing they chose to target in their comments. We found that comments addressed five main areas: the students' ideas, their control of form and mechanics, their ability to employ appropriate academic writing and research conventions, the processes of writing, and global issues, relating to the entire essay.

Table 18.3 shows that comments overwhelmingly addressed the ideational content of the writing. The teachers focused their praise mainly on ideas (64%) and were much less likely to praise either formal or academic aspects of the texts. Criticisms also tended to address this aspect and comprised over 43% of all the criticisms given, with less than a quarter concerned with formal issues. Suggestions focused evenly on ideas and academic concerns and slightly less on formal language-related issues. Although it may appear that the teachers were giving priority to meaning issues, it is worth reiterating that we focused only on the summary comments in this study and that a great deal of in-text feedback actually had a language focus (see Hyland, F., 1998).

TABLE 18.3 Focus of teacher feedback acts (%)

	Ideas	Form	Academic	Process	General	Totals
Praise	63.8	13.8	13.3	2.3	6.8	100
Criticism	43.8	23.5	30.1	2.6	0.0	100
Suggestion	36.3	23.4	36.3	4.0	0.0	100
Overall	50.7	19.2	24.3	2.8	3.0	100

Mitigation in teacher end comments

While it is an important pedagogic resource, teacher written feedback is also an evaluative genre in which student writing is judged and pronounced on. Evaluation always carries with it the seeds of potential friction because criticism can represent a direct challenge to a writer and undermine his or her developing confidence. Praise, too, also carries risks, for while it conveys support and interest, it can also damage an open relationship between the teacher and student as it implies a clear imbalance of authority. As every teacher knows, responding to student writing entails more than deciding whether to comment on form or content; it involves delicate social interactions that can enhance or undermine the effectiveness of the comment and the value of the teaching itself.

This is one reason why teachers may choose to compliment the ideas in a student essay or opt to present a criticism in the form of a suggestion. However, the response choices available to teachers are not limited to these broad macro-purposes. It is clear from our data that when expressing judgements, the two teachers were aware of the affective, addressee-oriented meanings their comments conveyed. As a result, baldly negative comments such as "Poor spelling" or "Referencing is inadequate" were rare. In fact, 76% of all criticism and 64% of suggestions were mitigated in some way. Praise was presented baldly 75% of the time but was itself widely used to tone down the negative effect of comments. Redressive strategies principally involved paired-patterns, hedges, personalisation, and interrogative syntax, and often an act included at least one of these strategies. The ways that these were distributed across different categories are shown in Table 18.4.

TABLE 18.4 Acts mitigated by different strategies

	Paired acts	Hedged	Personal	Interrogative	Unmitigated
Praise	68 (31.2)	26 (12.0)	5 (2.3)	2 (1.0)	117 (53.7)
Criticism	67 (43.8)	79 (51.6)	17 (11.1)	19 (12.4)	24 (15.7)
Suggestion	35 (28.2)	44 (35.5)	14 (11.3)	5 (4.0)	36 (29.0)
Overall	156 (31.5)	135 (27.3)	42 (4.9)	23 (4.7)	177 (35.8)

Acts can be mitigated by more than one strategy simultaneously.

Paired act patterns

One of the most obvious features of our data was the frequency with which these teachers combined their critical remarks with either praise (20%), suggestions (15%), or both (9%). One-fifth of the criticisms were accompanied by praise, the adjacency of the two acts serving to create a more balanced comment, slightly softening the negativity of the overall evaluation:

6. Vocabulary is good but grammar is not accurate and often makes your ideas difficult to understand.
   The order is OK, but the problem with this essay is the difficulty of finding the main idea.
   You used the information in the diagram well. Although you did not mention the expansion of the earth's crust.

Here then teachers seem to syntactically subordinate criticism to praise by preceding a negative comment with a positive one, a strategy we found later to be all too obvious to the students themselves.

Fifteen percent of criticisms were linked to suggestions, expanding what might be seen as a blunt criticism into a proposal for improvement, thus adding a more effective pedagogic and interactional dimension:

7. This conclusion is all a bit vague. I think it would be better to clearly state your conclusions with the brief reasons for them.
   This is a very sudden start. You need a more general statement to introduce the topic.
   This essay tends to waffle on a bit, I'm afraid. Try to make it much tighter and clarify your thoughts.

Once again, the full force of the criticism is assuaged by the second part of a pair. As the above examples indicate, suggestions could focus on the students' texts and proposals for revisions, but some given on both drafts and final versions of writing covered general principles and extended the suggestions to future writing behaviour, as in (8):

8. Maho, as I said on your first draft, a lot of this essay is about your learning history and, therefore, not directly relevant to the topic. At least you haven't shown how it is relevant. At university, you must answer the question you choose and keep on the topic.

This example is also interesting for its dialogic nature or intertextuality (Bakhtin, 1981) expressed through the back references to previous feedback. There were many examples in this data of feedback referring back to previous advice, to advice given orally during writing workshops, and also to points made by the teacher during classroom teaching sessions. Such examples illustrate the importance of context for a complete understanding of the analysis of feedback.

Following praise–criticism and criticism–suggestion, the most common pattern in our data was the praise–criticism–suggestion triad. This strategy serves to both mitigate the potential threat of the criticism and to move the students towards improving either their current text or their writing processes more generally in the longer term:

9. References very good. Two small problems. (1) Bibliography (at end of essay) — include initials of author. (2) Be careful about referencing inside the essay. Avoid said.
   Interesting content, but difficult to understand. I think you need to ask for help from flatmates, classmates, friends, me to read your writing and see if they can understand it.

For the most part, praise was less specific than criticism and frequently more cursory than developmental. In fact, previous work on compliments has suggested that these tend to be highly formulaic, in terms of both their grammatical structure and limited range of vocabulary (Johnson, 1992). However, while our data includes some comments which focused quite closely on what was good, suggestions tended to be pedagogically more useful and often functioned to narrow the issue considerably in praise–suggestion pairs:

10. Good movement from general to specific. But you need to make a clearer promise to the reader.
    This is a good essay but you have to expand your ideas.

In sum, criticisms of student writing in our data were frequently accompanied by either praise or suggestions which acted to fulfil both pedagogic and interpersonal functions for teachers.

Hedges

A second mitigation strategy was the use of hedges. Not surprisingly in such an evaluative genre, hedges were widely used to tone down criticisms and reflect a positive, sympathetic relationship with student-writers. While hedges have both an epistemic and affective function (Hyland, K., 1998), their principal purpose here was not to suggest probability, but to mitigate the interpersonal damage of critical comments. Our findings here differ considerably from Ferris' (1997) study of the feedback of one experienced ESL teacher, in which only 15% of comments contained hedges. This may be because we recognised more features as hedges, or it might relate to cultural or individual differences in feedback practices. In our data, hedges occurred frequently in end comments and comprised mainly modal lexical items, imprecise quantifiers, and usuality devices such as *often* and *sometimes*.

The following comment, for example, is structurally a criticism – suggestion pair, but the teacher has further drawn the sting from the comment by including a number of lexical softeners:

11. Some of the material seemed a little long-winded and I wonder if it could have been compressed a little.

Here the comment is made more tentative through the imprecise weakening devices *some* and a *little*, the lexical verbs *seemed* and *wonder*, the modal *could*, and the use of the hypothetical form. The response is weakened further by being couched in personal terms, a strategy we will discuss further below.

Hedges largely occurred in evaluations of content, where one might expect the interpersonal effects of direct criticism to be potentially most harmful, but they were also found in other categories of comment. They also appeared to be most used to mitigate the full force of stand-alone acts of criticism:

12. This is actually a little bit too long.
    Your conclusion was a bit weak.
    The essay is rather middle-heavy.
    There is possibly too much information here.

Hedges not only occurred in criticisms. Both teachers also used them to weaken the impositions which a direct suggestion may make on the reader:

13. Tomoko wrote well on this, you might ask to read hers.
    You could still make the promise to the reader clearer.
    You might also need to study the use of 'ing' and 'ed' forms in complex sentences.
    It might also be good to change the order of your paragraphs/ideas.

Perhaps surprisingly, hedges also often occurred in statements of praise. While common, praise was rarely fulsome and was at times very faint indeed:

14. Although parts of this essay have improved, they seem to be mainly surface level adjustments.
Fairly clear and accurate.
You show a reasonable understanding of this.
Mostly fairly good.

It seems here that praise is being used less to compliment the writer on an aspect of his or her essay than to call attention to some weakness. Hedged praise, in other words, might work for the writer as a rephrased criticism, simply prefacing a criticism or signaling a problem in a way which is less threatening to the teacher–student relationship.

Personal attribution

A third strategy to soften the force of criticisms was to signal it as reflecting a *personal opinion*. As we noted above, judging another's work is always an inherently unequal interaction because the power to evaluate is nonreciprocal and lies exclusively with the teacher. By expressing their commentary as a personal response, however, teachers can make a subtle adjustment to the interactional context and perhaps foreground a different persona. It allows them to relinquish some of their authority and adopt a less threatening voice. In other words, personal attribution allows teachers to react as ordinary readers, rather than as experts, and to slightly reposition themselves and their relationship to the student-writer. This can be seen in the following examples:

15. I'm sorry, but when reading the essay, I couldn't see any evidence of this really. Perhaps you should have given me your outline to look at with the essay.
I find it hard to know what the main point of each paragraph is.
My concern in this essay is that you introduce several terms in the introduction but do not provide a definition for any.

Specifying oneself as the source of an opinion can qualify its force by acknowledging that others may hold an alternative, and equally valid, view (Hyland, 2000a; Myers, 1989). Thus, in (15) there is a suggestion that another reader may see the required evidence or find the main point of each paragraph. The personal expression of criticism, then, reminds the reader that the comment carries only the view of one individual, thereby conveying the limitation of the criticism. The decision of writers to soften censure by rhetorically announcing their presence in the discourse was also employed to phrase suggestions. Representing a comment as their individual opinion rather than an uncontested necessity helped to mitigate what might otherwise be seen as a command:

16. I still believe a major effort to read English would improve your grammar. At university, I suggest you try to use a computer with a spelling checker because lecturers will not tolerate bad spelling.

Interrogative syntax

The final form of mitigation employed by these two teachers in their summary comments was to construct the criticism in *interrogative form*. Questions are a means of highlighting knowledge limitations and can be used to weaken the force of a statement by making it relative to a writer's state of knowledge. While they generally seek to engage and elicit a response from the reader, questions also express the writer's ignorance or doubt and, therefore, can mitigate the imposition of a suggestion or a criticism:

17. The first two paragraphs — do they need joining?
    Did you check your spelling carefully? Why not make a spelling checklist of words you often get wrong and use this before handing in your final? I think you have selected appropriate ideas to include. Why do you want to add more ideas?

In addition, questions are also useful when one wishes to protect oneself or one's reader from the full effects of what might be considered serious allegations. Interrogatives undermine a categorical interpretation of the underlying proposition, and because of this they can, in some cases, be seen as the teacher's attempt to withhold full commitment from the possible implications of a statement, an important consideration when the issue is one of plagiarism or the extent to which another person was involved in writing or editing the essay:

18. You only mention Ward once in the essay. Are all the other ideas your own? You need to make it clear which are yours and which are hers. Have you used quotations here? Some of it sounds like it might be. Did you get some help with the editing?

Mitigation: Motivations and miscommunications

Motivations

Why did the teachers use so many mitigation devices when giving feedback? We explored their motivations partly by examining the talk-aloud protocols they conducted when responding to the case study subjects' assignments and partly by examining the interviews carried out with both teachers at the beginning of the course to discover their beliefs and reported practices about feedback.

The two teachers took slightly different roles when responding to student texts. Joan reported that she sought to stress the "most important or most generalisable problem" in her feedback and that she tried not to overwhelm the students by criticising all their problems. She mentioned an experience on a previous course which had affected the way she gave feedback, making her less willing to be directly critical:

I had a Korean student who was kind of a fossilisation problem, I guess. And her writing was just full of errors and like you didn't even have paragraphs and it was very short. On the very first test, I think I made some criticisms. . .and she wrote in her journal that she found this very devastating and "please try and encourage me" and so after that I modified my feedback to try and be more positive. I mean, I had been positive but I felt it was my duty to point out that there were major problems here. I mean, it's hard sometimes to get a balance between being a realist and being positive. But once she told me that, I made a conscious effort.

Nadia first suggested that she viewed her role when responding to writing as "a reader rather than a know-it-all teacher." However, while this was her ideal, she was not always able to escape from a more authoritative role and was sometimes directive in her approach. Having elaborated her stance, she said to the interviewer, "Am I saying the right things or not? I'm probably going contrary to all kinds of research." She, therefore, seemed to be aware that there were more and less appropriate ways of giving feedback and that a too dominant teacher role might be considered unacceptable.

Nadia had previously worked on a first language writing course which had emphasised the importance of not appropriating student texts, and this agenda was obviously in her mind when both responding to interview questions and giving feedback. The topic of ownership of student texts has been an important and prominent topic in L1 writing research (see, for example, Brannon & Knoblauch, 1982; Onore, 1989). The arguments concerning appropriation are summarised by Knoblauch and Brannon (1984, p. 118), who have argued that writing can be "stolen" from a writer by the teacher's comments. They suggest that if students follow directive feedback too closely, they may develop neither their cognitive skills nor their writing abilities, but merely rewrite texts to reflect their teachers' concerns.

Teachers' fears of text appropriation have also surfaced in the ESL writing classroom. However, there has been some debate about their relevance and validity, especially for L2 writers. Reid (1994) has claimed that if they were overconcerned about appropriation, teachers might fail to give L2 students the direct and concrete help they needed. Nadia's stance suggests that she was aware of the issue of ownership of texts, and the mitigation of her feedback may have been a way of toning down what might be seen as over-directive interventions in the students' writing. By couching comments and criticisms as tentative suggestions, both teachers may have been attempting to intervene with clearer consciences, seeking to avoid demotivating students with negativity and removing the potential disquiet created by seeming to take over their texts.

Another reason for mitigating the force of comments was to minimise the possible threat which criticism carries for the "face" or public self-image of students. This was particularly evident in cases where teachers dealt with plagiarism, a very sensitive issue for feedback and something teachers are often unwilling to address directly. Teachers may also be aware of the literature that suggests that plagiarism is, at least partly, a western cultural concept (e.g., Pennycook, 1996; Scollon, 1994). Scollon (1994), for instance, discusses the ambiguity of the concept of plagiarism for nonnative writers unfamiliar with the individualistic authorial role expected in English academic writing which demands an inseparable link between ideas and those who first presented them. However, while students from cultures with more collectivist discourse practices may regard the nonattribution of authorship more positively than university handbooks, ESL teachers often feel that it is important to address these concerns when giving feedback. Not wishing to seem culturally insensitive, however, they often appear reluctant to do this directly.

Both teachers in this study disliked approaching students head-on over issues of plagiarism. In two think-aloud protocols conducted when giving feedback on assignments, Joan debated how to address this issue and make the students aware of plagiarism. She finally decided on indirectness:

It doesn't sound like her words — I hate accusing people of plagiarism, but when you think it is, what do you do?

Joan wrote as her feedback — *Where did you get this information? Have you used quotations?* — and considered this interrogative form "a subtle way of saying it." Unfortunately, the student's retrospective interview revealed that she failed to identify these as rhetorical questions and to detect the underlying criticism and the implicit suggestion. The offending text remained in her final draft.

Joan again felt uncomfortable with this issue in another case and debated how to tell Maho that she had plagiarised:

> Uh so, bother — do I accuse her of plagiarism? — I can ask her where she got this information from anyway.

Once again, she used the formula, "Where did you get this information from?" in her feedback rather than directly confronting the issue. Nadia, when confronted with a case of plagiarism of an entire essay, was also unwilling to be direct. She responded with a very indirect comment comprising a personalised and hedged criticism followed by a second considerably hedged criticism:

> I am afraid that this may not be your own work. You may have gotten some/ considerable help with it.

Perhaps unsurprisingly, this failed to open a dialogue with the student on the topic.

A final reason for the use of mitigation strategies is that teachers often see them as a means of helping to maintain or develop good relationships with their students. As we have seen, one way of taking the sting out of critical remarks was through the use of positive comments, but students did not always welcome empty remarks, and their reception of praise varied considerably. Some thought such feedback served no useful function. A student in Nadia's class, Zhang, for example, believed positive remarks were "useless" unless they were backed up by "serious" comments that he could act on:

> Because the teacher gives only a few words — It's OK or it's interesting — I think it's useless. [. . .] It is difficult for the teacher to be serious all the time, but in fact the teacher should try their best to be more serious. The students, they think very much about the teacher's feedback. If the teacher always not serious, the students will be feel very disappointed.

This view of positive comments was reinforced by another student in Nadia's class, Mei Ling, who said she discounted positive comments because "I want to know my weaknesses most." In her interview, she spoke against positive feedback as a waste of time; what she wanted was what she termed "negative" feedback highlighting her problems. For her, positive comments were insincere and, therefore, worthless:

> Sometimes maybe the teacher doesn't mean it, but they just try to encourage you. [. . .] Because there is always "but" after the positive. Sometimes the teacher just tries to find something good in my essay and then may be that strength is not the main point.

Her observation, incidentally, is an astute piece of informal discourse analysis, as our data shows that most of the praise–criticism pairs were linked by a contrastive connector.

However, the teachers were also right in their assumption that positive comments could be motivating for students. Maho emphasised the importance she attached to them:

> If teacher give me positive comments it means I succeed.

Nadia's student Samorn also described how a lack of positive comments affected both her attitudes to writing and her reception of feedback:

> . . .if feedback is not so good, I mean that teacher criticise many mistake I have, then I feel — "Oh I don't like writing." I am very interested in teacher's comments every time. I like to read it and when I read it and if it says "it's good but your problem is grammatical problem," then I will turn back to see how many mistakes I have. But if the comment is very bad and maybe not good enough, maybe I'll stop for a while and keep it and take it out and look at again later.

These student perspectives on positive comments suggest that they need to be used with care by teachers, rather than just included to make critical comments more palatable.

Miscommunications

While we did not initially set out to look for cases of miscommunication, it soon became clear during the student interviews that they were often unable to understand the teachers' mitigated comments. In fact, teachers did sometimes reflect aloud whether students would understand their more indirect statements. For example, Joan debated this possibility as she responded to an essay although she ultimately let the comment stand:

> *The main idea is clear — but the organisation of sentences in the introduction — and conclusion — is confusing? No. Is chaotic? Ha — but*

the organisation of sentences in the introduction and conclusion — is jumbled? Um could be improved? err — *could be improved —* sometimes students don't understand that as a criticism but he should.

While we have no quantifiable data to determine the full extent of this problem, each case study provided several examples where students failed to understand, or only partly understood, such mitigated comments. This study allows us to do little more than point to this as a potentially serious issue, but we believe that it opens interesting possibilities for more detailed research. In this final section, we draw on the transcripts of the retrospective interviews and the teachers' protocols. This preliminary data strongly suggests that teachers may sometimes forget that students are reading their feedback in a foreign language and that being more indirect and "subtle" may actually result in significant misunderstandings.

Our first example comes from Nadia's response to Zhang's assignment where she wanted him to define a term which he used throughout the essay. She drew his attention to this with a bald criticism followed by an interrogative suggestion:

My concern in this essay is that you introduce several terms in the introduction but do not provide a definition for any. I should like to know more about 'macroscopic' — is this what you are talking about in your conclusion?

However, in his final draft of the essay, the term was still being used without a definition, and Nadia commented that this aspect "still hasn't changed" in this version. Zhang was surprised and somewhat put out to read this comment and said that he "didn't understand what she wanted":

R   So why do you think your teacher made that comment?
Z   I think we have err, maybe I understand this word, no. I can't use this
      word correctly. The meaning I do not understand very well.
R   Is that what she's telling you? That she thinks you don't understand it?
Z   No.
R   What's she saying?
Z   She did not say anything about that.
R   She mentioned macroscopic — *know more about it.*
Z   Yes well it seems to me that this word means overall, but I don't know
      whether it's right or not.
R   Now here she's written *still hasn't changed* (final copy). What do you
      think she's saying there?
Z   I don't know.

> R   Do you think she wanted you to put in some definitions here of words like macroscopic?
> Z   I don't know, in fact I didn't understand what she meant.

Clearly, Zhang did not simply refuse to respond to this comment on purpose: "I didn't mean to ignore it, but I did not understand it. I think the teacher is always right."

Other examples of miscommunication were found in all the case studies. For example, Joan noted a problematic statement in the final paragraph of Keith's assignment and discussed this in her protocol:

> (Joan reading) "based on above data, we can perceive the higher sales revenue creates more profits" — of course. Now should I say something about this — I'll ask *isn't this statement obvious/always true?* — I hope he doesn't mind the suggestion.

The retrospective interview revealed that Keith failed to understand the negative implication of the question:

> R   Right — Why has she written isn't this statement obvious/always true?
> K   I don't know.
> R   Did you do anything with that comment?
> K   No.
> R   Do you know what she means by it?
> K   *Isn't this obvious/always true?* Just a question I think.
> R   She's just wondering?
> K   Yes wondering. But. . .I thought this is true.

In another part of the assignment, Joan felt that Keith was repeating himself and tried indirectly to alert him to the fact that certain information was superfluous with a hedged criticism:

> It seems funny to have this repeated — oh so this second part is like a summary — this just seems like a summary — it's not necessary. I might just write — *The second section seemed like a summary of the previous information.*

However, this comment was misinterpreted as a positive comment and served to reinforce this strategy:

> R   Why did you change this part?
> K   Because here — *Your second section seemed like a summary.* Because in my proposal, this includes two parts, so I think this is very necessary.

At the end of part one I make, like Joan says, a summary. So also I have put in another summary in here.

R   Right so you now have short summaries at the end of each part?

K   Yes.

A further example comes from Class B. During her protocol, Nadia gave the following comments on Samorn's conclusion to her essay comparing a free market and planned economy:

"In a free market economy, there are more productive efficiency than in a planned economy and consumers are happier for they can choose and get the goods they want." Ha she clearly knows which one she wants, good. . . . Still the conclusion is a bit abrupt. I might just write — *the conclusion may be a bit abrupt — you could re-state some of the main points*.

Samorn, however, misinterpreted this hedged criticism–hedged suggestion pair as a negative comment on her stance in the conclusion, as this extract from her interview shows:

R   Right, where it says here, "The conclusion may be a bit abrupt, you could restate some of the main points in the conclusion" — do you do that at all?

S   Yes, I think I do. In the conclusion, I change it from my first draft. My first draft I say that the free market economy is better than a planned economy, but this one I will not say like that, I will say I can not tell which is one is better. And I know that in fact I shouldn't have said that which one is better because I cannot say that.

We have already discussed Joan's reluctance to directly accuse Maho of plagiarism by couching her comment as an interrogative criticism: "Where did you get this information from?" This phrasing meant the meaning was impenetrable for Maho, however, who took the question literally as a request for a reference, as can be seen from her interview transcript:

R   "Where did you get this information from?" Did you change anything to show where you got the information from?

M   Not yet and I wonder where can I put all those information, references?

R   Right. So you weren't sure how to do the referencing?

M   Not quite sure.

R   So you've left it out?

M   Yes.

Miscommunication can occur even when teachers seek to address students' requests for particular kinds of help. Maho, for example, wrote on her essay cover sheet that she wanted feedback on her ideas and also some suggestions for new ideas. This was how Joan responded in her protocol:

> *There are* — no — you is more personal perhaps — *You have interesting ideas here and seem to have thought deeply about the ideas* — mmm — I think I might comment on — she wanted more ideas? — I might comment on that she has selected ideas well — *I think you have selected good* — maybe some of those actual examples weren't very good *appropriate ideas to include* — and I might just ask her *why do you want to add more ideas?*

Thus, Joan first considered a comment which gave hedged praise, moved to personalised and hedged praise, and finally ended up with a criticism mitigated by an interrogative. In an interview, Maho revealed that she was confused by the indirectness of this comment and reinterpreted it as a criticism of what she saw as her major problem, that of organisation:

R "You have some very interesting ideas here and seem to have thought deeply about the ideas. I think you have selected appropriate ideas to include. Why do you want to add any more ideas?" Why do you think your teacher made this comment?

M Because my ideas not concentrate on one point. Just this idea's about that one and this idea's about this one here. Something like that. I mean she means I think.

R Right so you think that she means you have different ideas in different places.

M Yes not well organised.

R From this comment you think that?

M Yes.

We have attempted to show from these extracts that while teachers often have laudable interpersonal and pedagogic reasons for mitigating their feedback, tentative comments have the very real potential to cloud issues and create confusion. While further research is needed to explore this in greater detail, it seems that mitigated criticism was most opaque to students and a source of particular confusion, especially where it was phrased interrogatively and not coupled with an explicit suggestion for revision. Indirectness frequently seems to be counter-productive to the aim of clearly conveying the point the teacher wishes to make and is often reinterpreted by students according to their own writing concerns and agendas. In other cases, failure to understand implied criticisms or toned down praise leads the student to revise aspects of the text which are not problematic. Our data clearly illustrate that indirectness can result in communication problems, and this is an aspect of feedback which needs to be investigated more directly by future studies.

Some conclusions and teaching implications

This paper focused on an important aspect of teacher feedback: the summary comments at the end of student assignments, and considered them in terms of their functions as praise, criticism, and suggestions. Critical analysts have consistently argued that patterns of language use contribute to the relationships which help structure social positions, and the two teachers discussed here seemed very aware that their responses to student writing had the potential to construct the kinds of relationships which could either facilitate or undermine a student's writing development. They recognised that offering praise and criticism expresses and confirms the teacher's right to evaluate a student's work and, as a result, sought to blur the impact of this dominance. By combining these acts into patterns of praise–criticism, criticism–suggestion, and praise–criticism–suggestion, and through use of hedges, question forms, and personal attribution, they sought to enhance their relationship, minimise the threat of judgement, and mitigate the full force of their criticisms and suggestions.

It is clear, however, that while responding to student writing is an important element of the teacher's role, it is also a practice that carries potential dangers and requires careful consideration. We have seen that despite the best intentions of these teachers and their desire to respond positively and effectively, the effect of their mitigation was often to make the meaning unclear to the students, sometimes creating confusion and misunderstandings. This result, incidentally, contrasts with Ferris' (1997) finding that hedged comments were more likely to lead to positive revisions than those without hedges. Presumably, the hedges did not directly encourage greater revisions, but were included in comments that were more directly usable. That is, teachers are more likely to hedge suggestions and criticisms (both of which implicitly or explicitly invite revision) than positive comments (many of which cannot be acted on). Future studies might investigate such links between "usable" hedged and nonhedged feedback and revision practices, using both analyses of student essays and interview data, so that we could learn more about the effects of hedges.

While our analysis represents the feedback comments of only two teachers and should, therefore, be treated cautiously, we believe these mitigation patterns are used more widely and, thus, our findings have potential application for L2 pedagogy. The kinds of miscommunication we have identified suggest that it may be a good idea for teachers to look critically at their own responses and to consider ways of making them clear to students. One aspect they could consider more closely is the number of mitigation devices they use. Although teachers need to guard against being over-directive, sometimes it may be necessary to deal with problems and possible solutions quite frankly. This is especially important with learners of low English proficiency since they may be less familiar with indirectness and fail to understand implied messages. Nor should we overestimate the ability of more advanced learners to recover the point of our remarks, as hedges are often invisible to L2 readers (Hyland, 2000b). Indirectness, in other words, can open the door to misinterpretation. More research that employs retrospective interview techniques to look at how students actually understand indirect feedback is needed.

Our data also strongly suggests that misinterpretations were common when dealing with plagiarism, due to the teachers' unwillingness to address this issue directly. However, while hedging our comments on such sensitive topics may cause us less discomfort and help preserve the student's face, we also have to think about his or her future participation in academic environments that invariably have exacting standards of attribution and referencing. Admonishments about plagiarism are common in style guides and academic handbooks, and most universities tend to treat the issue as a punishable offence. However, while we may be critical of these official practices or embarrassed by raising the issues in our comments on the work of particular students, shrouding the issue by forcing the reader through a maze of indirectness serves little purpose and fails to assist L2 learners seeking admission to further study. Like other cultural and rhetorical conventions of academic communities, we should address these issues clearly and directly, both in our classes and our feedback.

At the same time, our discussion has shown that students vary considerably in what they want from their teachers in the form of feedback. Some students value positive comments very highly while others simply discount them as merely mitigation devices, so we need to take care when making positive comments. In particular, these may need to be specific rather than formulaic and closely linked to actual text features rather than general praise. Most importantly, praise should be sincere. Students, as we have seen, are adept at recognising formulaic positive comments which serve no function beyond the spoonful of sugar to help the bitter pill of criticism go down.

In sum, when we pick up a pile of student essays to mark, we do not approach them with a tabula rasa. We have in our minds a stock of tried and tested phrases to choose from which relate in various ways to our own experiences of feedback, the kinds of teachers we are, and what we are trying to achieve. We alter these to fit specific students and their needs and personalities. This means that our comments go far beyond simple decisions to address form or content or to praise mechanics or criticise organisation. We are generally acutely aware of the importance of feedback in both providing helpful advice on our students' writing and in negotiating an interpersonal relationship which will facilitate its development. The ways that we frame our comments can transform students' attitudes to writing and lead them to improvements, but our words can also confuse and dishearten them. We hope that the description we have offered here may encourage teachers to reexamine their feedback to ensure it is clear and constructively helpful to students.

Appendix A. Teacher interview prompts

Attitudes to teaching writing

1 Could you describe the approach to teaching essay writing that you usually use when teaching ESL/EFL students?

2 What do you think is the biggest problem for EFL/ESL students when they try to write academic essays?

Approach and attitudes to giving feedback

1 How helpful do you think teacher written feedback is for improving students' writing?

2 What do you think is your main role when you respond in writing to a student's draft?

3 What do you think is your main role when you respond to a student's completed writing?

4 When you respond to drafts of student essays, are there any aspects of the texts which you focus on more than others? What are they?

5 Do you focus on the same aspects when responding to final drafts?

6 Do you think that teacher feedback is more helpful on a draft during the writing process or on completed writing? Why do you think so?

7 When you give feedback, which of your comments do you expect to be most useful to students to help them improve their writing?

8 Thinking specifically about language problems, what form of written feedback on language problems in drafts of writing do you usually give to students?

Attitudes to other forms of feedback

1 Do you think teacher feedback given orally, i.e., in an individual conference or while walking around the class, is more or less helpful than written feedback or are they about as helpful as each other? Why?

2 How useful do you think peer feedback is to ESL students in helping them improve a draft of their writing? Why do you think so?

Expectations of student behaviour

1 How do you expect students to use the written feedback you give them on their writing?

2 What would you expect students to do if they could not understand your comments or could not correct the mistakes in their writing after receiving feedback?

Reflections on own experiences of feedback

1 Could you describe an occasion where you felt that you have given a very successful feedback? This might be to an individual or to a whole class. It might involve one episode or a treatment carried on over a whole course.

2 If you were to pass one piece of advice about giving effective feedback to a new teacher of ESL writing, what would that be?

Appendix B. Student retrospective interview prompts

General overview of the revisions carried out

1 How long did you spend revising this draft?

2 Could you describe what you did as you revised? For example, did you read the feedback first or did you refer to the feedback as you revised?

3 What were the main changes you made to the draft?

4 What do you think was the most important change you made to the draft?

General overview of feedback use

1   What was the most useful feedback your teacher gave you on this draft?

2   Did you get feedback from any other source?

3   What use did you make of your peers' comments?

On global comments and changes

1   Why do you think your teacher made this comment?

2   What changes did you make to the writing after you read this comment?

3   Do you feel more satisfied with your writing now?

4   Why (not)?

On localised comments and changes

1   What do you think this comment is asking you to do?

2   What change did you make to your writing because of this comment?

3   Do you think your change has improved the writing?

4   How has it improved your writing?

On comments and corrections ignored

1   Why do you think your teacher made this comment?

2   Why did you not make any changes to the writing?

3   Do you think there is still a problem with the writing?

Student evaluation of their success in revising

1   When you look at your first and second drafts, do you feel satisfied with your revisions?

2   Do you feel that the essay has improved? How?

3   Is there anything about writing that you learned from writing this essay that you will remember and use in the future?

19

Specificity revisited: How far should we go now?

1 Introduction

In this paper I want to briefly revisit a concept central to ESP but which remains persistently contentious: the idea of *specificity*. Put most simply, this resolves into a single question: are there skills and features of language that are transferable across different disciplines and occupations, or should we focus on the texts, skills and language forms needed by particular learners? This question lies at the heart of what our profession is and what we do in our classrooms. Yet our inability to reach an answer weakens our potential effectiveness as teachers, causes uncertainty about our role, and creates confusion about the goals of ESP itself. With university ESP/EAP once again under attack from tighter budget constraints, it is critical that we give this issue discussion space. My aim here is to argue a case for specificity, that ESP involves teaching the literacy skills which are appropriate to the purposes and understandings of particular communities, and hopefully stimulate a debate through which we can critically examine our practices as teachers.

2 Specificity and literacy

Since the term first emerged in the 1960s, ESP has consistently been at the front line of both theory development and innovative practice in teaching

English as second/other language. Assisted by a healthy receptiveness to the understandings of different perspectives, ESP has consistently provided grounded insights into the structures and meanings of texts, the demands placed by academic or workplace contexts on communicative behaviours, and the pedagogic practices by which these behaviours can be developed. ESP is, in essence, research-based language education and the applied nature of the field has been its strength, tempering a possible over-indulgence in theory with a practical utility.

The success of this marriage of theory and practice has, I believe, been achieved because of clear emphasis on the idea of specificity, a concept fundamental to most definitions of ESP. It was central to Halliday, MacIntosh, and Strevens' (1964) ground-breaking work nearly 40 years ago, for example, and highlighted by Strevens (1988) who characterised ESP as centred on the language and activities appropriate to particular disciplines, occupations and activities and required by particular learners. By stressing students' target goals and the need to prioritise competencies, specificity clearly distinguishes ESP and general English, and has helped decouple university language teaching from the grammar or 'personal writing' approaches of earlier days.

Equally importantly, it has given ESP its heavy dependence on a strong research orientation which highlights the importance of target behaviours and which, in turn, has influenced the kinds of data we collect, the ways we collect it, and the theories we use to understand it. The imperative to inform classroom decisions with knowledge of the language features, tasks and practices of particular communities has led us to develop and sharpen concepts such as *genre, authenticity, discourse community, communicative purpose,* and *audience* which are now common coinage in applied linguistics. It has also meant the development of both ethnographic and text analytic research methods to help us get at what is going on in particular contexts.

But this practical orientation has also been a serious weakness, particularly in universities, where ESP is often regarded as a 'service activity', shunted off into special units, and marginalised as a remedial exercise designed to fix-up students' problems. The assumption underlying this practice is that there is a single literacy which students have failed to acquire, probably because of gaps in school curricula or the insufficient application of learners themselves. Students are seen as coming to their university studies with a deficit of literacy skills which can be topped up in a few English classes. Literacy can thus be taught to students as a set of discrete, value-free rules and technical skills usable in any situation. A more sophisticated version of this view recognises that ESL must not be taught in a vacuum and should prepare students for the language and skills they will be exposed to, but identifies these in terms of broad functional varieties called 'Academic English' or 'Business English' that can simply be applied in any relevant situation.

Conveniently for administrators, although rarely made explicit in the official rationales for the restructuring of university language programmes, such solutions are also often cheaper, logistically undemanding, and require less skilled staff to implement.

Both these positions, however, obviously disregard *specificity,* and in so doing, I believe, undermine our pedagogic effectiveness, weaken our academic role, and threaten our professionalism. Unfortunately, such positions seem to be gaining ground in many universities where, under the impact of increasing economic stringency and the banner of 'rationalisation', there have been moves away from specificity in university ESP classes towards teaching more 'generic' skills and language. In other words, back towards practices that are no longer ESP but closer to general English teaching.

3 General English for specific purposes?

There are four main reasons advanced for taking a general ESP approach, often referred to as a 'wide angle perspective'.

First, some ESL experts have expressed doubts about the possibility of identifying, and thus teaching, specific varieties at all, as Ruth Spack's (1988) paper (from which I borrow my title) illustrates. Spack's much discussed view is that language teachers lack the expertise and the confidence to teach subject specific conventions and that we should leave these to those who know best, the subject specialists themselves, and focus instead on general principles of inquiry and rhetoric. Second, there is the idea that LSP is simply too hard for students at lower levels of English proficiency who need to acquire a 'general English' suitable for all contexts before they can study LSP. Third, is the point made briefly earlier, that the systematic analyses of tasks and texts is an extravagant indulgence in times of cut-backs.

Last, and most important, the argument is often made that there are generic skills and forms of language that are the same across a range of disciplines, professions, or purposes. Hutchinson and Waters (1987, p. 165–166), for example, claim that there are insufficient variations in the grammar, functions or discourse structures of different disciplines to justify a subject-specific approach. The labels *English for General Academic Purposes* (EGAP) and *English for General Business Purposes* (EGBP), originally introduced by Blue (1988), seem to endorse the idea of broad literacy domains. Discussions of these terms in recent books offering overviews of the field by Dudley-Evans and St John (1998) and Jordan (1997) show how far the idea of non-specific ESP has crept in to our current thinking and practices.

In response, the position that specialist discourse should be left to subject specialists can be countered from several directions. It seems evident, for example, that subject teachers generally lack both the expertise and desire to teach literacy skills. Many subject specialists appear to believe that academic discourse conventions are largely self-evident and universal (e.g. Lea & Street, 1999) and are often content to simply assign grades to products without worrying too much how the product was arrived at (e.g. Braine, 1988). Nor is it entirely clear what the 'general principles of inquiry and rhetoric' we are counselled to teach actually are or, even if we could identify them, how these might help address students' urgent needs to operate effectively in specific disciplines (Johns, 1988). Perhaps most importantly, a decade on from Spack, we are now in a better position to describe the literacy cultures of different academic majors more precisely and with more confidence. This knowledge is related, moreover, to our professional responsibility to use these descriptions of target forms and tasks to best assist our students.

The second argument, that weak students need to control core forms before getting on to specific, and presumably more difficult, features of language is, quite simply, not supported by research in second language acquisition. Students do not learn in this step-by-step fashion according to some externally imposed sequence. They acquire features of the language as they need them, rather than incrementally in the order that teachers present them. Students may need to attend more to sentence-level features at lower proficiencies, and perhaps require remedial attention in some areas, but there is no need to ignore either discourse or discipline at any stage.

It is also worth noting that this position also neglects the fact that most ESP/EAP learners need to acquire competence in a range of subject specific communicative skills in addition to particular forms and genres. Most ESP courses place considerable emphasis on preparing students to engage effectively in their target communities, providing guidance on such activities as asking questions in tutorials, participating in meetings, writing on-line technical documentation, and so on. Participation in these activities rarely depends on students' full control of 'common core' grammar features and few ESP teachers would want to delay instruction in such urgently demanded skills while students perfected their command of, say, the article system or noun-verb agreement.

Third, it seems ironic that ESP has to defend itself against arguments that research-based teaching is uneconomical. Not only was cost-effectiveness one of the justifications given for the initial introduction and early impetus of ESP, but this is also a characteristic often unquestioned in corporate settings where the expertise of specialist practitioners is actively sought and highly valued. There are, of course, differences in corporate and academic contexts which make direct comparisons difficult, particularly as it is often harder to precisely specify target tasks and texts and to sustain student motivation in universities. By moving beyond limited occupational arenas and into the academy we also bring politics into the picture, as time spent on English runs up against other demands on students which are often fiercely resisted. I do not believe that these reasons are sufficient to force us into intuitively devised GESP courses however. On the contrary, we can reasonably respond that if there is some kind of general, universally useful, ESP, then it can equally, and more cost-effectively, be acquired together with the specific variety of the target discipline (e.g. Flowerdew & Peacock, 2001; Strevens, 1988).

Finally, there is the idea which underlies all the others: that ESP involves teaching general skills and forms that are transferable across contexts and purposes. This is what Bloor and Bloor (1986) call the *common core hypothesis*, the idea that "many of the features of English are found in all, or nearly all, varieties" (Leech & Svartvik, 1994). Most ESP and study-skills textbooks are obviously based on this idea, and there are numerous courses organised around 'core' themes such as 'business writing' and 'oral presentations', and 'core' topics like 'persuasive language', 'expressing cause and effect', and so on. Bloor and Bloor argue that a major weakness of the common core is that it focuses on a formal system and ignores the fact that any form has many possible meanings depending on the context in which it is used. Defining what is common is perhaps relatively straightforward when dealing with grammatical forms that comprise a finite set, but gets unwieldy when we introduce meaning and use. By incorporating meaning into the common core then, we are led to the notion of specific varieties, and to the inescapable consequence that learning should take place within these varieties.

More seriously perhaps, is the problem of identifying exactly what comprises a core. Ann Johns (1997, p.58–64) has sought to elaborate such common principles in 'general expository academic prose' by drawing on the work of various composition theorists. Her list includes features such as explicitness, intertextuality, objectivity, emotional neutrality, correct social relations, appropriate genre requirements, use of metadiscourse and hedging, and display of a disciplinary vision. However, these familiar, apparently self-evident, features of academic writing are only 'core' in a very general sense and give the misleading impression of uniform disciplinary practices. As Johns goes on to point out, each of these points is further refined and developed differently within each discipline, so that some fields, such as literature, may actually subscribe to none of them. Thus while they may be useful in providing learners with a basis for challenging unreflective stereotypes or examining specific practices in their own fields, such 'core' features offer an inadequate foundation for understanding disciplinary conventions or developing academic writing skills.

4 Different strokes for different folks

The discourses of the academy do not form an undifferentiated, unitary mass but a variety of subject-specific literacies. Disciplines have different views of knowledge, different research practices, and different ways of seeing the world, and as a result, investigating the practices of those disciplines will inevitably take us to greater specificity.

The idea of professional communities, each with its own particular practices, genres, and communicative conventions, thus leads us towards a specific role for ESP. But this is not to deny that students also cross boundaries. They inhabit complex academic and social worlds, moving outside their disciplines to take elective courses, discussing problems and assignments with peers, lecturers and advisors, and engaging in a disparate range of spoken and written genres. We have to recognise, of course, that our students need to function in numerous social environments and that our courses should equip them with the necessary skills to do so. Such epistemological, ontological, social and discoursal border-crossings pose enormous challenges for students and teachers alike, but a good starting point is to recognise the literacy practices that help mark off these borders. The notion of specificity thus provides learners with a way of understanding the diversity they encounter at university. It shows them, in other words, that literacy is relative to the beliefs and practices of social groups and to the purposes of their individual members in accomplishing their goals.

The principle of specificity receives strong theoretical endorsement from the philosophical perspective of social constructionism (e.g. Bruffee, 1986; Rorty, 1979) and the critiques and extensions of it (e.g. Bizzell, 1992; Blyler & Thralls, 1993). This stresses that disciplines and professions are largely created and maintained through the distinctive ways that members jointly construct a view of the world through their discourses. We work within communities in a particular time and place, and these communities are created by our communicative practices; so writing is not just another aspect of what goes on in the professions or disciplines, it is seen as actually *producing* them. The model is of "independent creativity disciplined by accountability to shared experience" (Richards, 1987, p. 200). As a result, the teaching of specific skills and rhetoric cannot be divorced from the teaching of a subject itself because what counts as convincing argument, appropriate tone, persuasive interaction, and so on, is managed for a particular audience (Berkenkotter & Huckin, 1985; Hyland, 2000a, 2001a).

Equally persuasively, we can also turn to a large, and very diverse, body of research evidence to back up this view of discipline and profession-specific variation. Once again, I will draw on academic writing to illustrate some of this research, but the point I want to make applies more widely to ESP taught in university contexts.

First, there is a considerable collection of survey results which show that the writing tasks students have to do at university are specific to discipline and related to educational level. In the humanities and social sciences, for example, analysing and synthesising multiple sources is important, while in science and technology, activity-based skills such as describing procedures, defining objects, and planning solutions are required (Casanave & Hubbard, 1992). In post-graduate programmes it seems that engineers give priority to describing charts, while business studies faculty require students to compare ideas and take a position (Bridgeman & Carlson, 1984). In undergraduate classes, questionnaire data suggests that lab reports are common in chemistry, program documentation in computer science, and article surveys in maths (Wallace, 1995).

More interestingly, these differences begin to multiply when we move beyond the classifications of questionnaire designers. Genre categories blur when actual assignment handouts and essay scripts are considered, for example, and the structure of common formats such as the experimental lab report can differ completely across different technical and engineering disciplines (Braine, 1995). Ethnographic case studies of individual students and courses reinforce this picture, revealing marked diversities of task and texts in different fields (e.g. Candlin & Plum, 1999; Prior, 1998). It is not difficult to imagine how complicated this can become for students in joint degrees or interdisciplinary studies like business studies, for example, where a student may have to produce texts in fields as diverse as accountancy and corporate planning.

Overall then, this literature points to the fact that different disciplines identify different types of writing as features of academic literacy and that terms like *lab reports, lectures,* or *memos* imply neither homogeneity nor permanence. Members' (or folk) taxonomies can, of course, be a useful first step into describing target contexts and are helpful in identifying what insiders see as similar and different (e.g. Swales, 1990, p. 54). It is, however, easy to be misled into believing there is greater similarity in the communicative resources of different communities than is actually the case. The expectations that a genre label calls to mind in one field may be very different to those it evokes in another, and we should hesitate before regarding such identifiers as objective and invariable descriptions of the ways members organise their communicative practices.

This view of multiple literacies is reinforced by text analysis research. While academic genres are often identified by their conventional surface features, they are actually forms of social action designed to accomplish socially recognised purposes with some hope of success. While such purposes are influenced by personal factors and subject to individual choices, these choices are likely to be relatively limited in practice. This is because successful academic writing depends on the individual writer's projection of a shared context. We are more likely to achieve our disciplinary purposes if we frame our messages in ways which appeal to appropriate culturally and institutionally legitimated relationships (e.g. Dillon, 1991; MacDonald, 1994). This is why, for example, the 'common core' features of academic prose listed earlier often differ considerably in their frequency, expression and function across disciplines. Simply, the ways that writers present their arguments, control their rhetorical personality, and engage their readers reflect preferred disciplinary practices (Hyland, 2000a).

One major reason for these differences in disciplinary discourses then, is that texts reveal generic activity (Berkenkotter & Huckin, 1995; Swales, 1990). They build on the writer's knowledge of prior texts, and therefore, exhibit repeated rhetorical responses to similar situations with each generic act involving some degree of innovation and judgement. This kind of typification not only offers the individual writer the resources to manage the complexities of disciplinary engagement, but also contributes to the stabilisation and reproduction of disciplines. This directs us to the ways disciplinary texts vary, not only in their content but in different appeals to background knowledge, different means of establishing truth, and different ways of engaging with readers.

In sum, this research shows that scholarly discourse is not uniform and monolithic, differentiated merely by specialist topics and vocabularies. It has to be seen as an outcome of a multitude of practices and strategies, where argument and engagement are crafted within specific communities that have different ideas about what is worth communicating, how it can be communicated, what readers are likely to know, how they might be persuaded, and so on (e.g. Faigley, 1985).

5 Putting the 'S' back into ESP

Essentially ESP rests on the idea that we use language to accomplish purposes and engage with others as members of social groups. It is concerned with communication rather than language and with the processes by which texts are created and used as much as with texts themselves. What this means is that the field seeks to go beyond intuitive laundry lists of common core features and the autonomous views of literacy that such lists assume, to the practices of real people communicating in real contexts. A *specific* conception of ESP thus recognises that while generic labels such as 'academic English' or 'scientific English' may be a convenient shorthand for describing general varieties, they conceal a wealth of discursive complexity.

Unfortunately, however, such labels disguise variability and tend to misrepresent academic literacy as a naturalised, self-evident and non-contestable way of participating in academic communities. This in turn encourages the idea that there is one general 'academic English' (or 'business English', etc.) and one set of strategies for approaching reading and writing tasks that can be applied, in a painting-by-numbers fashion, across disciplines. By divorcing language from context, such an autonomous view of academic literacy misleads learners into believing that they simply have to master a set of rules which can be transferred across fields. As I noted above, many subject specialists also subscribe to this view, taking academic writing conventions to be unproblematically universal and unreflectively available. Because they are rarely provided with a means of conceptualising the varied epistemological frameworks of the academy, students (and teachers) are often unable to see the consequences these have for communication or distinguish differences in the disciplinary practices they encounter at university (Plum & Candlin, 2001).

By ignoring specificity, moreover, we also run the risk of creating an unbridgeable gulf between the everyday literacies that students bring with them from their homes and those that they find in the university. In such circumstances it is easy for both learners and teachers to reify these powerful academic and professional literacy practices; to see them as autonomous, abstract and beyond their control. Acquisition of disciplinary knowledge involves an encounter with a new and dominant literacy, and because academic ability is frequently evaluated in terms of competence in this literacy, students often find their own literacy practices to be marginalised and regarded as failed attempts to approximate these standard forms. So, by detaching academic literacy from its social consequences, it is easy to see communication difficulties as learners' own weaknesses and for ESP to become an exercise in language repair. The only way to counter this is to bring these practices back to earth by targeting specific contexts and drawing on the experiences of our learners. Only by taking the notion of specificity seriously can ESP find ways to undermine a 'single literacy' view and to replace 'remedial' approaches to teaching with those that address students' own perceptions and practices of writing.

I am aware this is not an easy task. Putting specificity into practice can involve considerable challenges, not least in balancing students' needs (assessed in terms of the specific discursive practices of their fields) with competing departmental demands (particularly as to time and learning priorities), and with institutional constraints (often ultimately specified in terms of a viable group size). Nor have I mentioned the wishes of individual students, who typically have their own personal agendas and pedagogic preferences. I acknowledge that students have little time for abstractions or linguistic theory and I am sympathetic to the comment of one ESPJ reviewer that they are frequently looking for a quick return and concrete, understandable teaching points. There is nothing about specific teaching however which prevents us for delivering on these demands, in fact it suggests ways of targeting them more effectively.

A major problem of heterogeneous classes is actually finding enough common ground among students, but one solution is to exploit the specificity of their circumstances through the opportunities that such classes offer to contrast their disciplinary experiences and expectations (cf. Swales & Feak, 2000). This kind of rhetorical consciousness raising not only helps satisfy students' demands for personal relevance, but also reveals to them the multi-literate nature of the academy. Becoming literate in one's discipline essentially means developing an awareness of the functions of texts and how these functions are conventionally accomplished. By making contact with those outside their field, students may more easily come to see that communication does not entail adherence to a set of universal rules but involves making rational choices based on the ways texts work in specific contexts.

Moving beyond the classroom, specificity is not only central to our teaching and the ways we perceive the disciplines and professions, it is also critical to how we move forward as a field of inquiry and practice. Placing specificity at the heart of our role means that we are less likely to focus on decontextualised forms, to see genres as concrete artefacts rather than interactive processes, or to emphasise a one-best-way approach to genre and interaction. It also means that ESP might become less vulnerable to claims of 'accommodationism' to dominant political and institutional orders (e.g. Benesch, 2001; Pennycook, 1994). A discipline-specific view of literacy makes it easier for both teachers and students to see the complex ways in which discourse is situated in unequal social relationships and how its meanings are represented in social ideologies. Clearly there is a real need for us to be more flexible in our pedagogies, more wide-ranging in our research, and more critical in our professional practices. But we need to hold fast to those things we have got right: a commitment to revealing the workings of other communicative worlds to our students by grounding pedagogical decisions in an understanding of target texts and practices.

Together, all this leads to the important conclusion that expertise in a subject means being able to use its discourses in the specific ways that one's readers are likely to find effective and persuasive. While we may often talk about reports, memos, oral presentations, and so on as overarching genres and universal skills, these take on meaning only when they are situated in real contexts of use. Put simply, students do not learn in a cultural vacuum: their disciplinary activities are a central part of their engagement with others in their disciplines and they communicate effectively only by using its particular conventions appropriately.

ESP therefore involves developing new kinds of literacy, equipping students with the communicative skills to participate in particular academic and professional cultural contexts. Establishing exactly what are the specific language, skills, and genres of particular groups on which we need to base learning priorities may well be expensive, time consuming and skill-intensive. But it is this research which both makes our teaching effective and our practices professional, and we should not give these up easily. There is, then, only one possible response to the question posed in the title of this paper: effective language teaching in the universities involves taking specificity seriously. It means that we must go as far as we can.

Commentary on Part V

Ann Johns

One could surmise from reading the other sections of this collection that Ken Hyland is best known for his research, significant because it asks interesting or unusual questions of language and texts, generally employing corpus linguistics as well as other approaches, such as expert interviews. That is true. However, for several years, I have been using some of his research-based texts in writing classes for students across the disciplines at various academic levels; and I am certain that others have done the same. There are several reasons for this. First, Ken's articles make evidenced-based arguments for variation in language use among the disciplines. Since most undergraduate students and their instructors seem to believe that there is one academic language that can be taught to students across the disciplines, I can point to 'Specificity revisited...', in this collection or an online article on the Writing-across-the-Curriculum website, '10 disciplines and discourse: Social interactions in the construction of knowledge', among others, to argue against the 'general academic language' view. For teachers of K-12 students, I use Ken's work to make a similar argument, as evidenced in my chapters focusing on the Common Core, national standards mandated in a majority of the states in the United States (Johns 2016a,b).

Graduate students, particularly those who are native speakers of languages other than English, seem to be savvier than younger students and their teachers. They know from their research studies that there are specific ways that language is used in their disciplines. However, having in most cases studied traditional grammar and vocabulary building, the latter of which is generally based upon the 'bible of vocabulary lists' (the AWL), created by Avril Coxhead, even these students tend to be unaware of Hyland's work. Thus, in my classes, the more sophisticated students examine the findings from Hyland's 'Stance and engagement ...' article from *Discourse Studies* (2005 and Chapter 6 this volume) and their favourite text, 'As can be seen: Lexical bundles and disciplinary variation (2008 and chapter 15 this volume)'. My international Fulbright students completed studies and presentations on research papers in their disciplines based on the 2008 article, sometimes varying or renaming Ken's categories for their purposes. Then they rewrote the Beatles' 'Eleanor Rigby', with words taken from 'As can be seen ...' to summarize their findings. So Ken's argument for variation, in addition to his classification schemes, is valuable to students at every level of instruction – as well as to their instructors.

There are other reasons for using Hyland's EAP work in classes at various academic levels: Ken's writing is accessible, certainly, and the structures of his texts are conventional for an applied linguist as well as clearly marked with appropriate metadiscourse. So, for example, my students examine the research results in Hyland and Tse (2004: 169 and Chapter 7 this volume). Then, using questions from the classic Swales and Feak volume (2012), they examine one or more of their disciplines' research articles for moves and metadiscourse.

Needless to say, reading Hyland's work has revolutionized my EAP teaching over the years, providing approaches which both students at several academic levels appreciate. And though I speak as one instructor, undoubtedly many others have benefited from his work, as well. Now I turn to the chapters in this section.

Chapter 16: I remember Ken's amazement when the *Journal of Second Language Writing* accepted this manuscript 'without even editing!' introducing many readers of this prestigious journal to genre approaches to teaching. More than ten years before this article appeared, John Swales's groundbreaking volume (1990) was circulating through the world; and later (1997), I had attempted to bring genre pedagogies to teachers, particularly of multilingual undergraduates. Those in the 'Sydney School' (e.g. Martin 1993; Macken-Horarik 2002) had been particularly active as theorists and practitioners, as well. However, because Ken presented genre and its resources as a response to teaching 'process', noting that 'writing cannot be distilled down to a set of cognitive processes', this paper undoubtedly affected a large, previously uninformed academic population, many of whom were located in the United States. A recent count of the citations of this text (26 February 2017) is 734. My guess is that this number does not indicate appropriately the number of teachers who have been influenced by Ken's careful, sustained argument for genre pedagogies.

Chapter 17: Born in 1951, Ken was in his 40s when he completed his PhD, having taken the typical British EFL (English as a Foreign Language) route of teaching internationally for many years, in his case, in Sudan[1], Saudi Arabia, Malaysia, Papua New Guinea and elsewhere. For teachers, this is a real plus, since it means that with a rich practitioner background, he is able to give

---

[1] What is it about Sudan? John Swales also taught there.

them explicit, useful assistance. This piece, written soon after he finished his dissertation on hedging, demonstrates Ken's continuing argument that even texts in science, often considered 'objective', are not neutral or template driven. In this article, he focuses on appropriate hedging, employed to 'present [writer's] claims cautiously, accurately, and modestly' (p. 417, this volume). Ken notes this linguistic feature's importance by saying:

> The appropriate use of hedges helps writers develop academic arguments and establish a relationship with their readers. (p. 409, this volume)

Ken then goes on to present a particularly rich list of 'Pedagogic Tasks' for the productive use of headings, expert informant involvement and addressing audiences.

Chapter 18: In working towards their volume entitled *Feedback in second language writing: Contexts and issues* (2006), Ken and Fiona Hyland produced this article, focusing on what was of most interest to them: the interactions between teachers and students, particularly the ways in which indirectness in teacher feedback 'has a real potential for incomprehension and miscommunication'. Providing many examples of written feedback as well as results of interviews with both teachers and students, the authors warn of the 'potential dangers' entailed in the language we use to evaluate student writing. This is a useful and thought-provoking article. But this is my problem: like every teacher I know, I dislike grading papers, so I often work at night with a glass of wine to fortify me. Not surprisingly, the results from this research made me cringe. Have my students ever understood and made good use of my written feedback?

Chapter 19: Appropriate as the final EAP text in the collection, this article makes a lengthy argument, begun by the famous Peter Strevens in 1988, that ESP should always be distinguished from TENOR (Teaching English for No Obvious Reason), because I must be specific to context. Here, Ken insists that

> we need to hold fast to those things we [ESP teachers and researchers] got right: a commitment to revealing the workings of other communicative worlds to our students by grounding pedagogical decisions on an understanding of target texts and practices. (p. 467, this volume)

This comment points to what EAP should probably move towards next: taking a page from Rhetorical Genre Studies theory to make the literacy context even more central to our understanding of academic texts and pedagogies.

References

Acknowledgements: Gratitude that grates [Editorial]. (1996). *The Economist*, 340, 83.

Adams Smith, D. (1984). Medical discourse: Aspects of author's comment. *English for Specific Purposes*, 3, 25–36.

Adler, E. S., Gent, C. E. and Overmeyer, C. B. (1998). The home style homepage: Legislator use of the world wide web for constituency contact. *Legislative Studies Quarterly*, vol 23 (4), 585–95.

Aguinis, H., Nesler, M. S., Quigley, B. M., Suk-Jae-Lee and Tedeschi, J. T. (1996). Power bases of faculty supervisors and educational outcomes for graduate students. *The Journal of Higher Education*, 67 (3), 267–97.

Ahmad, U. (1995). *Academic Language and Culture: Some Observations on Scientific Malay and scientific English*. Paper presented at the RELC Conference. Singapore.

Allwright, R., Woodley, M. and Allwright, J. (1988). Investigating reformulation as a practical strategy for the teaching of academic writing. *Applied Linguistics*, 9(3), 236–56.

Altenberg, B. (1998). On the phraseology of spoken English: The evidence of recurrent word combinations. In A. Cowie (ed.), *Phraseology: Theory, Analysis and Applications*, 101–22. Oxford: Oxford University Press.

Anderson, J. (1980). The lexical difficulties of English medical discourse for Egyptian students. *English for Specific Purposes (Oregon State University)*, 37(4).

Anson, C. (1989). Response styles and ways of knowing. In C. Anson (ed.), *Writing and Response*, 332–65. Urbana, IL: NCTE.

Anstey, M. and Bull, G. (2004). *Literacy as Social Practice: The Literacy Labyrinth*, 2nd edn. Sydney: Pearson.

Anthony, L. (2011). AntConc 3.4.3. http://www.laurenceanthony.net/software.html

Arnaud, P. and Bejoint, H., eds (1992). *Vocabulary and applied linguistics* London: Macmillan.

Arnaudet, M. and Barrett, M. (1984). *Approaches to Academic Reading and Writing*. Englewood Cliffs, NJ: Prentice Hall.

Arnold, J. and Miller, H. (2001). Breaking away from grounded identity? Cyberculture and gendered academic identities on the web. Paper presented at constructing cyberculture(s): Performance, pedagogy and politics in online spaces, College Park, MD.

Atkinson, D. (1996). The philosophical transactions of the royal society of London, 1675–1975: A sociohistorical discourse analysis. *Language in Society*, 25(03), 333–71.

Atkinson, D. (1999). *Scientific Discourse in Sociohistorical Context: The Philosophical Transactions of the Royal Society of London, 1675–1975*. Mahwah, NJ: Lawrence Erlbaum.

Atkinson, D. (2002). Comments on Ryoko Kabuta's 'Discursive construction of the image of US classrooms'. *TESOL Quarterly*, 36, 79–84.

Aull, L. (2015). *First-year University Writing: A Corpus-based Study with Implications for Pedagogy*. Basingstoke, UK: Palgrave Macmillan.

Aull, L. L. and Lancaster, Z. (2014). Linguistic markers of stance in early and advanced academic writing a corpus-based comparison. *Written Communication*, 31(2), 151–83.

Azar, M. (1997). Concession relations as argumentation. *Text-Interdisciplinary Journal for the Study of Discourse*, 17(3), 301–16.

Bakhtin, M. M. (1986). *The Dialogic Imagination: Four Essays.* Translated by C. Emerson and M. Holquist. Edited by M. Holquist. Austin: University of Texas Press.

Bakhtin, M. (1986). *Speech Genres and other Late Essays*. Austin: University of Texas Press.

Barber, C. (1988). Some measurable characteristics of modern scientific prose. In J. Swales (ed.), *Episodes in ESP*, 1–16. Cambridge: Cambridge University Press. (Reprinted from *Contributions to English Syntax in and Philology*, 21–43, edited by F. Behre, 1962, Gothenburg, Sweden: Almqvist & Wiksell.)

Bartholomae, D. (1986). Inventing the university. *Journal of Basic Writing*, 5(1), 4–23.

Barton, D (1994). *The Social Basis of Literacy: An Introduction to the Ecology of Written Language*. Oxford: Blackwell.

Barton, D. and Hamilton, M. (1998). *Local Literacies*. London: Routledge.

Barton, E. L. (1995). Contrastive and non-contrastive connectives metadiscourse functions in argumentation. *Written Communication*, 12(2), 219–39.

Bates, L., Lane, J. and Lange, E. (1993). *Writing Clearly: Responding to ESL Composition*. Boston: Heinle and Heinle.

Bauer, L. and Nation, P. (1993). Word families. *International Journal of Lexicography*, 6(4), 253–79.

Baynham, M. (2006). Performing self, family and community in Moroccan narratives of migration and settlement. In A. De Fina, D. Schiffrin and M. Bamberg (eds), *Discourse and Identity*, 376–97. Cambridge: Cambridge University Press.

Bazerman, C. (1984). Modern evolution of the experimental report in physics: Spectroscopic articles in Physical Review, 1893–1980. *Social Studies of Science*, 14(2), 163–96.

Bazerman, C. (1988). *Shaping Written Knowledge*. Madison: University of Wisconsin Press.

Bazerman, C. (1993). Foreword. In N. Blyler and C. Thralls (eds), *Professional Communication: The Social Perspective*, vii–x. Newbury Park, CA: Sage.

Bazerman, C. (1994). Systems of genres and the enactment of social intentions. In A. Freedman and P. Medway (eds), *Genre and the New Rhetoric*, 79–101. London: Taylor and Francis.

Bazerman, C. (1994). *Constructing Experience*. Carbondale: Southern Illinois University.

Beauvais, P. J. (1989). A speech act theory of metadiscourse. *Written Communication*, 6(1), 11–30.

Becher, T. (1989). *Academic Tribes and Territories: Intellectual Inquiry and the Cultures of Disciplines*. Buckingham: Open University Press.

Becher, T. and Trowler, P. (2001). *Academic Tribes and Territories: Intellectual inquiry and the cultures of disciplines*. Milton Keynes: SRHE and Open University Press.

Beebe, L. and Takahashi, T. (1989). Sociolinguistic variation in face-threatening speech acts. In M. Eisenstein (ed.), *The Dynamic Interchange: Empirical Studies in Second Language Variation*, 199–218. New York: Plenum.

Belcher, D. D. (1994). The apprenticeship approach to advanced academic literacy: Graduate students and their mentors. *English for Specific Purposes*, 13, 23–34.

Belcher, D. D. (1997). An argument for nonadversarial argumentation: On the relevance of the feminist critique of academic discourse to L2 writing pedagogy. *Journal of Second Language Writing*, 6(1), 1–21.

Belcher, D. D. (2007). Seeking acceptance in an English-only research world. *Journal of Second Language Writing*, 16(1), 1–22.

Ben-Ari, E. (1987). On acknowledgements in ethnographies. *Journal of Anthropological Research*, 43(1), 63–84.

Benesch, S. (2001). *Critical English for Academic Purposes: Theory, Politics and Practice*. Mahwah, NJ: Lawrence Erlbaum.

Benwell, B. and Stokoe, E. (2006). *Discourse and Identity*. Edinburgh: Edinburgh University Press.

Berger, P. and Luckman, T. (1967). *The Social Construction of Reality*. London: Allen Lane.

Berkenkotter, C. and Huckin, T. (1995). *Genre knowledge in disciplinary communication*. Hillsdale, NJ: Lawrence Erlbaum.

Berliner, D. (2003). Educational research: the hardest science of all. *Educational Researcher*, 32, 18–20.

Besnier, N. (1990). Language and affect, *Annual Review of Anthropology*, 19, 419–51.

Bhatia, V. K. (1999). Integrating products, processes and participants in professional writing. In C. N. Candlin and K. Hyland (eds), *Writing: Texts, Processes and Practices*, 21 39. London: Longman.

Bhatia, V. K. (2002). A generic view of academic discourse. In J. Flowerdew (ed.), *Academic Discourse*, 21–39. London: Longman.

Biber, D. (1988). *Variation Across Speech and Writing*. Cambridge: Cambridge University Press.

Biber, D. (2006). Stance in spoken and written university registers. *Journal of English for Academic Purposes*, 5(2), 97–116.

Biber, D. (2006). *University Language: A Corpus-based Study of Spoken and Written Registers*. Amsterdam: Benjamin.

Biber, D., Connor, U. and Upton, T. (2007). *Discourse on the Move: Using Corpus Analysis to Describe Discourse Structure*. Amsterdam: John Benjamins.

Biber, D., Conrad, S. and Cortes, V. (2004). If you look at ...: Lexical bundles in university teaching and textbooks. *Applied Linguistics*, 25(3), 371–405.

Biber, D. and Finegan, E. (1989). Styles of stance in English: lexical and grammatical marking of evidentiality and affect. *Text*, 9(1), 93–124.

Biber, D. and Gray, B. (2016). *Grammatical Complexity in Academic English: Linguistic Change in Writing*. Cambridge: Cambridge University Press.

Biber, D., Johansson, S., Leech, G., Conrad, S. and Finegan, E., eds (1999). *Longman Grammar of Spoken and Written English*. London: Longman.

Bizzell, P. (1982). Cognition, convention and certainty: what we need to know about writing. *Pre/Text*, 3, 213–41.

Bizzell, P. (1989). 'Cultural criticism': A social approach to studying writing. *Rhetoric Review*, 7(2), 224–30.

Bizzell, P. (1992). *Academic Discourse and Critical Consciousness*. Pittsburgh: University of Pittsburgh Press.

Blakeslee, A. M. (1997). Activity, context, interaction, and authority learning to write scientific papers in situ. *Journal of Business and Technical Communication*, 11(2), 125–69.

Bloor, M. and Bloor, T. (1986). *Language for Specific Purposes: Practice and Theory*. In CLCS occasional papers. Dublin: Centre for Language & Communication Studies, Trinity College.

Bloor, M. and Bloor, T. (1991) Cultural expectations and socio-pragmatic failure in academic writing. In B. Heaton and P. Adams (eds), *Socio-cultural Issues in English for Academic Purposes*, Review of ELT 1(2), pp. I–12.

Bloor, T. (1996). Three hypothetical strategies in philosophical writing. In E. Ventola and A. Mauranen (eds), *Academic Writing: Intercultural and Textual Issues*, 19–43. Amsterdam: John Benjamins.

Blue, G. (1988). Individualising academic writing tuition. In P. Robinson (ed.), *Academic Writing: Process and Product*, 95–9. ELT Documents 129.

Blyler, N. R. and Thralls, C. (1993). *Professional Communication: The Social Perspective*. Newbury Park, CA: Sage.

Bondi, M. (1999). *English across Genres: Language Variation in the Discourse of Economics* Modena: Edizioni Il Fiorino.

Bondi, M. (2004). 'If you think this sounds very complicated, you are correct': awareness of cultural difference in specialized discourse. In C. Candlin and M. Gotti (eds), *Intercultural Aspects of Specialised Discourse*, 53–78, Bern: Peter Lang.

Bondi, M. and Mauranen, A. (2003). Editorial: Evaluative language use in academic discourse. *Journal of English for Academic Purposes*, 2(4), 269–71.

Bonnardel, N., Piolat, A. and Le Bigot, L. (2010). The impact of colour on website appeal and users' cognitive processes. *Displays*, 32(2), 69–80.

Booth, V. (1985) *Communicating in Science Writing and Speaking*. Cambridge: Cambridge University Press.

Bourdieu, P. (1977). *Outline of a Theory of Practice*. Cambridge: Cambridge University Press.

Bourdieu, P. (1980). The production of belief: contribution to an economy of symbolic goods. *Media, Culture and Society*, 2, 261–93.

Bourdieu, P. (1991). *Language and Symbolic Power*. Oxford: Polity Press.

Bourdieu, P. and Passeron, J.-C. (1996). Introduction: language and relationship to language in the teaching situation. In P. Bourdieu, J.-C Passeron and M. de Saint Martin (eds), *Academic Discourse*, 1–34. The Hague: Mouton.

Braine, G. (1988). A reader reacts (commentary on Ruth Spack's 'Intiating ESL students into the academic discourse community: how far should we go?'). *TESOL Quarterly*, 22(4), 702.

Braine, G. (1995). Writing in the natural sciences and engineering. In D. Belcher and G. Braine (eds), *Academic Writing in a Second Language: Essays on Research and Pedagogy*, 113–34. New York: Ablex.

Brandt, D. and Howarth, P. (1990) *Literacy as Involvement: The Acts of Writers, Readers and Texts.* Carbondale, IL: Southern Illinois University Press.

Brannon, L. and Knoblauch, C. H. (1982). On students' rights to their own texts: A model of teacher response. *College Composition and Communication*, 33, 157–66.

Brenton, F. (1996). Rhetoric in competition: The formation of organizational discourse in conference on college composition and communication abstracts. *Written Communication*, 13(3), 355–84.

Brett, P. (1994). A genre analysis of the results section of sociology articles. *English for Specific Purposes*, 13(1), 47–59.

Bridgeman, B. and Carlson, S. (1984). Survey of academic writing tasks. *Written Communication*, 1, 247–80.

Brophy, J. (1981). Teacher praise: A functional analysis. *Review of Educational Research*, 51, 5–32.

Brown, J., Collins, A. and Duguid, P. (1989). Situated cognition and the culture of learning. *Educational Researcher,* 18, 32–42.

Brown, P. and Levinson, S. (1987) *Politeness: Some Universals in Language Usage.* Cambridge: Cambridge University Press.

Brown, R. (2004). Self-composed: rhetoric in psychology personal statements. In *Written Communication*, 21(3), 242–60.

Bruffee, K. (1986). Social construction: language and the authority of knowledge. A bibliographical essay. *College English*, 48, 773–9.

Bunton, D. (1999). The use of higher level metatext in PhD theses. *English for Specific Purposes*, 18, S41–56.

Butler, J. (1990) *Gender Trouble: Feminism and the Subversion of Identity.* London: Routledge.

Butt, D., Fahey, R., Feez, S., Spinks, S. and Yalop, C. (2000). *Using Functional Grammar: An Explorer's Guide*, 2nd edn. Sydney: NCELTR.

Cadman, K. (1997). Thesis writing for international students: a question of identity? *English for Specific Purposes*, 16(1), 3–14.

Caesar, T. (1992). On acknowledgements. *New Orleans Review*, 19(1), 85–94.

Cameron, D. (1992). *Feminism and Linguistic Theory*. London: MacMillan.

Cameron, D. (1995). *Verbal Hygiene*. London: Routledge.

Cameron, D. (2000). *Good to Talk? Living and Working in a Communication Culture.* London: Sage.

Cameron, D. (2006a). *The Myth of Mars and Venus*. Oxford, UK: Oxford University Press.

Cameron, D. (2006b). *On Language and Sexual Politics*. London: Routledge.

Campbell, C. (1990). Writing with others' words: using background reading text in academic compositions. In B. Droll (ed.), *Second Language Writing*, 211–30. Cambridge: Cambridge University Press.

Campion, M. E. and Elley, W. B. (1971). *An Academic Vocabulary List* Wellington, New Zealand: New Zealand Council for Educational Research.

Canagarajah, A. S. (2002). *Critical Academic Writing and Multilingual Students*. Ann Arbor, MI: University of Michigan Press.

Canagarajah, S. (2013). *Translingual Practice: Global Englishes and Cosmopolitan Relations*. New York: Routledge.

Candlin, C. N. and Plum, G. A. (1999). Engaging with challenges of interdiscursivity in academic writing: researchers, students and teachers. In

C. N. Candlin and K. Hyland (eds), *Writing: Texts, Processes and Practices*, 193–217. London: Longman.

Cardelle, M. and Corno, L. (1981). Effects on second language learning of variations in written feedback on homework assignments. *TESOL Quarterly*, 15(3), 251–61.

Carter, R. (1998). *Vocabulary: Applied Linguistics Perspectives*. London: Routledge.

Casanave, C. P. (1995). Local interactions: Constructing contexts for composing in a graduate sociology program. In G. Braine and D. Belcher (eds), *Academic Writing in a Second Language: Essays on Research and Pedagogy*, 83–110. Norwood, NJ: Ablex.

Casanave, C. P. and Vandrick, S., eds (2003). *Writing for Scholarly Publication*. Mahwah, NJ: Lawrence Erlbaum.

Casanave, C. and Hubbard, P. (1992). The writing assignments and writing problems of doctoral students: faculty perceptions, pedagogical issues and needed research. *English for Specific Purposes*, 11, 33–49.

Chafe, W. (1985). Linguistic differences produced by differences between speaking and writing. In D. Olson, N. Torrance and A. Hildyard (eds), *Literature, Language and Learning: The Nature and Consequences of Reading and Writing*, 105–23. Cambridge: Cambridge University Press.

Chafe, W. (1986). Evidentiality in English conversation and academic writing. In W. Chafe and J. Nichols (eds), *Evidentiality: The Linguistic Coding of Epistemology*. Norwood, NJ: Ablex.

Chandler, D. and Roberts-Young, D. (1998). The construction of identity in the personal homepages of adolescents. http://www.aber.ac.uk/media/ Documents/short/strasbourg.html.

Chandler, D. (1998). Personal homepages and the construction of identities on the web. Paper presented at Aberystwyth Post-International Group conference on linking theory and practice: Issues in the politics of identity. University of Wales (9–11 September). http://www.aber.ac.uk/media/

Chang, Y.-Y. and Swales, J. (1999). Informal elements in English academic writing: Threats or opportunities for advanced non-native speakers? In C. N. Candlin and K. Hyland (eds), *Writing: Texts, Processes and Practices*, 145–67. London: Longman.

Channell, J. (1994). *Vague Language*, Oxford: Oxford University Press.

Chargaff, E. (1974). Building the tower of Babel. *Nature*, 248: 778.

Charles, M. (2006). The construction of stance in reporting clauses: A cross-disciplinary study of theses. *Applied Linguistics*, 27, 492–518.

Charles, M. (2007). Reconciling top-down and bottom-up approaches to graduate writing: Using a corpus to teach rhetorical functions. *Journal of English for Academic Purposes*, 6, 289–302.

Charles, M. (2012). Proper vocabulary and juicy collocations: EAP students evaluate do-it-yourself corpus-building. *English for Specific Purposes*, 31, 93–102.

Charles, M. (2014). Getting the corpus habit: EAP students' long-term use of personal corpora. *English for Specific Purposes*, 35, 30–40.

Cheng, X. and Steffensen, M. (1996). Metadiscourse: A technique for improving student writing. *Research in the Teaching of English*, 30(2), 149–81.

Cherry, R. (1988). Ethos vs persona: self-representation in written discourse. *Written Communication*, 5, 251–76.

Chin, E. (1994). Redefining 'context' in research on writing. *Written Communication*, 11(4), 445–82.

Cho, S. (2004). Challenges of entering discourse communities through publishing in English: Perspectives of nonnative-speaking doctoral students in the united states of America. *Journal of Language, Identity, and Education*, 3(1), 47–72.

Choi, Y. H. (1988). Text structure of Korean speakers' argumentative essays in English. *World Englishes*, 7(2), 129–37.

Christie, F. (1987). Genres as choice. In I. Reid (ed.), *The Place of Genre in Learning: Current Debates*, 22–34. Deakin: Deakin University Press.

Christie, F. (1996). The role of a functional grammar in development of a critical literacy. In G. Bull and M. Anstey (eds), *The Literacy Lexicon*, 46–57. Sydney: Prentice-Hall.

Christie, F. and Martin, J. R., eds (1997). *Genre in Institutions: Social Processes in the Workplace and School*. New York: Continuum.

Clark B. (1987). *The Academic Life: Small Worlds, Different Worlds*. Princeton, NJ: The Carnegie Foundation for the Advancement of Teaching.

Clyne, M. (1987). Cultural differences in the organisation of academic texts. *Journal of Pragmatics*, 11, 211–47.

Čmejrková, S. (1996). Academic writing in Czech and English. In E. Ventola and A. Mauranen (eds), *AcademicVwwriting: Intercultural and Textual Issues*, 137–52. Amsterdam: Benjamins.

Coates, J. (1987). Epistemic modality and spoken discourse. *Transactions of the Philological Society*, 85(1), 110–31.

Coates, J. 1983. *The Semantics of the Modal Auxiliaries*. Beckenham: Croom Helm.

Cobb, T. (2004). The online corpus builder. Available from http://www.er.uqam.ca/

Coe, R. M. (2002). The new rhetoric of genre: Writing political briefs. In A. M. Johns (ed.), *Genre in the Classroom*, 195–205. Mahwah, NJ: Erlbaum.

Cohen, A. D. and Cavalcanti, M. C. (1990). Feedback on compositions: Teacher and student verbal reports. In B. Kroll (ed.), *Second Language Writing: Research Insights for the Classroom*, 155–77. Cambridge: Cambridge University Press.

Cohen, A., Glasman, H., Rosenbaum-Cohen, P. R., Ferrara, J. and Fine, J. (1988). Reading English for specialized purposes: Discourse analysis and the use of standard informants. In P. Carrell, J. Devine, D. Eskey (eds), *Interactive Approaches to Second Language Reading*, 152–67. Cambridge: Cambridge University Press.

Cohen, M. and Manion, L. (1994). *Research Methods in Education*, 4th edn. London: Croom Helm.

Cohen, M., Manion, L. and Morrison, K. (2000). *Research Methods in Education*, 5th edn. London: Routledge.

Connor, U. (1996). *Contrastive Rhetoric*. Cambridge: Cambridge University Press.

Connor, U. and Upton, T. (2004). The genre of grant proposals: a corpus linguistic analysis. In U. Connor and T. Upton (eds), *Discourse in the Professions: Perspectives from Corpus Linguistics*, 235–56. Amsterdam: John Benjamins.

Council of Biology Editors. 1978. *Scientific Style and Format the CBE Manual for Authors, Editors, and Publishers*, 4th edn. Cambridge: Cambridge University Press.

Connors, R. J. and Lunsford, A. (1993). Teachers' rhetorical comments on student papers. *College Composition and Communication*, 44, 200–23.

Conrad, S. and Biber, D. (2000). Adverbial marking of stance in speech and writing. in S. Hunston and G. Thompson (eds), *Evaluation in Text*, 56–73. Oxford: Oxford University Press.

Cooper, M. M. (1989). Why are we talking about discourse communities? Or, foundationalism rears its ugly head once more. In M. M. Cooper and M. Holzman (eds), *Writing as Social Action*, 203–20. Portsmouth, New Hampshire: Boyton/Cook.

Cope, B. and Kalantzis, M., eds (1993). *The Powers of Literacy: A Genre Approach to Teaching Writing*. Bristol, PA: Falmer Press.

Cortes, V. (2004). Lexical bundles in published and student disciplinary writing: Examples from history and biology. *English for Specific Purposes*, 23, 397–423.

Coxhead, A. (2000). A new academic word list. *TESOL Quarterly*, 34, 213–38.

Coxhead, A. (2002). The academic word list: A corpus-based word list for academic purposes. In B. Ketterman and G. Marks (eds), *Teaching and Language Corpora (TALC) Conference Proceedings*, 73–89. Atlanta, GA: Rodopi.

Coxhead, A. (2011). The Academic Word List 10 years on: Research and teaching implications. *TESOL Quarterly*, 45, 355–62.

Coxhead, A. (2016). Reflecting on Coxhead (2000), 'A new academic word list'. *TESOL Quarterly*, 50, 181–5.

Coxhead, A. and Nation, I. S. P. (2001). The specialized vocabulary of English for academic purposes. In J. Flowerdew and M. Peacock (eds), *Research Perspectives on English for Academic Purposes*, 252–67. Cambridge: Cambridge University Press.

Crismore, A. (1989). *Talking with Readers: Metadiscourse as Rhetorical Act* New York: Peter Lang.

Crismore, A. and Farnsworth, R. (1989). Mr. Darwin and his readers: Exploring interpersonal metadiscourse as a dimension of ethos, *Rhetoric Review*, 8(1), 91–112.

Crismore, A. and R. Farnsworth. 1990. Metadiscourse in popular and professional discourse. In W. Nash (ed.): *The Writing Scholar: Studies in the Language and Conventions of Academic Discourse*, 118–36. Newbury Park, CA: Sage.

Crismore, A., R. Markkanen and M. Steffensen. 1993. Metadiscourse in persuasive writing: A study of texts written by American and Finnish university students. *Written Communication*, 10(1), 39–71.

Cronin, B. (1995). *The Scholar's Courtesy: The Role of Acknowledgements in the Primary Communication Process*. London: Taylor Graham.

Cronin, B. and Overfelt, K. (1994). The scholar's courtesy: A survey of acknowledgement behaviour. *Journal of Documentation*, 50(3), 165–96.

Cronin, B., McKenzie, G. and Rubio, L. (1993). The norms of acknowledgement in four humanities and social sciences disciplines. *Journal of Documentation*, 49(1), 29–43.

Cronin, B., McKenzie, G. and Stiffler, M. (1992). Patterns of acknowledgement. *Journal of Documentation*, 48(2), 227–39.

Daiker, D. (1989). Learning to praise. In C. Anson (ed.), *Writing and Response*, 103–. Urbana, IL: NCTE.

Danath, J. and Boyd, B. (2004). Public displays of connection. *BT Technology Journal*, 22(4), 1–11.

Dant, T. (1991). *Knowledge, Ideology and Discourse*. London: Routledge.

Davies, B. and Harre, R. (1990). Positioning: The discursive production of selves. *Journal for the Theory of Social Behaviour*, 20, 43–63.

Davis, E. and Schmidt, D. (1995). *Using the Biological Literature A Practical Guide*, 2nd edn. New York: Dekker.

Day, D., Antaki, C., ed., and Widdicombe, S., ed. (1998). Being ascribed, and resisting, membership of an ethnic group. In C. Antaki and S. Widdicombe (eds), *Identities in Talk*, 151–70. Sage Publications, Incorporated.

Day, R. (1994). *How to Write and Publish a Scientific Paper*. Phoenix, AZ: Oryx Press.

de Fina, A., Schiffrin, D. and Bamberg, M., eds (2006). *Discourse and Identity*. Cambridge: Cambridge University Press.

Delpit, L. (1988). The silenced dialogue: Power and pedagogy in educating other people's children. *Harvard Educational Review*, 58, 280–98.

Dillon, G. (1991). *Constructing Texts: Elements of a Theory of a Composition and Style*. Bloomington: Indiana University Press.

Dillon, G. (1991). *Contending rhetorics: writing in academic disciplines*. Bloomingdale: Indiana University Press.

Ding, H (2007). 'Genre analysis of personal statements: Analysis of moves in application essays to medical and dental schools'. In *English for Specific Purposes*, 26, 368–92.

Dixon, J. (1987). The question of genres. In I. Reid (ed.), *The Place of Genre in Learning: Current Debates*, 9–21. Geelong, Australia: Deakin University.

Doheny-Farina, S. and Odell, L. (1985). Ethnographic research on writing: Assumptions and methodology. In L. Odell and D. Goswami (eds), *Writing in Non-Academic Settings*. New York: Guilford.

Dragga, S. (1986, March). Praiseworthy grading. A teacher's alternative to editing error. Paper presented at the Conference on College Composition and Communication, New Orleans.

Dressen, D. (2003). 'Geologists' implicit persuasive strategies and the construction of evaluative evidence. *Journal of English for Academic Purposes*, 2, 273–90.

Dubois, B. (1988). Citation in biomedical journal articles. *English for Specific Purposes*, 7, 181–94.

Dudley-Evans, T. (1992) Socialisation into the academic community: Linguistic and stylistic expectations of a PhD thesis as revealed by supervisor comments. In Heaton, B., Howarth, P. and Adams, P. (eds), *Sociocultural Issues in English for Academic Purposes*, 41–51. Review of ELT.

Dudley-Evans, T. and St John, M. J. (1998). *Developments in English for Specific Purposes*. Cambridge: Cambridge University Press.

Duguid, P. (2005). The art of knowing: Social and tacit dimensions of knowledge and the limits of the community of practice. *The Curriculum Journal*, 21, 109–18.

Dumont, K. and Frindte, W. (2005). Content analysis of the homepages of academic psychologists. *Computers in Human Behavior*, 21(1), 73–83.

Edwards, A. (2005). Let's get beyond community and practice: The many meanings of learning by participating. *The Curriculum Journal*, 16, 49–65.

Einstein, A. (1934). *Essays in Science*. New York: The Philosophical Library.

Elbow, P. (1991). Reflections on academic discourse: How it relates to freshmen and colleagues. *College English*, 53, 135–55.

Emig, J. (1983). *The Web of Meaning.* Upper Montclair, NJ: Boynton/Cook.

Erni, J. (2001). Like a postcolonial culture: Hong Kong re-imagined. *Cultural Studies*, 15(3/4), 389–418.

Ervin, E. (1993). Interdisciplinarity or 'an elaborate edifice built on sand'? Rethinking rhetoric's place. *Rhetoric Review*, 12, 84–105.

Fahnestock, J. (1986) Accommodating science: The rhetorical life of scientific facts. *Written Communication*, 3(3), 275–96.

Faigley, L. (1985). Non-academic writing: the social perspective. In L. Odell and D. Goswami (eds), *Writing in Non-Academic Settings*, 231–48. New York: Guildford Press.

Faigley, L. (1986). Competing theories of process: A critique and a proposal. *College Composition and Communication*, 48, 527–42.

Fairclough, N. (1992). *Discourse and Social Change.* Cambridge: Polity Press.

Fairclough, N. (1995). Critical Discourse Analysis. London: Longman.

Farrell, P. (1990). *Vocabulary in ESP: A Lexical Analysis of the English of Electronics and a Study of Semi-Technical Vocabulary.* (CLCS Occasional Paper No. 25). Dublin, Ireland: Trinity College, Centre for Language and Communication Studies.

Feez, S. (2002). Heritage and innovation in second language education. In A. M. Johns (ed.), *Genre in the Classroom*, 47–68. Mahwah, NJ: Erlbaum.

Ferris, D. R. (1995). Student reactions to teacher response in multiple-draft composition classrooms. *TESOL Quarterly*, 29(1), 33–53.

Ferris, D. R. (1997). The influence of teacher commentary on student revision. *TESOL Quarterly*, 31, 315–39.

Ferris, D. R., Pezone, S., Tade, C. R. and Tinti, S. (1997). Teacher commentary on student writing: Descriptions and implications. *Journal of Second Language Writing*, 6, 155–82.

Firth, J. R. (1951). *Modes of Meaning. Essays and Studies (The English Association)*, 118–49.

Flowerdew, J. (1993). Concordancing as a tool in course design. *System*, 21, 231–44.

Flowerdew, J. (1999). Problems in writing for scholarly publication in English: The case of Hong Kong. *Journal of Second Language Writing*, 8(3), 243–64.

Flowerdew, J. (2001). Attitudes of journal editors toward nonnative speaker contributions. *TESOL Quarterly*, 35, 121–50.

Flowerdew, J. and Peacock, M. (2001). Issues in EAP: a preliminary perspective. In J. Flowerdew and M. Peacock (eds), *Research Perspectives on English for Academic Purposes*, 8–24. Cambridge: Cambridge University Press.

Flowerdew, L. (2005). An integration of corpus-based and genre-based approaches to text analysis in EAP/ESP: Countering criticisms against corpus-based methodologies. *English for Specific Purposes*, 24, 321–32.

Foucault, M. (1972) *The Archaeology of Knowledge* London: Tavistock.

Foucault, M. (1974). *The order of things.* London: Tavistock Press.

Foucault, M. (1981). The order of discourse. In R. Young (ed.), *Untying the Text: A Post-Structuralist Reader*, 48–78. Boston: Routledge.

Fraser, B. (1980). Conversational mitigation. *Journal of Pragmatics*, 4, 341–50.

Fraser, B. and Nolan, W. (1981). The association of deference with linguistic form. *International Journal of the Sociology of Language*, 27, 93–109.

Freadman, A. (2002). *Uptake. The Rhetoric and Ideology of Genre: Strategies for Stability and Change.* Hampton Press.

Freedman, A. and Medway, P., eds. (1994). *Genre and the New Rhetoric*. London: Taylor & Francis.

Gambier, Y., ed (1998). Discoun professionnels en francais. New York: Peter Lang.

Garfield, E. (1998). Random thoughts on citationology. *Scientometrics*, 43, 69–76.

Gardner, D. and Davies, M. (2014). A new academic vocabulary list. *Applied Linguistics*, 35(3), 305–27.

Gee, J. (1999). *An Introduction to Discourse Analysis*. London: Routledge.

Gee, J. (2004). *Situated Language and Learning: A Critique of Traditional Schooling*. London: Routledge.

Gee, T. C. (1972). Students' responses to teacher comments. *Research in the Teaching of English*, 6, 212–21.

Geertz, C. (1983). *Local Knowledge: Further Essays in Interpretive Anthropology*. New York: Basic Books.

Geertz, C. (1988) *Words and Lives: The Anthropologist as Author*. Palo Alto, CA: Stanford University Press.

Gergen, K. (1991). *The Saturated Self: Dilemmas of ildentity in Contemporary Life*. New York: Basic Books.

Gergen, K. J. and Thatchenkery, T. J. (1996). Organisational science as social construction: postmodern potentials. *The Journal of Applied Behavioral Science*, 32(4): 356–77.

Giannoni, D. S. (2002). Worlds of gratitude: A contrastive study of acknowledgment texts in English and Italian research articles. *Applied Linguistics*, 23(1), 1–31.

Giddens, A. (1991). *Modernity and Self-Identity: Self and Society in the Late Modern Age*. Cambridge: Polity Press.

Gilbert, G. (1977). Referencing as persuasion. *Social Studies of Science*, 7, 113–22.

Gilbert, G. (1976). The transformation of research findings into scientific knowledge. *Social Studies of Science*, 6, 281–306.

Gilbert, G. and M. Mulkay (1984). *Opening Pandora's Box: A Sociological Analysis of Scientific Discourse*. Cambridge: Cambridge University Press.

Gilbert, S. F. (1995). Introduction: Postmodernism and science. *Science in Context*, 8, 559–61.

Goffman, E. (1967). *Interaction Ritual*. Garden City, NY: Anchor Books.

Goffman, E. (1971). *The Presentation of Self in Everyday Life*. Harmondsworth: Penguin Books.

Gosden, H. (1993). Discourse functions of subject in scientific research articles. *Applied Linguistics*, 14(1), 56–75.

Grabe, W. (1984) *Towards Defining Expository Prose Within a Theory of Text Construction*, Unpublished doctoral dissertation. Los Angeles: University of Southern California.

Granger, S. (1994) New insights into the learner lexicon: A preliminary report from the international corpus of English. In L. Flowerdew and K. K. Tong (eds), *Enrering Text*, 102–13. Hong Kong: Hong Kong University of Science and Technology.

Gray, B. and Biber, D. (2012). Current Conceptions of Stance. In K. Hyland and C. Sancho-Guinda (eds), *Stance and Voice in Written Academic Genres*, 15–33. London: Palgrave.

Greene, M. (2013). History of the journal Nature. doi:10.1038/nature06243

Grice, H. (1975). Logic and conversation. In P. Cole and J. Morgan (eds), *Syntax and Semantics, Vol. 3: Speech Acts*. New York: Academic Press.

Groom, N. (2005). Pattern and meaning across genres and disciplines: an exploratory study. *Journal of English for Academic Purposes*, 4(3), 257–77.

Gunnarsson, B.-L. (1995). Studies of language for specific purposes: A biased view of a rich reality. *International Journal of Applied Linguistics*, 5(1), 111–34.

Hall, S. (1996). Introduction: Who needs identity? In S. Hall and P. Du Gay (eds), *Questions of Cultural Identity*, 1–17. London: Sage.

Halliday, M. and Martin J. (1993). *Writing Science. Literacy and Discursive Power*. London: Falmer Press.

Halliday, M. A. K. (1978). *Language as a Social Semiotic: The Sociological Interpretation of Language and Meaning*. London: Edward Arnold.

Halliday, M. A. K. (1988). On the language of physical science. In M. Ghadessey (ed.), *Registers of Written English*, 162–78. London: Pinter.

Halliday, M. A. K. (1994). *An introduction to Functional Grammar*, 2nd edn. London: Edward Arnold.

Halliday, M. A. K. (1998). Things and relations: Regrammaticising experience as technical knowledge. In J. R. Martin and R. Veel (eds), *Reading Science*, 185–235. London: Routledge.

Halliday, M., MacIntosh, A. and Strevens, P. (1964). *The Linguistic Sciences and Language Teaching*. London: Longman.

Halliday, M. A. K. (2004). The Language of Science. In J. Webster (ed.). *5th volume of the Collected Works of M.A.K. Halliday*. London/New York: Continuum

Halloran, S. (1984). The birth of molecular biology: An essay in the rhetorical criticism of scientific discourse. *Rhetoric Review*, 3, 70–83.

Hammond, J. and Macken-Horarik, M. (1999). Critical literacy: Challenges and questions for ESL classrooms. *TESOL Quarterly*, 33, 528–44.

Hanania, E.A S. and Akhtar, K. (1985) Verb form and rhetorical function in science writing: A study of MS theses in biology, chemistry, and physics. *English for Specific Purpose*, 4, 49–58.

Hanlein, H. (1998). *Studies in Authorship Recognition – A Corpus-Based Approach*. Frankfort, KY: Peter Lang.

Harris, J. (1989). The idea of a discourse community in the study of writing. *College Composition and Communication*, 40, 11–22.

Hasan, R. (1996). Literacy, everyday talk and society. In R. Hasan and G. Williams (eds), *Literacy in Society*, 377–424. London: Longman.

Haswell, R. (1991). *Gaining Ground in College Writing: Tales of Development and Interpretation*. Dallas: Southern Methodist University Press.

Hatch, J., Hill, C. and Hayes, J. (1993). When the messenger is the message. *Written Communication*, 10, 569–98.

Hawking, S. (1993). *Black Holes and Baby Universes and Other Essays*. New York: Bantam Books.

Hedgcock, J. and Lefkowitz, N. (1994). Feedback on feedback: Assessing learner receptivity to teacher response in L2 composing. *Journal of Second Language Writing*, 3(2), 141–63.

Herbert, A. (1965). *The Structure of Technical English*. London: Longman.

Herbert, R. (1990). Sex-based differences in compliment behaviour. *Language in Society*, 19, 201–24.

Hess, M. (2002). A nomad faculty: English professors negotiate self-representation in university web space. *Computers and Composition*, 19(2), 171–89.

Hinds, J. (1983) Contrastive rhetoric: Japanese and English. *Text*, 3(2), 183–95.

Hinkel, E. (1997). Indirectness in L1 and L2 academic writing. *Journal of Pragmatics*, 27(3), 360–86.

Hoey, M. (1991). *Patterns of Lexis in Text*. Oxford: Oxford University Press.

Hoey, M. (2001). *Textual Interaction: An Introduction to Written Text Analysis*. London: Routledge.

Hoey, M. (2005). *Lexical Priming: A New Theory of Words and Language*. London: Routledge.

Hoey, M. (1988). Writing to meet the reader's needs: Text patterning and reading strategies. *Tronheim Papers in Applied Linguistics*, 4, 51–73.

Holmes, D. I. (1994). Authorship attribution. *Computers and the Humanities*, 28(2), 87–106.

Holmes, J. (1982). Expressing doubt and certainty in English. *RELC Journal*, 13, 19–28.

Holmes, J. (1984) Modifying illocutionary force. *Journal of Pragmatics*, 8, 345–65.

Holmes, J. (1988). Doubt and certainty in ESL textbooks. *Applied Linguistics*, 91, 20–44.

Holmes, J. (1995). *Women, Men and Politeness*. London: Longman.

Hood, S. (2010). *Appraising Research: Evaluation in Academic Writing*. London: Palgrave.

Howell, J. and Memering, D. (1986) *Brief Handbook for Writers*. Englewood Cliffs, NJ: Prentice Hall.

Huang, J. C. (2010). Publishing and learning writing for publication in English: Perspectives of NNES PhD students in science. *Journal of English for Academic Purposes*, 9(1), 33–44.

Huddleston, R. (1988). Review article. *Journal of Linguistics*, 24, 137–74.

Hundt, M. and Mair, C. (1999). "Agile" and "Uptight" Genres: The Corpus-based Approach to Language Change in Progress. *International Journal of Corpus Linguistics*, 4(2), 221–42

Hunston, S. (1993). Evaluation and ideology in scientific writing. In M. Ghadessy, *Register Analysis: Theory and Practice*, 57–73. London: Pinter.

Hunston, S. (1994). Evaluation and organisation in a sample of written academic discourse. In M. Coulthard (ed.), *Advances in Written Text Analysis*, 191–218. London: Routledge.

Hunston, S. (2002). *Corpora in Applied Linguistics*. Cambridge: Cambridge University

Hunston, S. and Thompson, G., eds (2000). *Evaluation in Text: Authorial Stance in the Construction of Discourse*. Oxford: Oxford University Press.

Hutchinson, T. and Waters, A. (1987). *English for Specific Purposes*. Cambridge: Cambridge University Press.

Hyland, F. (1998). The impact of teacher written feedback on individual writers. *Journal of Second Language Writing*, 7(3), 255–86.

Hyland, K. (1994) Hedging in academic writing and EAP textbooks. *English for Specific Purposes*, 13(3), 239–55.

Hyland, K. (1996a). Writing without conviction? Hedging in science research articles. *Applied Linguistics*, 17(4), 433–54.

Hyland, K. (1996b). Talking to the academy: Forms of hedging in science research articles. *Written Communication*, 13, 251–81.

Hyland, K. (1997). Scientific claims and community values: Articulating an academic culture. *Language and Communication*, 17(1), 19–31.

Hyland, K. (1998a). *Hedging in Scientific Research Articles*. Amsterdam: John Benjamins.

Hyland, K. (1998b). Exploring corporate rhetoric: Metadiscourse in the CEO's letter. *Journal of Business Communication*, 35(2), 224–45.

Hyland, K. (1998c). Persuasion and context: the pragmatics of academic metadiscourse. *Journal of Pragmatics*, 30, 437–55.

Hyland, K. (1999). Disciplinary discourses: Writer stance in research articles. In C. N. Candlin and K. Hyland (eds), *Writing: Texts, Processes and Practices*, 99–121. London: Longman.

Hyland, K. (2000a). *Disciplinary Discourses: Social Interactions in Academic Writing*. London: Longman.

Hyland, K. (2000b). Hedges, boosters and lexical invisibility: Noticing modifiers in academic texts. *Language Awareness*, 9(4), 179–97.

Hyland, K. (2001a). Bringing in the reader: Addressee features in academic articles. *Written Communication*, 18(4), 549–74.

Hyland, K. (2001b). Humble servants of the discipline? Self mention in research articles. *English for Specific Purposes*, 20(3), 207–26.

Hyland, K. (2002a). Directives: Argument and engagement in academic writing. *Applied Linguistics*, 23(2), 215–39.

Hyland, K. (2002b). Specificity revisited: How far should we go now? *English for Specific Purposes*, 21(4), 385–95.

Hyland, K. (2002c). Genre: Language, context and literacy. *Annual Review of Applied Linguistics*, 22, 113–35.

Hyland, K. (2002d). What do they mean? Questions in academic writing. *Text*, 22, 529–57.

Hyland, K. (2002e). Authority and invisibility: Authorial identity in academic writing. *Journal of Pragmatics*, 34, 1091–12.

Hyland, K. (2003). *Second Language Writing*. New York: Cambridge University Press.

Hyland, K. (2004a). *Disciplinary Discourses: Social Interactions in Academic Writing*. Ann Arbor, MI: University of Michigan Press.

Hyland, K. (2004b). Disciplinary interactions: Metadiscourse in L2 postgraduate writing. *Journal of Second Language Writing*, 13, 133–51.

Hyland, K. (2004c). Graduates' gratitude: The generic structure of dissertation acknowledgements. *English for Specific Purposes*, 23(3), 303–24.

Hyland, K. (2005a). *Metadiscourse: Interactions in Writing*. London: Continuum.

Hyland, K. (2005b). Stance and engagement: A model of interaction in academic discourse. *Discourse Studies*, 6(2), 173–191.

Hyland, K. (2006). *English for Academic Purposes*. London: Routledge.

Hyland, K. (2008a). Academic clusters: Text patterning in published and postgraduate writing. *International Journal of Applied Linguistics*, 18(1), 41–62.

Hyland, K. (2008b). As can be seen: Lexical bundles and disciplinary variation. *English for Specific Purposes*, 27(1), 4–21.

Hyland, K. (2009). *Academic Discourse*. London: Continuum.

Hyland, K. (2011a). The presentation of self in scholarly life: Identity and marginalization in academic homepages. *English for Specific Purposes*, 30, 286–97.

Hyland, K. (2011b). Projecting an academic identity in some reflective genres. *Ibérica*, 21, 9–30.

Hyland, K. (2012a). *Disciplinary Identities*. Cambridge: Cambridge University Press.

Hyland, K. (2012b). Undergraduate understandings: Stance and voice in final year reports. In K. Hyland and C. Sancho Guinda (eds), *Stance and Voice in Written Academic Genres*, 134–50. Basingstoke, UK: Palgrave Macmillan.

Hyland, K. (2015). *Academic Publishing: Issues and Challenges in the Construction of Knowledge*. Oxford: OUP.

Hyland, K. (2016). General and specific EAP. In K. Hyland and P. Shaw (eds), *The Routledge Handbook of EAP*, 17–29. London: Routledge.

Hyland, K. and Diani, G., eds (2009). *Academic Evaluation: Review Genres in University Settings*. Basingstoke, UK: Palgrave Macmillan.

Hyland, K. and Hyland, F., eds (2006). *Feedback in Second Language Writing: Contexts and Issues*. Cambridge: Cambridge University Press.

Hyland, K. and Jiang, F. (2016). We must conclude that...: A diachronic study of academic engagement. *Journal of English for Academic Purposes*, 24, 29–42.

Hyland, K. and J. Milton (1997). Qualification and certainty in L1 and L2 students' writing. *Journal of Second Language Writing*, 16(2), 183–205.

Hyland, K. and Sancho Guinda, C., eds (2013). *Stance and Voice in Written Academic Genres*. Basingstoke, UK: Palgrave Macmillan.

Hyland, K. and Tse, P. (2004). Metadiscourse in academic writing: A reappraisal. *Applied Linguistics*, 25(2), 156–77+288.

Hyland, K. and P. Tse (2005a). Hooking the reader: a corpus study of evaluative that in abstracts. *English for Specific Purposes*, 24, 123–39.

Hyland, K. and Tse, P. (2005b). Evaluative that constructions: Signalling stance in research abstracts. *Functions of Language*, 12(1), 39–63.

Hyland, K. and Tse, P. (2007). 'Is there an "academic vocabulary"?' *TESOL Quarterly*, 41(2), 235–54.

Hyland, K. and Tse, P. (2012). 'She has received many honours': Identity in article bio statements. *Journal of English for Academic Purposes*, 11(2), 155–65.

Hyon, S. (1996). Genre in three traditions: Implications for ESL. *TESOL Quarterly*, 30, 693–722.

Ivani, R. (1998). *Writing and Identity: The Discoursal Construction of Identity in Academic Writing*. Amsterdam: John Benjamins.

Ivani, R. and Simpson, J. (1992). Who's who in academic writing? In N. Fairclough (ed.), *Critical Language Awareness*, 141–73. London: Longman.

Jacoby, S. and Gonzales, P. (1991). The constitution of expert-novice in scientific discourse. *Issues in Applied Linguistics*, 2, 149–81.

Jespersen, O. (1924). *The Philosophy of Grammar*. London: Allen & Unwin.

Jewitt, C. (2005). Multimodality, reading, and writing for the 21st century. *Discourse: Studies in the Cultural Politics of Education*, 26, 315–31.

Johns, A. (1990). Coherence as a cultural phenomenon: Employing ethnographic principles in the academic milieu. In U. Connor and A. Johns (eds), *Coherence in Writing: Research and Pedagogical Perspectives*, 209–26. Alexandria, VA: TESOL.

Johns, A. (1993). Written argumentation for real audiences: Suggestions for teacher research and classroom practice. *TESOL Quarterly*, 27(I), 75–90.

Johns, A. M. (1988). Another reader reacts (commentary on Ruth Spack's "Intiating ESL students into the academic discourse community: how far should we go?"). *TESOL Quarterly*, 22(4), 705–06.

Johns, A. M. (1991) Faculty assessment of ESL student literacy skills: Implications for writing assessment. In. L. Hamp-Lyons (ed.), *Assessing*

Second Language Writing in Academic Contexts, 167–79. Norwood, NJ: Ablex.

Johns, A. M. (1997). *Text, Role and Context: Developing Academic Literacies*. Cambridge: Cambridge University Press.

Johns, A.M. (2002). Genre and ESL/EFL composition instruction. In B. Kroll (ed.), *Exploring the Dynamics of Second Language Writing*. New York: Cambridge University Press.

Johns, A. M. (2016a). The Common Core in the United States: A major shift in standards and assessments. In K. Hyland and P. Shaw (eds), *The Routledge Handbook of English for Academic Purposes*, 461–76. London: Routledge.

Johns, A. M. (2016b). Reader-culture-text mergence: Seven pedagogical principles. In L.C. de Oliveira (ed.), *The Common Core State Standards for Literacy in History/Social Studies, Science, and Technical Subjects for English Language Learners* (Grades 6-12). Alexandria, VA: TESOL Press.

Johns, A. and Swales, J. (2002). Literacy and disciplinary practices: Opening and closing perspectives. *Journal of English for Academic Purposes*, 1, 13–28.

Johns, T. and King, P. eds (1991). *Classroom Concordancing*. Birmingham: Centre for English Language Studies.

Johnson, D. M. (1992). Compliments and politeness in peer review texts. *Applied Linguistics*, 13, 52–71.

Jones, W. P. and Keene, M. L. (1981). *Writing Scientific Papers and Reports*, 8th edn. Dubuque, Iowa: W C Brown.

Jordan, R. (1997). *English for Academic Purposes*. Cambridge: Cambridge University Press.

Jordon, R. (1990) *Academic Writing Course*. London: Collins.

Judson, H. (1995). *The Eighth Day of Creation: The Makers of the Revolution in Biology*. Harmondsworth: Penguin Books.

Juola, P. and Baayen, R. H. (2005). A controlled-corpus experiment in authorship identification by cross-entropy. *Literary and Linguistic Computing*, 20(1), 59–67.

Kassirer, J. and Angell, M. (1991). On authorship and acknowledgements. *The New England Journal of Medicine*, 325(21), 1510–12.

Kaufer, D. and Geisler, C. (1989). Novelty in academic writing. *Written Communication*, 6(3), 286–311.

Keh, C. (1990). Feedback in the writing process: A model and methods for implementation. *ELT Journal*, 44(4), 294–305.

Kellenberger, E. (1989). Origins of molecular biology. *Plant Molecular Biology Reporter*, 7, 231–4.

Kennedy, G. (1998). *An Introduction to Corpus Linguistics*. London: Longman.

Kent, T. (1991). On the very idea of a discourse community. *College Composition and Communication*, 42(4),425–45.

Khamkhien, A. (2014). Linguistic features of evaluative stance: Findings from research article discussions. *Indonesian Journal of Applied Linguistics*, 4, 54–69.

Killingsworth, M. J. (1992). Discourse communities local and global. *Rhetoric Review*, 11, 110–22.

Killingsworth, M. J. and Gilbertson, M. K. (1992). *Signs, Genres and Communication in Technical Communication*. Amityville, New York: Baywood.

Knoblauch, C. H. and Brannon, L. (1984). *Rhetorical Traditions and the Teaching of Writing*. Upper Montclair, NJ: Boynton/Cook.

Knorr-Cetina, K. (1981). *The Manufacture of Knowledge*. Oxford: Pergamon Press.

Kolb, D. A. (1981). Learning styles and disciplinary differences. In A. Chickering, *The Modem American College*, 232–55. San Francisco: Jossey Bass.

Koppel, M., Argamon, S. and Shimoni, A. R. (2002). Automatically categorizing written texts by author gender. *Literary and Linguistic Computing*, 17, 401–12.

Koutsantoni, D. (2004). Attitude, certainty and allusions to common knowledge in scientific research articles. *Journal of English for Academic Purposes*, 3, 163–82.

Koutsantoni, D. (2007). *Developing Academic Literacies: Understanding Disciplinary Communities' Culture and Rhetoric*. Oxford: Peter Lang.

Kraus, W. (2000). Making identity talk. On qualitative methods in a longitudinal study. FQS, 1, 2. <http://www.qualitative-research.net/index.php/fqs/article/view/1084>.

Kress, G. (1989). *Linguistic Processes in Sociocultural Practice*. Oxford: Oxford University Press.

Kress, G. (2003). *Literacy in the New Media Age*. London: Routledge.

Kress, G. (2010). *Multimodality: A Social Semiotic Approach to Contemporary Communication*. London: Routledge.

Kress, G. R. and Hodge, R. I. V. (1993). *Language as Ideology*. London: Routledge.

Kress, G. and van Leuwen, L. (1996). *Reading Images: The Grammar of Visual Design*. London: Routledge.

Kuhn, T. (1970). *The Structure of Scientific Revolutions*, 2nd edn. Chicago: University of Chicago Press.

Kuo, C.-H. (1993). Problematic issues in EST materials development. *English for Specific Purposes*, 12, 171–81.

Kuo, C.-H. (1999). The use of personal pronouns: Role relationships in scientific journal articles. *English for Specific Purposes*, 18(2), 121–38.

Labov, W. (1984) 'Intensity.' In D. Schiffrin (ed.) *Meaning, Form and Use in Context: Linguistic Applications*, 43–70. Washington: Georgetown University Press.

Lachowicz, D. (1981). On the use of the passive voice for objectivity, author responsibility and hedging in EST. *Science of Science*, 2(6), 105–15.

Lage, H., Heitmann, D., Cingolani R., Grambow, R, and Ploog, K. (1991). Center-of mass quantization of excitons in GaAs quantum-well wires. *Physics Review Letters*, B44 6550-3.

Lakoff, G. (1972) Hedges: A study in meaning criteria and the logic of fuzzy concepts. *Chicago Linguistic Society Papers*, 8, 183–228.

Lancaster, Z. (2014). Exploring valued patterns of stance in upper-level student writing in the disciplines. *Written Communication*, 31, 27–57.

Lannon, J. (1986). *The Writing Process, a Concise Rhetoric*, 2nd edn. Boston: Little Brown.

Lantolf, J. P. (1999). Second culture acquisition: Cognitive considerations. In E. Hinkel (ed.), *Culture in Second Language Teaching and Learning*, 28–46. Cambridge: Cambridge University Press.

Latour, B. and Woolgar, S. (1979). *Laboratory Life: The Social Construction of Scientific Ffacts*. Beverly Hills: Sage.

Lave, J. and Wenger, E. (1991). *Situated Learning: Legitimate Peripheral Participation*. Cambridge University Press.

Lawler, S. (2008). *Identity: Sociological Perspectives*. Cambridge: Polity Press.

Lea, M. and Street, B. (1999). Writing as academic literacies: Understanding textual practices in higher education. In C. N. Candlin and K. Hyland (eds), *Writing: Texts, Processes and Practices*, 62–81. London: Longman.

Lee, E. and Norton, B. (2003). Demystifying publishing: A collaborative exchange between graduate student and supervisor. In C. Casanave and S. Vandrick (eds), *Writing for Publication: Behind the Scenes in Language Education*, 17–38). Mahwah, NJ: Lawrence Erlbaum Associates.

Lee, J. J. and Deakin, L. (2016). Interactions in L1 and L2 undergraduate student writing: Interactional metadiscourse in successful and less-successful argumentative essays. *Journal of Second Language Writing*, 33, 211–34.

Lee, D. and Swales, J. (2006). A corpus-based EAP course for NNS doctoral students: Moving from available specialized corpora to self-compiled corpora. *English for Specific Purposes*, 25, 56–75.

Leech, G. and Svartvik, J. (1994). *A Communicative Grammar of English*, 2nd edn. London: Longman.

Leki, I. (1990). Coaching from the margins: Issues in written response. In B. Kroll (ed.), *Second Language Writing: Research Insights for the Classroom*, 57–68. Cambridge: Cambridge University Press.

Lemke, J. (1995). *Textual Politics: Discourse and Social Dynamics*. London: Taylor and Francis.

Lemke, J. L. (2002). Travels in hypermodality. *Visual Communication*, 1, 299–325.

Lerat, P. (1995). Les langues specialisees. Paris: Presses universitaires de France.

Lester, J. D. (1993). *Writing Research Papers*, 7th edn. New York: HarperCollins.

Lewis, M. (1997). *Implementing the Lexical Approach*. Hove: Language Teaching Publications.

Li, S. L. and Pemberton, R. (1994). An investigation of students' knowledge of academic and sub-technical vocabulary. *Proceedings of the Joint Seminar on Corpus Linguistics and Lexicology*, 83–196. Hong Kong: Hong Kong University of Science and Technology.

Li, Y. Y. (2006). A doctoral student of physics writing for publication: A sociopolitically-oriented case study. *English for Specific Purposes*, 25(4), 456–78.

Li, Y. Y. (2014). Seeking entry to the north American market: Chinese management academics publishing internationally. *Journal of English for Academic Purposes*, 13, 41–52.

Li, Y. Y. and Flowerdew, J. (2007). Shaping Chinese novice scientists' manuscripts for publication. *Journal of Second Language Writing*, 16, 100–17.

Lillis, T. and Curry, M. J. (2010). *Academic Writing in a Global Context*. London: Routledge.

Lillis, T. (2001). *Student Writing: Access, Regulation, Desire*. London: Routledge.

Lincoln, Y. S. and Guba, E. G. (1985). *Naturalistic Inquiry*. Beverley Hills, CA: Sage.

Lindsay, D. (1984). *A Guide to Scientific Writing*. Melbourne: Longman.

Liu, C., Arnett, K., Capella, L. and Beatty, R. (1997). Web sites of the Fortune 500 companies: Facing customers through home pages. *Information & Management*, 31(6), 335–45.

Liu, Y. and O'Halloran, K. L. (2009). Intersemiotic texture: Analyzing cohesive devices between language and images. *Social Semiotics*, 19(4), 367–87.

Luke, A. (1996). Genres of power? Literacy education and the production of capital. In R. Hasan and A. G. Williams (eds), *Literacy in Society*, 308–38. London: Longman.

Lundquist, L., Picht, H. and Qvistgaard, J., eds (1998). *LSP: Identity and Interface. Research, Knowledge and Society*. Copenhagen: LSP Centre, Copenhagen Business School.

Luukka, M.-R. (1995). Puhuttua ja kirjoitettua tiedetta. Funktionaalinen ja yhteisollinen nakokulma tieteen kielen interpersonaalisiin piirteisiin. *Jyvaskyla Studies in Communication*, 4.

MacDonald, S. (1994). *Professional Academic Writing in the Humanities and Social Sciences*. Carbondale: Southern Illinois University Press.

Macken-Horarik, M. (2002). 'Something to shoot for': A systemic functional approach to teaching genre in secondary school science. In A. M. Johns (ed.), *Genre in the Classroom*, 21–46. Mahwah, NJ: Erlbaum.

Malcolm, L. (1987). What rules govern tense usage in scientific articles?. *English for Specific Purposes*, 6, 31–44.

Mangan, K. (2012). Social networks for academics proliferate, despite some doubts. The Chronicle of Higher Education. http://www.bing.com/search?q=So cial+Networks+for+Academics+Proliferate%2C+Despite+Some+Doubts6csr c=IETopResult&:FORM=IEioTR

Manual on scientific writing (1993). Collingwood, VIC: Tafe Publications.

Mao, L. R. (1993). I conclude not: Toward a pragmatic account of metadiscourse. *Rhetoric Review*, 11(2), 265–89.

Markkanen, R. and Schroder, H. (1989). Hedging as a translation problem in scientific texts. In C. Lauren and M. Nordman (eds), *Special Language: From Humans Thinking 10 Thinking Machines*, 171–9. Clevedon: Multilingual Matters.

Markus, H. and Nurius, P. (1986). Possible selves. *American Psychologist*, 41(9): 954–68.

Martin, J. (1993). Genre and literacy – modelling context in educational linguistics. *Annual Review of Applied Linguistics*, 13, 141–72.

Martin, J. R. (2001). Beyond exchange: APPRAISAL systems in English. In S. Hunston and G. Thompson (eds), *Evaluation in Text*, 142–75. Oxford: Oxford University Press.

Martin, J. R. (1992). *English Text System and Structure*. Amsterdam: Benjamins.

Martin, J. R. and D. Rose. (2003). *Working with Discourse Meaning Beyond the Cclause*. London: Continuum.

Martin, J. R. and White, P. (2005) *The Language of Evaluation: Appraisal in English*. London: Palgrave.

Martin, J., Christie, F. and Rothery, J (1987). Social processes in education: A reply to Sawyer and Watson (and others). In I. Reid (ed.), *The Place of Genre in Learning: Current Debates*, 35–45. Geelong, Australia: Deakin University Press.

Massey University. (2004). The headwords of the Academic Word List. Retrieved 13 March 2007, from http://language.massey.ac.nz/staff/awl/headwords.shtml.

Mauranen, A. (2001). Reflexive academic talk: Observations from MICASE. In R. Simpson and J. Swales (eds), *Corpus Linguistics in North America*, 165–78. Ann Arbor, MI: University of Michigan Press.

Mauranen, A. (2006). Speaking the discipline: Discourse and socialisation in ELF and LI English. In K. Hyland and M. Bondi (eds), *Academic Discourse Across Disciplines*, 271–94. Frankfort: Peter Lang.

Mauranen, A. (1993). Contrastive ESP rhetoric: Metatext in Finnish-English economics texts. *English for Specific Purposes*, 12(1), 3–22.

Mauranen, A. (1993). *Cultural Differences in Academic Rhetoric*. Frankfort: Peter Lang.

McCain, K. W. (1991). Communication, competition and secrecy: The production and dissemination of research-related information in genetics. *Science, Technology, & Human Values*, 16, 491–516.

McGrath, L. and Kuteeva, M. (2012). Stance and engagement in pure mathematics research articles: Linking discourse features to disciplinary practices. *English for Specific Purposes*, 31, 161–73.

McMillan, V. 1997. *Writing Papers in the Biological Sciences*, 2nd edn. Boston: Bedford Books.

Miller, C. (1984). Genre as social action. *Quarterly Journal of Speech*, 70, 157–78.

Miller, H. (1995). The presentation of self in electronic life: Goffman on the internet. http://www.psicopolis.com/psicopedia/selfweb.htm (accessed 28 April 2011).

Miller, T. (1998). Visual persuasion: A comparison of visuals in academic texts and the popular press. *English for Specific Purposes*, 17(1), 29–46.

Mills, G. and Water, J. A. (1986). *Technical Writing*, 5th edn. Fort Worth: HBJ.

Montgomery, S. (1996). *The Scientific Voice*. New York: The Guildford Press.

Moore, T. (2002). Knowledge and agency: A study of 'metaphenomenal discourse' in textbooks from three disciplines. *English for Specific Purposes*, 21(4), 347–66.

Moreno, A., Rey-Rocha, J., Burgess, S., Lopez-Navarro, I., Sachdev, I. (2012). Spanish researchers' perceived difficulty writing research articles for English medium journals: The impact of proficiency in English versus publication experience. *Ibérica*, 24, 157–84.

Mulkay. M. (1979). *Science and the Sociology of Knowledge*. London: Allen and Unwin.

Muller, G. (1985) *The American College Handbook of Contemporary English*. New York: Harper & Row.

Mur Duenas, P. (2012). Getting research published internationally in English: An ethnographic account of a team of finance Spanish scholars, struggles. *Ibérica*, 24, 139–56.

Murray, D. (1989). *Write to Learn*. New York: Holt, Rinehart and Winston.

Myers, G. (1985). Texts as knowledge claims: The social construction of two biology articles. *Social Studies of Science*, 15(4), 593–630.

Myers, G. (1989). The pragmatics of politeness in scientific articles. *Applied Linguistics*, 10, 1–35.

Myers, G. (1990). *Writing Biology: Texts in the Social Construction of Scientific Knowledge*. Madison, WI: University of Wisconsin Press.

Myers, G. (1991). Lexical cohesion and specialized knowledge in science and popular science texts. *Discourse Processes*, 14(1), 1e26.

Nagy, W. E., Anderson, R., Schommer, M., Scott, J. and Stallman, A. (1989). Morphological families in the internal lexicon. *Reading Research Quarterly*, 24, 263–82.

Namsaraev, V. (1997). Hedging in Russian academic writing in sociological texts. In R. Markkanen and H. Schroder (eds), *Hedging and Discourse: Approaches to the Analysis of a Pragmatic Phenomenon*, 64–79. Berlin: de Gruyter.

NASH, W. (1990) Introduction: The stuff these people write. In W. Nash (ed.), *The Writing Scholar: Studies in Academic Discourse*, 10. Newbury Park, CA: Sage.

Nation, I. S. P. (1990). *Teaching and Learning Vocabulary*. New York: Newbury House.

Nation, I. S. P. (2001). *Learning Vocabulary in Another Language*. New York: Cambridge

Nation, I. S. P. (2002). RANGE [computer software]. Available from http://www.vuw.ac.nz/lals/publications/software.aspx

Nattinger, J. and DeCarrico, J. (1992). *Lexical Phrases and Language Teaching*. Oxford: OUP.

Norton, B. (2000). *Identity and Language Learning: Gender, Ethnicity and Educational Change*. Harlow: Pearson Educational.

Nwogu, K. (1991). Structure of science popularizations: a genre analysis approach to the schema of popularized medical texts. *English for Specific Purposes*, 10, 111–23.

Nystrand, M. (1992). Social interactionism versus social constructionism: Bakhtin, Rommetveit, and the semiotics of written text. In A. H. Wold (ed.), *Theory of Language and Mind*, 157–73. Oslo: Scandinavian University Press.

Nystrand. M. (1989). A social interactive model of writing. *Written Communication*, 6, 66–85.

Ochs, E. and B. Schieffelin (1989). Language has a heart. *Text*, 9, 7–25.

Olsson, J. (2004). *Forensic Linguistics: An Introduction*. London: Continuum.

Onore, C. (1989). The student, the teacher, and the text: Negotiating meanings through response and revision. In: C. Anson (ed.), *Writing and Response*, 231–57). Urbana, IL: NCTE.

Pagel, W. J., Kendall, F. E. and Gibbs, H. R. (2002). Self-identified publishing needs of nonnative English-speaking faculty and fellows at an academic medical institution. *Science Editor*, 25(4), 111–4.

Palmer, F. (1990). *Modality and the English Modals*, 2nd edn. London: Longman.

Palmer, F.R. (1986) *Mood and Modality*. Cambridge: Cambridge University Press.

Paltridge, B. (2001). Genre, text type and the English for Academic Purposes (EAP) classroom. In A. M. Johns (ed.), *Genre in the Classroom*, 69–88. Mahwah, NJ: Erlbaum.

Paltridge, B. (2017). *The Discourse of Peer Review: Reviewing Submissions to Academic Journals*. Basingstoke, UK: Palgrave Macmillan.

Paltridge, B. and Wang, W. (2011). Contextualizing ESP research: Media discourses in China and Australia. In D. Belcher, A. Johns and B. Paltridge (eds), *New Directions in English for Specific Purposes Research*, 25–43. Ann Arbor, MI: University of Michigan Press.

Pare, A. (1993). Discourse regulations and the production of knowledge. In R. Spilka (ed.), *Writing in the Workplace: New Research Perspectives*, 111–23. Carbondale: Southern Illinois University Press.

Parks, M. and Archley-Landas, T. (2003). Communicating self through personal homepages: is identity more than screen deep? Paper presented at the annual conference of the International Communication Association, San Diego, CA.

Pellechia, M. (1997). Trends in science coverage: A content analysis of three US newspapers. *Public Understanding of Science*, 6, 49–68.

Pennycook, A. (1994). *The Cultural Politics of English as an International Language*. London: Longman.

Pennycook, A. (1994). The politics of pronouns. *ELT Journal*, 48(2), 173–8.

Pennycook, A. (1996). Borrowing others' words: Text, ownership, memory and plagiarism. *TESOL Quarterly*, 30(2), 201–30.

Plum, G. D. and Candlin, C. N. (2001). Becoming a psychologist: student voices on academic writing in psychology. In C. Baron, N. Bruce and D. Nunan (eds), *Knowledge and Discourse: Towards an Ecology of Language*, 238–66. London: Pearson.

Podgorecki, A. (1997). *Higher Faculties Cross National Study of University Culture*. Westport, CT: Praeger.

Poole, R. (2016). A corpus-aided approach for the teaching and learning of rhetoric in an undergraduate composition course for L2 writers. *Journal of English for Academic Purposes*, 21, 99–109.

Porter, J. (1992). *Audience and Rhetoric: An Archaeological Composition of the Discourse Community*. Englewood Cliffs, NJ: Prentice-Hall.

Powell, M. (1985). Purposive vagueness: An evaluation dimension of vague quantifying expressions. *Journal of Linguistics*, 21, 31–50.

Praninskas, J. (1972). *American University Word List*. London: Longman.

Prelli, L. (1989) *A Rhetoric of Science Inventing Scientific Discourse*. Columbia: University of South Carolina Press.

Prince, E., Frader, J. and Bosk, C. (1982) On hedging in physician-physician discourse. In R. D. Pietro (eds), *Linguistics and the Professions*, 83–97). Hillsdale, NJ: Ablex.

Prior, P. (1998). *Writing/Disciplinarity: A Sociohistoric Account of Literate Activity in the Academy*. Mahwah, NJ: Lawrence Erlbaum.

Quirk, R., Greenbaum, S., Leech, G. and Svartvik, J. (1972). *A Grammar of Contemporary English*. Harlow: Longman.

Quirk, R., S. Greenbaum, G. Leech and J. Svartvik (1985). *A Comprehensive Grammar of the English Language*. London: Longman.

Raimes, A. (1991). Out of the woods: Emerging traditions in the teaching of writing. *TESOL Quarterly*, 25, 407–30.

Ramanathan, V. and Atkinson, D. (1999). Individualism, academic writing, and ESL writers. *Journal of Second Language Writing*, 8, 45–75.

Rauch, Y M. (1992). The rhetoric of the probable in scientific commentaries: the debate over the species status of the Red Wolf. *Technical Communication Quarterly*, 6, 91–104.

Rawlins, J. (1980). What's so wrong with "In my opinion"? *College English*, 41, 670–4.

Reid, J. (1994). Responding to ESL students' texts: The myths of appropriation. *TESOL Quarterly*, 28(2), 273–94.

Reynolds, M. and Dudley-Evans, T., eds (1999). *Genre Analysis: Current Perspectives, Applications and Contributions*. Cambridge: CUP

Rheingold, H. (1995). *The Virtual Community: Finding Connection in a Computerized World*. London: Minerva.

Richards, J. (1993) Beyond the textbook: The role of commercial materials in language teaching. *RELC Journal*, 24(l), 1–14.

Richards, S. (1987). *Philosophy and Sociology of Science: An Introduction*, 2nd edn. Oxford: Blackwell.

Rogoff, B. (1990) *Apprenticeship in Thinking: Cognitive Development in Social Context*. New York, NY: Oxford University Press.

Rorty, R. (1979). *Philosophy and the Mirror of Nature*. Princeton: Princeton University Press.

Rothstein, E. (1996). Technology: Connections. Can Twinkies think, and other ruminations on the Web as a garbage depository. *New York Times*. March 4.

Rounds, P. (1982). *Hedging in Written Academic Discourse: Precision and Flexibility*. Mimeo, Michigan: The University of Michigan.

Rowntree, K. (1991). *Writing for Success*. Auckland: Longman Paul.

Sacks, H., Schlegloff, E. and Jefferson, G. (1974). A simplest systematics for the organization of turn-taking in conversation. *Language*, 50, 697–735.

Salager-Meyer, F. (1994) Hedges and textual communicative function in medical English written discourse. *English for Specific Purposes*, 13(2), 1499170.

Salager-Meyer, F. M., Ariza, A. A. and Zambrano, N. (2003) The scimitar, the dagger and the glove: Intercultural differences in the rhetoric of criticism in Spanish, French and English medical discourse (1930-1995), *English for Specific Purposes*, 22, 223–47.

Sarangi, S. and Roberts, C., eds (1999). *Talky Work and Institutional Order: Discourse in Medical, Mediation and Management Settings*. Berlin: Mouton de Gruyter.

Scarcella, R. and Brunak, J. (1981). On speaking politely in a second language. *International Journal of the Sociology of Language*, 1981(27), 59–76.

Schiffrin, D. (1980). Metatalk: Organisational and evaluative brackets in discourse. *Sociological Inquiry Language and Social Interaction*, 50, 199–236.

Schmitt, D. and Schmitt, N. (2005). *Focus on Vocabulary: Mastering the Academic Word List*. London: Longman.

Schröder, H. (1991). Linguistic and text-theoretical research on Language for specific purposes. In H. Schroder (ed.), *Subject-oriented Texts: Language for Special Purposes and Text Theory*, 1–48. Berlin: de Gruyter.

Schröder, H. (1995). Der Stil wissenschaftlichen Schrcibens zwischen Disziplin, Kultur und Paradigma. In G. Stickel (*Stilfragen*). Berlin: de Gruyter.

Scollon, R. (1994). As a matter of fact: The changing ideology of authorship and responsibility in discourse. *World Englishes*, 13(1), 333–46.

Scollon, R. and Scollon, S. (1995). *Intercultural Communication*. Oxford: Blackwell.

Scott, M. (1996). *Wordsmith Tools 4*. Oxford University Press.

Scott, M. (2004). *Wordsmith Tools (version 4)*. Oxford, UK: Oxford University Press.

Scott, M. and Tribble, C. (2006). *Textual Patterns*. Amsterdam: Benjamin.

Searle, D. and Dillon, D. (1980). The message of marking: Teacher written responses to student writing at intermediate grade levels. *Research in the Teaching of English*, 14(3), 233–42.

Semino, E. and Short, M. H. (2004). *Corpus Stylistics*. London: Longman.

Seone, E. (2013). On the conventionalisation and loss of pragmatic function of the passive in Late Modern English scientific discourse. *Journal of Historical Pragmatics*, 14(1), 70–99

Severino, C. (1993). The sociopolitical implications of response to second language and second dialect writing. *Journal of Second Language Writing*, 2(3), 181–201.

Shannon, C. and Weaver, W. (1963). *Mathematical Theory of Communication*. Illinois: University of Illinois Press.

Shapin, S. (1984). Pump and circumstance: Robert Boyle's literary technology. *Social Studies of Science*, 14, 481–520.

Shaw, P. (1992). Reasons for the correlation of voice, tense and sentence function in reporting verbs. *Applied Linguistics*, 13, 302–19.

Sinclair, J. (1991). *Corpus, Concordance and Collocation*. Oxford: Oxford University

Sinclair, J. (1999). A way with common words. In: H. Hasselgard and S. Oksefjell, (eds), *Out of Corpora*, 157–79) Amsterdam: Rodopi.

Skelton, J. (1988) Comments in academic articles. In Grunwell, P. (ed.), *Applied Linguistics in Society*. London: CILTIBAAL.

Smith, H. (1985). *Readable Writing, Revising for Style*. Belmont, CA: Wadsworth.

Spack, R. (1988). Initiating ESL students into the academic discourse community: How far should we go? *TESOL Quarterly*, 22(1), 29–52.

Spencer, C. and Arbon, B. (1996). *Foundations of Writing: Developing Research and Academic Writing Skills*. Lincolnwood, IL: National Textbook Co.

Sperber, D. and Wilson, D. (1986). *Relevance: Communication and Cognition*. Oxford: Blackwell.

Sperling, M. (1994). Constructing the perspective of teacher – as reader: A framework for studying response to writing. *Research in the Teaching of English*, 28, 175–203.

Starfield, S. and Ravelli, L. J. (2006). "The writing of this thesis was a process that I could not explore with the positivistic detachment of the classical sociologist": Self and structure in new humanities research theses. *Journal of English for Academic Purposes*, 5(3), 222–43.

Stern, S. R. (2004). Expressions of identity online: Prominent features and gender differences in adolescents' world wide web home pages. *Journal of Broadcasting & Electronic Media*, 48(2), 218–43.

Storer, N. and Parsons, T. (1968). The disciplines as a differentiating force, in E. B. Montgomery (ed.), *The Foundations of Access to Knowledge*, 101–21. Syracuse: Syracuse University Press.

Street, B. (1984). *Literacy in Theory and Practice*. Cambridge, UK: Cambridge University Press

Street, B. (1995). *Social Literacies*. London: Longman.

Strevens, P. (1988). ESP after twenty years: A reappraisal. In M. Tickoo (ed.), *ESP: State of the Art*, 1–13. Singapore: SEAMEO Regional Language Centre.

Strunk, W. J. and White, E. (1959) *The Elements of Style*. NY: Macmillan.

Stubbs, M. (1996). *Text and Corpus Analysis*. Oxford: Blackwell.

Stubbs, M. (2005). Conrad in the computer: Examples of quantitative stylistic methods. *Language and Literature*, 14(1), 5–24.

Sullivan, D. (1996). Displaying disciplinarity. *Written Communication*, 13(2), 221–50.

Swales, J. (1987) Utilising the literatures in teaching the research paper. *TESOL Quarterly*, 21(l), 41–68.

Swales, J. (1990). *Genre analysis: English in Academic and Research Settings*. Cambridge: Cambridge University Press.

Swales, J. (1993). Genre and engagement. *Revue Beige de Philologie et Histoire*, 71(3), 689–98.

Swales, J. (1996). Occluded genres in the academy: The case of the submission letter. In E. Ventola and A. Mauranen (eds), *Academic Writing: Intercultural and Textual Issues*, 45–58. Amsterdam: John Benjamins.

Swales J. (1998) *Other Floors, Other Voices: A Textography of a Small University Building*. Mahwah, NJ: Lawrence Erlbaum

Swales, J. M. (2004). *Research Genres: Explorations and Applications*. Cambridge: Cambridge University Press.

Swales, J. (2009). *Incidents in an Educational Life: A Memoir (of sorts)*. Ann Arbor: University of Michigan Press.

Swales, J. and Malczewski, B. (2001). Discourse management and new episode flags in MICASE. In R. Simpson and. J. Swales (eds), *Corpus Linguistics in North America*, 145–64. Ann Arbor, MI: University of Michigan Press.

Swales, J. and Feak, C. (2000). *English in Today's Research World: A Writing Guide*. Ann Arbor: University of Michigan Press.

Swales, J. and Feak, C. (2004) *Academic Writing for Graduate Students: Essential Tasks and Skills*, 2nd edn. Ann Arbor: University of Michigan Press.

Swales, J. M. and Feak, C. B. (2012). *Academic Writing for Graduate Students: Essential Tasks and Skills*, 3rd edn. Ann Arbor: University of Michigan Press.

Swales, J., Ahmad, U., Chang, Y., Chavez, D., Dressen, D. and Seymour, R. (1998). Consider this: The role of imperatives in scholarly writing. *Applied Linguistics*, 19, 97–121.

Taavitsainen, I. (1999). Metadiscursive practices and the evolution of early English medical writing (1375–1550). In J. M. Kirk (ed.), *Corpora Galore Analyses and Techniques in Describing English*. 191–207. Amsterdam: Rodopi.

Tadros, A. (1993) The pragmatics of text averral and attribution in academic texts. In M. Hoey (ed.), *Data Description, Discourse*, 98–114. London: Harper Collins.

Tang, J., Zhang, D. and Yao, L. (2007). Social network extraction of academic researchers. In Seventh IEEE international conference on data mining. http://keg.cs.tsinghua.edu.cn./../ICDM07-Tang-et-al-Academic-Network-Extraction.pdf. Accessed 28.04.11.

Tang, L. (2007). *Coping with Separation Chinese Seafarer–Partners in Cyberspace*. Unpublished PhD thesis, University of Cardiff. <http://www.sirc.cf.ac.uk/Nippon%20Fellows/Theses/Tang.pdf>.

Tannen, D. (1990). *You Just Don't Understand: Women and Men in Conversation*. New York: Ballantine.

Tardy, C. (2004). The role of English in scientific communication: Lingua franca or tyrannosaurus rex? *Journal of English for Academic Purposes*, 3, 247–69.

Tardy, C. (2015). *Beyond Convention: Genre Innovation in Academic Writing*. Ann Arbor: MI: University of Michigan Press.

Taylor, W. F. and Hoedt, K. C. (1966). The effect of praise upon quantity and quality of creative writing. *Journal of Educational Research*, 60, 80–3.

Thetela, P. (1997). Evaluated entities and parameters of value in academic research articles. *English for Specific Purposes*, 16, 101–18.

Thomas, J. (1983) Cross-cultural pragmatic failure. *Applied Linguistics*, 4(2), 91–112.

Thomas, S. and Hawes, T. (1994). Reporting verbs in medical journal articles. *English for Specific Purposes*, 13, 129–48.

Thompson, D. (1993) Arguing for experimental "facts" in science: a study of research article results sections in biochemistry. *Written Communication*, 8(I), 106–28.

Thompson, G. (1996). Voices in the text: discourse perspectives on language reports. *Applied Linguistics*, 17, 501–30.

Thompson, G. (2001). Interaction in academic writing: Learning to argue with the reader. *Applied Linguistics*, 22(1), 58–78.

Thompson, G. and Thetela, P. (1995). The sound of one hand clapping: the management of interaction in written discourse. *Text*, 15, 103–27.

Thompson, G. and Ye, Y. (1991). Evaluation of the reporting verbs used in academic papers. *Applied Linguistics*, 12, 365–82.

Thompson, J. B. (1984). *Studies in the Theory of Ideology*. Cambridge: Polity Press.

Thoms, L. and Thelwall, M. (2005). Academic home pages: Reconstruction of the self. First Monday, 10(12). <http://www.firstmonday.org/issues/issue10_12/thoms/>.

Todeva, E. (1999). Variability in academic writing: Hedging. *AAAL Conference*. New York, 6–9 March.

Toulmin, S. (1958). *The Uses of Argument*. Cambridge: CUP.

Toulmin, S. (1972). *Human Understanding*, vol. 1. Oxford: Clarendon Press.

Trimble, L. (1985). *English for Science and Technology: A Discourse Approach*. Cambridge: Cambridge University Press.

Tse, P. and Hyland, K. (2009). Discipline and gender: Constructing rhetorical identity in book reviews. In K. Hyland and G. Diani (eds), *Academic Evaluation: Review Genres in University Settings*, 05–121. London: Palgrave-MacMillan.

Tucker, P. (2003). Evaluation in the art-historical research article. *Journal of English for Academic Purposes*, 2, 291–312.

Turkle, S. (1996). *Life on the Screen: Identity in the Age of the Internet*. London: Weidenfeld & Nicolson.

Turner, J. C. (1987). Rediscovering the social group: A self-categorization theory. London: Blackwell.

Uccelli, P., Dobbs, C. L. and Scott, J. (2013). Mastering academic language organization and stance in the persuasive writing of high school students. *Written Communication*, 30, 36–62.

Valero-Garces, C. (1996). Contrastive ESP rhetoric: Metatext in Spanish-English Economics texts. *English for Specific Purposes*, 15(4), 279–94.

Vande Kopple, W. (1985). Some exploratory discourse on metadiscourse. *College Composition and Communication*, 36, 82–93.

Vande Kopple, W. (2002). 'Metadiscourse, discourse, and issues in composition and rhetoric.' In E. Barton and G. Stygall (eds), *Discourse Studies in Composition*, 91–113. Cresskill, NJ: Hampton Press.

Vel, D. O., Corney, M., Anderson, A. and Mohay, G. (2002, August 6–8). Language and gender author cohort analysis of e-mail for computer forensics. In *Second Digital Forensics Research Workshop*, Syracuse, NY.

Ventola, E. (1992) Writing scientific English: overcoming cultural problems. *International Journal of Applied Linguistics*, 2(2), 191–220.

Vygotsky, L. (1978). *Mind in Society: The Development of Higher Psychological Processes*. Cambridge, MA: Harvard University Press.

Wallace, R. (1995). *English for Specific Purposes in ESL Undergraduate Composition Classes: Rationale*. Unpublished doctoral dissertation, Illinois State.

Wang, K. M.-T. and Nation, P. (2004). Word meaning in academic English: Homography in the academic word list. *Applied Linguistics*, 25, 291–314.

Ward, J. (2009). EAP reading and lexis for Thai engineering undergraduates. *Journal of English for Academic Purposes*, 8(4), 294–301.

Wasko, M. M. and Faraj, S. (2000). It is what one does: Why people participate and help others in electronic communities of practice. *The Journal of Strategic Information Systems*, 9, 155–73.

Watson, J. (1968). *The Double Helix*. Harmondsworth: Penguin.

Weimer, W. (1977). Science as a rhetorical transaction: toward a nonjustificational conception of rhetoric. *Philosophy and Rhetoric*, 10, 1–29.

Wenger, E. (1998). *Communities of Practice: Learning, Meaning, and Identity*. Cambridge: Cambridge University Press.

Wertsch, J. 1991. *Voices of the Mind: A Sociocultural Approach to Mediated Action*. Cambridge, MA: Harvard University Press.

West, M. (1953). *A General Service List of English Words*. London: Longman.

White, P. (2003). Beyond modality and hedging: A dialogic view of the language of intersubjective stance. *Text*, 23(2): 2594–8.

Whitley, R. (1984). *The Intellectual and Social Organisation of the Sciences*. Oxford: Clarendon Press.

Wilkinson, A. M. (1992). Jargon and the passive voice: Prescriptions and proscriptions for scientific writing. *Journal of Technical Writing and Communication*, 22, 319–25.

Williams, J. (1981). *Style Ten Lessons in Clarity and Grace*. Boston: Scott Foresman.

Willis, D. (2003). *Rules, Patterns and Words: Grammar and Lexis in English Language Teaching*. Cambridge: Cambridge University Press.

Wilson, D. (2002). *The Englishization of Academe: A Finnish Perspective*. Jyväskylä, Finland: University of Jyväskylä Language Centre.

Winkler, A. and McCuen, J. (1989) *Writing the Research Paper: A Handbook*, 3rd edn. Sydney: Harcourt Brace Jovanovich.

Wortham, S. (2001). Interactional positioning and narrative self-construction. *Narrative Inquiry*, 10, 157–84.

Wortham, S. and Gadsden, V. (2006) 'Urban fathers positioning themselves through narrative: An approach to narrative self-construction.' In A. de Fina, D. Schiffrin and M. Bamberg (eds), *Discourse and Identity*, 315–41. Cambridge, UK: Cambridge University Press.

West, M. (1953). *A General Service List of English Words*. London: Longman.

Worthington, D. and Nation, I. S. P. (1996). Using texts to sequence the introduction of new vocabulary in an EAP course. *RELC Journal*, 27, 1–11.

Wray, A. and Perkins, M. (2000). The functions of formulaic language. *Language and Communication*, 20, 1–28.

Wu, M. H. (1992) Towards a contextual lexico-grammar: An application of concordance analysis in EST teaching. *RELC Journal*, 23(2), 18–34.

Wynn, E. and Katz, J. E. (1997). Hyperbole over cyberspace. Self-presentation and social boundaries in internet home pages and discourse. *The Information Society*, 13(4), 297–328.

Xue, G. and Nation, I. S. P. (1984). A university word list. *Language Learning and Communication*, 3, 215–99.

Yang, H. (1986). A new technique for identifying scientific/technical terms and describing science texts. *Literacy and Linguistic Computing*, 1, 93–103.

Yang, R. Y. and Allison, D. (2003). Research articles in applied linguistics: Moving from results to conclusions. *English for Specific Purposes*, 22(4), 365–85.

Yarber, R. (1985). *Writing/or College*. Glenview, IL: Scott Foresman.

Ylonen, S., Neuendorff, D. and Effe, G. (1993) Zur kontrastiven Analyse von medizinischen Fachtexten. Eine diachrone Studie. In C. Lauren and M. Nordman (eds), *Special Language: From Humans Thinking to Thinking Machines*. Clevedon: Multilingual Matters.

Yorio, C. (1989). Idiomaticity as an indicator of second language proficiency. In K. Hyltenstam and K. Obler (eds), *Bilingualism Across the Lifespan*, 55–72. Cambridge: Cambridge University Press.

Zamel, V. (1983). The composing processes of advanced ESL students: Six case studies. *TESOL Quarterly*, 17, 165–87.

Zamel, V. (1985). Responding to student writing. *TESOL Quarterly*, 19(1), 79–101.

Ziman, J. (1984). *An Introduction to Science Studies: The Philosophical and Social Aspects of Science and Technology*. Cambridge: Cambridge University Press.

Zinsser, W. (1976) *On Writing Well*. NY: Harper & Row.

Zuck, J. G. and Zuck, L. V. (1986) Hedging in newswriting. In A. M. Cornu, J. Van Parjis, M. Delahaye and L. Baten (eds), *Beads or Bracelets? How do we Approach LSP?* Selected papers from the 5th European Symposium on LSP, 172–80. Oxford: Oxford University Press.

Index